# Swords and Sword Makers of England and Scotland

This book is dedicated to my daughter, Ann Elizabeth Bezdek-Rumney, who was a tremendous help in preparing this book.

# Swords and Sword Makers of England and Scotland

## Richard H. Bezdek

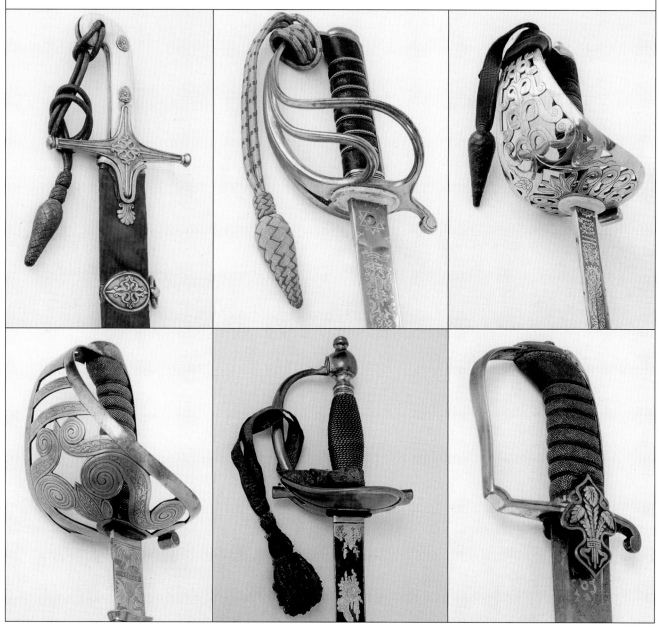

Paladin Press • Boulder, Colorado

*Sword and Sword Makers of England and Scotland*
by Richard H. Bezdek

Copyright © 2003 by Richard H. Bezdek

ISBN 1-58160-399-1
Printed in the United States of America

Published by Paladin Press, a division of
Paladin Enterprises, Inc.
Gunbarrel Tech Center
7077 Winchester Circle
Boulder, Colorado 80301 USA
+1.303.443.7250

Direct inquiries and/or orders to the above address.

PALADIN, PALADIN PRESS, and the "horse head" design
are trademarks belonging to Paladin Enterprises and
registered in United States Patent and Trademark Office.

All rights reserved. Except for use in a review, no
portion of this book may be reproduced in any form
without the express written permission of the publisher.

Neither the author nor the publisher assumes
any responsibility for the use or misuse of
information contained in this book.

Visit our Web site at www.paladin-press.com

# Table of Contents

Introduction · 1

## Chapter 1
Rulers of England · 3

## Chapter 2
Cavalry Swords Made at the Enfield Royal Small Arms Manufactory · 7

## Chapter 3
The Sword and Blade Making Center of Hounslow Heath · 9

## Chapter 4
The Sword and Blade Making Center of Shotley Bridge (The Hollow Sword Blade Company) · 19

## Chapter 5
German Edged Weapon Makers and Retailers Who Had Offices or Agents in London · 27

## Chapter 6
English Cutlers Who Were Appointed Sword Cutler to the Board of Ordnance · 29

## Chapter 7
English Sword Blade Markings · 31

## Chapter 8
The Cutlers Company of London · 33

## Chapter 9
English Sword Makers, Cutlers, Dealers, and Craftsmen Who Mounted Swords · 35

## Chapter 10
English Sword Makers, Cutlers, and Dealers Who Exported Swords to the United States · 217

## Chapter 11
English Armourers · 221

## Chapter 12
Rulers of Scotland · 227

## Chapter 13
Scottish Words Pertaining to Swords and Sword Guards · 229

## Chapter 14
Scottish Royal Armourers and Cutlers · 231

## Chapter 15
Scottish Sword Makers, Dealers, Cutlers, and Craftsmen Who Mounted Swords · 233

English Sword Photo Section · 269

Scottish Sword Photo Section · 361

Bibliography and Reference Material on English and Scottish Swords and Sword Makers · 387

# Contributors

I must express my appreciation to several organizations and people who contributed reference material to this project. Without their help, this book would be incomplete in several areas.

**John Arlett** (deceased), editor of *The Swordsman* and Senior Sword Advisor for Wilkinson Sword, London, provided historical information on Wilkinson swords.

**Colin Armstrong**, Executive Secretary, The Newcomen Society, provided old magazine articles on Shotley Bridge and the Hollow Sword Blade Company.

**Claude Blair** provided articles on Hounslow.

**Sue Charlin**, Editorial Assistant, The Oxford University Press, helped with "Notes and Queries" articles.

**Mr. T.R. Fattorini**, Thomas Fattorini Ltd., provided historical information about the company.

**Firmin & Sons PLC**, Birmingham, provided historical information on their company.

**James Forman** provided historical information on Scottish sword makers.

**S. James Gooding**, Publisher, *Arms Collecting* magazine, provided many articles from old back issues.

**Edmund Greenwood**, former Secretary, and A.B.L. Dove, Honorable Secretary, The Arms and Armour Society, provided many articles from old back issues of their journal.

**Neil Robert Grigg** provided the map of Hounslow and additional information.

**Hobson & Sons**, London, provided historical information on their company.

**Andy May**, good friend and fellow sword collector, worked very hard to obtain sword photos for this book.

**Stuart C. Mowbray**, Editor and Art Director, *Man at Arms* magazine, provided two volumes of *The History of the Cutlers Company of London* by Charles Welch.

**Newcastle City Library** provided information on Shotley Bridge.

**Anthony North**, Victoria & Albert Museum, helped with old magazine articles on sword makers.

**Henry Poole & Co.**, London, provided historical information on their company.

**W.G.M. Ragg** provided historical company information.

**Sheffield Local Studies Library** helped with articles on Shotley Bridge.

**Mrs. A.L. Thompson**, Local Studies Librarian, Durham City Reference Library, helped me locate articles on Shotley Bridge.
**University of Durham Library** helped with information on Shotley Bridge.
**John Tofts White** contributed large amounts of information on the sword and blade making center of Hounslow.

A special thanks to the people who contributed photos to this book:

**Martin Boswell** (English swords)
**Andrea Cameron**, Hounslow Library (Hounslow swords)
**Hennie Coetzee** (English sword)
**Mervyn Emms** (English swords)
**James Forman** (Scottish swords)
**Vanda Foster**, Deputy Curator, and **Andrea Cameron**, Senior Officer, Heritage Services, Gunnersbury Park Museum (Hounslow swords)
**Neil Robert Grigg** (Hounslow swords)
**Don F. Hamilton** (English swords)
**Geoffrey Jenkinson** (Scottish swords)
**J.G. Kirton** (English swords)
**Neil McGregor** (English swords)
**Robert B. Miller**, LionGate Arms & Armour (English swords)
**George C. Neumann** (English swords)
**Sheperd Paine** (English swords)
**Sam Saladino** (English and Hounslow swords)
**John Spooner**, **Richard Ratner**, and **Phillip Spooner** of West Street Antiques (English swords)
**Patrick R. Tougher**, Scottish Sword & Shield (Scottish swords)
**John Tofts White** (Hounslow swords)
**Bridgette Winsor** and **Sophia Popham**, Archivists, Birmingham City Archives (sword hilts of Matthew Boulton)
**Jan Zajac** (English swords)

# Preface

ollectors and researchers of English and Scottish swords will appreciate the huge amount of information available in this book. It is designed to make it easy for the reader to obtain specific information quickly. In Chapters 9 and 15, all material is presented in alphabetical and chronological order. The occupations of all artisans and merchants are shown in the left margin. All family and company names and addresses, including changes over the years, are listed, as is information about types of swords manufactured by each individual or company. Other chapters contain a large amount of data about a number of related areas of interest, including such early blade and sword making centers as Hounslow Heath and Shotley Bridge, early armourers in England and Scotland, the Cutlers Company of London, German sword makers and dealers with offices in England, and English sword makers and cutlers who exported swords to the United States of America. The last section of the book features many photographs of rare and unique English and Scottish swords.

# Introduction

 have collected antique English and Scottish cavalry swords for more than 25 years, and interest in these swords has grown tremendously during that time. One of the reasons for this is the many English and American movies, television programs, and historical documentaries that have appeared on wars fought from the sixteenth to the nineteenth centuries, including the Thirty Years War (1618–1648), English Civil War (1642–1651), American Revolutionary War (1775–1783), Napoleonic Wars (1793–1815), American War of 1812 (1812–1815), and Crimean War (1854–1856) as well as wars fought between Scottish patriots and the English crown. Swords were featured prominently in all of these films. Also during this period, there has been a huge renewal of interest in antiques in general. Therefore, there is now a large demand for military antiques, especially swords.

Unfortunately, the reference material about English and Scottish swords has not grown with the increasing interest in them. Although a few books have been published that identify English swords, there have been no new books about English sword makers, cutlers, and dealers since 1970, and no books have been published about Scottish sword makers and related fields for more than 20 years. A tremendous amount of new information on English and Scottish sword makers has come to light during that time. This book will help fill this gap in English and Scottish sword and sword maker reference material.

# Chapter 1: Rulers of England

| Dates of Reign | Ruler | Succession | Born | Died |
|---|---|---|---|---|
| 828–839 | Egbert | King of West Saxon (from 802); with submission of Mercia (828), first to bring all England under one rule | 775? | 839 |
| 839–858 | Ethelwulf | Son of Egbert | ? | 858 |
| 858–860 | Ethelbald | Son of Ethelwulf | ? | 860? |
| 860?–866? | Ethelbert | Second son of Ethelwulf | ? | 866? |
| 866?–871 | Ethelred I | Third son of Ethelwulf | ? | 871 |
| 871–899 | Alfred the Great | Fourth son of Ethelwulf | 849 | 899 |
| 899–924 | Edward the Elder | Son of Alfred | 870 | 924 |
| 924–940 | Athelstan | Eldest son of Edward the Elder | 895 | 940 |
| 940–946 | Edmund I | Half brother of Athelstan; murdered | 922? | 946 |
| 946–955? | Edred | Youngest son of Edward the Elder | 925? | 955? |
| 955?–959 | Edwy | Son of Edmund I | 939? | 959 |
| 959–975 | Edgar | Second son of Edmund I | 944 | 975 |
| 975–978 | Edward the Martyr | Son of Edgar; assassinated | 963? | 978 |
| 978–1016 | Ethelred II the Unready | Half brother of Edward the Martyr | 968? | 1016 |
| 1016–1016 | Edmund II Ironside | Eldest son of Ethelred II | 990 | 1016 |

## DANISH KINGS

| Dates of Reign | Ruler | Succession | Born | Died |
|---|---|---|---|---|
| 1016–1035 | Canute (Cnut, Knut) | By conquest and election | 994? | 1035 |
| 1035–1040 | Harold I (Harefoot) | Illegitimate son of Canute | ? | 1040 |
| 1040–1042 | Hardecanute | Son of Canute | 1018? | 1042 |

| Dates of Reign | Ruler | Succession | Born | Died |
|---|---|---|---|---|
| | | **SAXON KINGS** | | |
| 1042–1066 | Edward the Confessor | Son of Ethelred II | 1002? | 1066 |
| 1066–1066 | Harold II | Earl of Wessex and brother-in-law of Edward; elected by witan; killed at Battle of Hastings | 1022? | 1066 |
| | | **NORMAN KINGS** | | |
| 1066–1087 | William I the Conqueror | Second cousin of Edward the Confessor; gained throne by conquest | 1027 | 1087 |
| 1087–1100 | William II Rufus | Second surviving son of William I | 1056? | 1100 |
| 1100–1135 | Henry I Beauclerc | Youngest son of William I | 1068 | 1135 |
| 1135–1154 | Stephen | Third son of Count of Blois and Adela, daughter of William I. (Matilda, daughter of Henry I, was chosen "Lady of the English" in 1141 and ruled, although not crowned, for a few months.) | 1097? | 1154 |
| | | **THE PLANTAGENETS** | | |
| 1154–1189 | Henry II | Son of Geoffrey Plantagenet and Matilda | 1133 | 1189 |
| 1189–1199 | Richard I the Lion-Hearted | Eldest surviving son of Henry II | 1157 | 1199 |
| 1199–1216 | John | Youngest son of Henry II | 1167? | 1216 |
| 1216–1272 | Henry III | Eldest son of John | 1207 | 1272 |
| 1272–1307 | Edward I | Eldest son of Henry III | 1239 | 1307 |
| 1307–1327 | Edward II | Eldest surviving son of Edward I | 1284 | 1327 |
| 1327–1377 | Edward III | Eldest son of Edward II | 1312 | 1377 |
| 1377–1399 | Richard II | Eldest son of Edward III | 1367 | 1400 |
| 1399–1413 | Henry IV | Son of John of Gaunt, the fourth son of Edward III | 1362 | 1413 |
| 1413–1422 | Henry V | Eldest son of Henry IV | 1387 | 1422 |
| 1422–1461 1470–1471 | Henry VI | Only son of Henry V; died in Tower of London | 1421 | 1471 |
| | | **HOUSE OF YORK** | | |
| 1461–1470 1471–1483 | Edward IV | Grandson of Richard, son of Edmund, fifth son of Edward III | 1442 | 1483 |
| 1483–1483 | Edward V | Eldest son of Edward IV; murdered? | 1470 | 1483 |
| 1483–1485 | Richard III | Younger brother of Edward IV; killed at Battle of Bosworth Field | 1452 | 1485 |
| | | **HOUSE OF TUDOR** | | |
| 1485–1509 | Henry VII | Son of Edmund, eldest son of Owen Tudor and Catherine, widow of Henry V; his mother was a great-granddaughter of John of Gaunt | 1457 | 1509 |
| 1509–1547 | Henry VIII | Only surviving son of Henry VII | 1491 | 1547 |
| 1547–1553 | Edward VI | Son of Henry VIII and Jane Seymour | 1537 | 1553 |
| 1553–1558 | Mary I (Mary Tudor) | Daughter of Henry VIII and Catharine of Aragon | 1516 | 1558 |
| 1558–1603 | Elizabeth I | Daughter of Henry VIII and Anne Boleyn | 1533 | 1603 |

| Dates of Reign | Ruler | Succession | Born | Died |
|---|---|---|---|---|
| | | **HOUSE OF STUART** | | |
| 1603–1625 | James I (James VI of Scotland) | Son of Mary Queen of Scots and Lord Darnley; great-grandson of Margaret, daughter of Henry VII; king of Scotland (as James VI) from 1567 | 1566 | 1625 |
| 1625–1649 | Charles I | Only surviving son of James I; executed | 1600 | 1649 |
| | | **COMMONWEALTH AND PROTECTORATE** | | |
| 1649–1658 | Oliver Cromwell, Lord Protector | Put in power by the army | 1599 | 1658 |
| 1658–1659 | Richard Cromwell, Lord Protector | Third son of Oliver Cromwell | 1626 | 1712 |
| | | **HOUSE OF STUART RESTORED** | | |
| 1660–1685 | Charles II | Eldest surviving son of Charles I | 1630 | 1685 |
| 1685–1688 | James II | Second son of Charles I, expelled by Revolution of 1688 | 1633 | 1701 |
| 1689–1702 | William III | Son of William (Prince of Orange) and Mary, daughter of Charles I; installed by Revolution | 1650 | 1702 |
| 1689–1694 | Mary II | Eldest daughter of James II | 1662 | 1694 |
| 1702–1714 | Anne | Second daughter of James II; queen of Great Britain following union of England and Wales with Scotland in 1707 | 1665 | 1714 |
| | | **HOUSE OF HANOVER** | | |
| 1714–1727 | George I | Son of the Elector of Hanover and Sophia, who was daughter of Elizabeth, daughter of James I | 1660 | 1727 |
| 1727–1760 | George II | Son of George I | 1683 | 1760 |
| 1760–1820 | George III | Grandson of George II; insane from 1810, when Prince of Wales (later George IV) acted as regent | 1738 | 1820 |
| 1820–1830 | George IV | Eldest son of George III | 1762 | 1830 |
| 1830–1837 | William IV | Third son of George III | 1765 | 1837 |
| 1837–1901 | Victoria | Daughter of Edward, Duke of Kent, fourth son of George III | 1820 | 1901 |
| | | **HOUSE OF SAXE – COBURG – GOTHA** | | |
| 1901–1910 | Edward VII | Eldest son of Victoria and Prince Consort Albert of Saxe-Coburg-Gotha. | 1841 | 1910 |
| | | **HOUSE OF WINDSOR** (name changed in 1917) | | |
| 1910–1936 | George V | Second son Edward VII | 1865 | 1936 |
| 1936–1936 | Edward VIII | Eldest son of George V; abdicated | 1894 | 1936 |
| 1936–1952 | George VI | Second son of George V (created Duke of Windsor, Dec. 12) | 1895 | 1952 |
| 1952–2003 | Elizabeth II | Eldest daughter of George VI | 1926 | 2003 |

# Chapter 2

# Cavalry Swords Made at the Enfield Royal Small Arms Manufactory

The Enfield Royal Armoury was founded in 1811. It was located 11 miles north of London. In 1854, the name changed to the Enfield Royal Small Arms Manufactory.

### Pattern 1821 Light Cavalry Sword

In 1821, the pattern 1821 light cavalry sword was put into production. Six-thousand swords were made by March 1825, when production ceased in order to start production on the pattern 1821 heavy cavalry sword.

### Pattern 1821 Heavy Cavalry Sword

Production of the pattern 1821 heavy cavalry sword was originally deferred because there were 34,000 of the pattern 1796 heavy cavalry swords in stock. The chief storekeeper of the tower complained that the pattern 1821 heavy cavalry sword blade was designed too light, being even thinner than the pattern 1821 light cavalry sword. It was agreed that the sword blade was to be made thicker and stiffer. Two-thousand. pattern 1821 heavy cavalry swords were made at the Enfield Armoury, and the production was completed by November 1827.

In 1828 and 1835, complaints about pattern 1821 light and heavy cavalry swords prompted the armoury to request reports on the swords from the regiments to which they were issued. The reports said the scabbards for both were too weak, the blade fullers too deep, and the hilts too restrictive. In December 1835, an order for 1,000 heavy cavalry swords was given to the Enfield Armoury; the swords were given heavier blades and scabbards. The light cavalry swords that were made had larger pommels and the extended guard branches.

### Pattern 1853 Cavalry Sword

The pattern 1852 cavalry sword replaced both the pattern 1821 light and heavy cavalry swords and was issued to all regiments of cavalry. It was designed with sword maker Charles Reeves' patented tang, which was an extension of the blade (and as wide as the blade) as opposed to a separate piece welded to the blade. The grip was designed with two checkered leather strips riveted on to each side of the tang. The blade was curved and had a spear point to allow for both cutting and thrusting.

## Pattern 1864 Cavalry Sword

In March 1864, the pattern 1864 cavalry sword was adopted. The blade was basically the same, but the hilt's two-branch design was replaced with a sheet-metal guard pierced with a Maltese cross design.

## Patterns 1882 Long and Short, 1885, and 1890 Cavalry Swords

The patterns 1882 long and short, 1885, and 1890 cavalry swords all had basically the same type blade as the pattern 1864 sword. The sheet-metal guard, however, had turned-down edges to prevent damage to the cavalryman's uniform (there were complaints about uniform damage with the 1864 pattern). The scabbards on these four patterns were changed to allow for saddle mounting. The two loose carrying rings mounted on the top and middle of the scabbard were replaced with two stationary rings mounted on the top and on each side of the scabbard

## Pattern 1899 Cavalry Sword

The Enfield Armoury produced a small number of the pattern 1899 cavalry sword. The basic differences between the 1899 and 1890 patterns were that the 1899 sheet-metal guard was much larger and did not have a Maltese cross cut into it.

In 1900 and 1901, Enfield produced two experimental wooden cavalry sword scabbards and, in 1901, an experimental leather cavalry sword scabbard. In 1902, the armoury made 1,000 experimental wooden cavalry sword scabbards. In 1904, it made 75 experimental cavalry swords with leather scabbards, shaped leather grips, and straight blades.

## Pattern 1908 Cavalry Sword

The pattern 1908 sword had a large bowl-type guard and a very long, checkered, leather shaped grip. The straight blade straight narrowed at the tip. It was designed for thrusting only. Enfield produced a small number of these swords.

# Chapter 3

## The Sword and Blade Making Center of Hounslow Heath

In 1621, King James I declared that he needed military swords for 12,000 men (for the Thirty Years War, 1618–1648). Previously, he had to buy most of his swords from foreign sources, mostly from swordsmiths from Solingen, Prussia, a German kingdom. At that time, the Greenwich Royal Armouries were providing very few swords.

The king hoped to enlarge England's arms making capacity and provide employment for his subjects. In early 1621, he granted Thomas Murrey (cutler and secretary to the Prince of Wales; probably the wardrobe supplier to the prince) a patent for the sole manufacture of sword and rapier blades in England.

On June 13, 1621, by order of the court of the Cutlers Company, master John Porter, past master Thomas Chesshire, and royal cutler Robert South inspected an engine (machine for the making of sword blades, i.e., a blade mill or factory) and were to report their findings to the court at the next meeting. It was probably Thomas Murrey's blade mill.

In July 1621, Thomas Murrey presented his first group of sword blades to the Cutlers Company for inspection. The company rejected them, saying they needed much more work to come to "perfection," and the expense to make them was too large.

In 1624, King James I contracted with the Cutlers Company to provide 5,000 swords with hangers and girdles a monthly "to be used by the Earle of South Hampton and other honorable persons employed beyond the seas." The company purchased blades from bladesmiths in Solingen, Prussia, which were sold at a fixed rate to the members making the swords when the blades arrived (eight months later). The first purchase was for 48 dozen blades on April 29, 1624. They were sold to 16 members in groups of 24 at 5 pounds per 24. Over the next three years, blades were also purchased from merchants in London and Birmingham. All swords were delivered to the Tower of London.

In 1629, King Charles I and Sir John Heyden (Lieutenant of the Ordnance) decided to bring more German swordsmiths to England. Heyden was the younger brother of William Heyden, previous Lieutenant of Ordnance who was killed in 1623. Many German armourers and swordsmiths who had immigrated to England earlier worked in the Greenwich Royal Armouries (in operation 1511–1644). When Charles was Prince of Wales, his father, King James I, presented him with a sword made by Clemens Horn of Solingen, Prussia, so Charles knew of the quality of German sword blades.

It all started when Sir John Heyden, while on a diplomatic mission in Holland (probably Rotterdam) on behalf of King Charles I, encountered some German swordsmiths. The Germans were supposedly refugees fleeing from the terrors of the Thirty Years War. He persuaded some of them to immigrate to England and work under royal patronage. These swordsmiths were members of several sword-related crafts from Solingen, including *Schwertschmeides* (swordsmiths), *Klingenschmieldes* (bladesmiths/blade forgers), *Schwertschleifer* (sword/blade grinders), *Schwertfegers* (sword/blade polishers), and *Schwertharters* (sword/blade hardeners).

The route to England from Solingen went through the Netherlands and coastal Holland to Rotterdum and then to London. That is why many documents of the time referred to the Solingen Germans who immigrated to England as Dutch and why they called their blades "Dutch" blades.

Solingen is located on the southern edge of the Ruhr Valley on the Wupper River. The Wupper is a tributary of the larger Rhine River, which runs through Germany and the Netherlands to Rotterdam, Holland's largest seaport. The German swordsmiths could easily travel the Wupper River to the Rhine, down to Rotterdam, and by boat to London.

The German swordsmiths were settled at Hounslow Heath, a fairly flat farming area about 12 miles from London on the Staines Road (part of the old London road to Bath). They built their homes, workshops, and blade mills along the New Cutt River (later called the Duke of Northumberland River), which was actually a man-made canal. The river provided fast running water to turn waterwheels, which ran the equipment in their mills such as bellows and trip hammers.

The leading Germans who set up blade mills were bladesmiths of some stature in Solingen (i.e., guild members) who employed other Germans. The following German bladesmiths (probably blade mill owners) signed their blades:

- Peter Munsten the Younger (changed name to Peter English), c. 1629–1642
- Johann Kindt (Kinndt, Kennett), c. 1629–1659
- Johannes Hoppe (Hoppie) the Younger, c. 1633–1642
- Caspar Karn (Carnis), c. 1629–1642
- Clemens (Clamas) Meigen, c. 1629–1642
- Caspar Fleiseh, c. 1629–1642
- Johannes Dell (Bell), c. 1649–1685

Other known German swordsmiths and bladesmiths working in Hounslow were:

- Johann Konigs (Connyne), c. 1629–1642
- Clemens Horn the Younger, c.1629–1642
- Ceile Herder, c. 1649–1659
- Peter Henekels (Henkell), c. 1660–1685
- Johannes Meigen, c. 1629–1642
- Heinrich (Henry) Hoppe (Hoppie) the Elder, c. 1629–1642
- Joseph Hoppe Hoppie, c. 1629–1642

Benjamin Stone, a London cutler, played a large part in the sword and blade making center of Hounslow. Born c. 1591, Stone was the son of yeoman John Stone of Arundel, Sussex. He apprenticed to well-known London cutler William Bals (Balls) from 1604–1613. Stone became a freeman (allowed to sell his wares in London) of the Cutlers Company of London in 1613. His London shop was on Bartholemew Lane. Stone's identified apprentices are:

- Robert Salisbury, 1614
- Simon Connell, 1614
- William Handiday, 1620
- William Holmes, 1620–1628
- Ellia Browne, 1623
- Joseph Roger, 1626
- Robert Cooper, 1628
- William Hall, 1629
- John Mashrother, 1631
- John Hester, 1631
- James Hagan, 1638
- Thomas Hunt, 1647

Stone became a large purveyor of swords and blades to the Office of Ordnance. As time went on, he had many conflicts with the Cutlers Company and the royal cutler Robert South about the way he conducted his business. He was fined many times and even jailed over such things as striking other cutlers, putting incorrect marks on his products, buying knives made outside of London, swearing at officers of the company, and keeping unregistered apprentices. He also made enemies among cutlers in the Cutlers Company of London by complaining that their swords were of poor quality.

During this time (1613–1628), Stone was a typical London cutler, i.e, an assembler of swords and knives. He would buy finished blades from local bladesmiths and bladesmiths from Solingen, Prussia, Venice and Milan, Italy, and Toledo, Spain. He then fitted hilts (i.e., handle and guard) and leather scabbards, which he made or bought himself, to the blades. Many of his hilts were of cast brass, so Stone must have had a brass foundry. He also made or bought sword belts.

In 1621, Stone bought 30 swords from cutler Jacob Fulwater, son of Henry Fulwater. On June 27, 1628, he delivered 800 swords to the Office of Ordnance for the arming of troops recruited for Buckingham's proposed expedition to relieve La Rochelle. Of these swords, 350 were Italian swords from Venice and 450 were "Dutch" swords from Solingen. All of them had Irish (basket) hilts. On June 30, 1628, Stone sold 150 blades to the Office of Ordnance.

Knowing the king's need for military swords and blades and the availability of rough blades from the Germans who the king had brought to Hounslow, Stone set up a sword mill in Hounslow Heath in 1629 by converting an existing mill on the New Cutt River. His operation was actually a sword finishing mill—he bought rough-forged blades from the Germans (who mostly used imported German steel, which was better than English steel), grinded and polished them, then fit them with hilts and leather scabbards. The mill was powered by a waterwheel that operated the grinding stones and polishing wheels. Most swords and blades were sold to the Office of Ordnance and delivered to the Tower Armoury. He also refitted and refurbished swords for the Office of Ordnance and sold many blades to London cutlers who assembled swords. The German bladesmiths at Hounslow could obtain German steel through the port of London, and Stone could ship blades and finished swords from London to other English ports or overseas.

Hounslow Heath was a rather open farming area located approximately 12 miles from London in the county of Middlesex. It was on the Staines Road, a section of the old Roman road that led from London to Bath. The road went through the towns of Hammersmith, Staines, Bedfont, and Horton (Old Cohnbrook) before finally ending in Bath. Bedfont and Horton were in Hounslow Heath, as was the town of Isleworth. The Isleworth River (later called the Crane River) ran through Hounslow Heath. Joseph Jenks built a sword blade mill in Isleworth in 1639.

Hounslow Heath was a perfect spot for Stone to set up a sword mill. It was far enough away from London so the Cutlers Company of London and the Royal Cutler could not interfere with his production. (The rules and regulations of the Cutlers Company didn't apply to this suburban area.) Brass hilts, which were cheaper and easier to make than steel hilts, could be used at Hounslow, whereas the Cutlers Company did not allow London cutlers to use brass, at that time considered a poor and soft metal. Many brass hilts could be cast from the same mold, while steel hilts had to be made one at a time (cast steel had not been invented).

On June 20, 1632, Stone signed a lease for the land where his sword mill was located. (He probably rented the land between 1629–1632.) He leased three acres, which included a mill (probably a grain mill) on the New Cutt River, from George (Lord) Berkeley (Mowbray, Segrave and Bruce) at the east end of the town of Bedfont, parish of East Bedfont Cum [and] Hatton, county of Middlesex. Stone's sword mill straddled the New Cutt River, which ran between the Colne and Isleworth Rivers. The mill was located near the confluence of the New Cutt and Isleworth Rivers near the Baber Bridge. The Staines Road ran over the Baber Bridge, which straddled the Isleworth River.

Stone's lease began on the feast of Michaelmas (September 29, 1632) and ran for 21 years. It stated that he could not sublease the land or change the sword mill to any other kind of mill. He was to pay Lord Berkeley 32 pounds a year in four payments of 8 pounds each, to be paid at the feasts of St. Thomas the Apostle, the Anunciation of the Blessed Virgin Mary, St. John the Baptist, and St. Michael the archangel (Michaelmes). Stone was to pay all taxes on the property and properly maintain the land and mill.

Stone converted Lord Berkeley's mill into a sword mill. It had grinding and polishing stones powered by a waterwheel on the New Cutt River. He provided steel (probably German) to the German bladesmiths, who took the material to their workshops for shaping. (They probably also obtained German steel themselves.) Their sword blade mills, also located along the New Cutt River, included blade forges (furnaces) with bellows and trip hammers run by a waterwheel. The rough blades were taken to Stone's sword mill. Stone paid the Germans by the piece, then finished the blades by grinding and polishing, attached hilts, and fitting each with a leather scabbard. As well as English workers, Stone had German *Schliefers* (grinders) and *Fegers* (polishers) working for him. He sold blades to London cutlers (assemblers), but most finished swords were sold to the Office of Ordnance. Stone's Hounslow sword mill was now in direct competition with the cutlers of London for sword contracts with the Office of Ordnance.

Besides purchasing rough sword blades from the Hounslow German bladesmiths, Stone bought blades from Birmingham bladesmiths and imported blades from Solingen bladesmiths. He did not forge any blades at his mill. The German bladesmiths of Hounslow did not like to share their blade-making techniques with the English. The guilds in Solingen had harsh penalties for any bladesmith and his family who allowed German blade-making techniques to be shown to foreigners, including jail and death.

Around 1630, the king, in desperate need of military swords, granted John Kirke, an armourer and sword maker with the Greenwich Royal Armoury, a patent to make sword and rapier blades.

In 1630, when William Bals (Balls), Stone's previous master, died, Stone petitioned the Cutlers Company for Bals' mark (a bunch of grapes). The court delayed its approval until he paid a 10 shilling fine that he owed the Company. Stone obtained the mark in 1631.

In 1631, the royal cutler Robert South was advanced 100 pounds by the Cutlers Company of London to investigate the making of sword and rapier blades "for the good of the company and the kingdom."

On July 18, 1631, Stone delivered 4,356 swords (at 6 shillings each) with Irish (basket) hilts to the Office of Ordnance. They were to be used to arm the force raised by the Marquis of Hamilton to assist the king in his intervention in Sweden during the Thirty Years War.

In 1636, Stone advertised as follows:

> *I have perfected the art of blade making and my factory at Hounslow Heath could produce blades as good and cheap as any to be found in the Christian world. The price being fortified by the long experience and quality on the part of my German blade makers.*

On July 1, 1636, Stone petitioned the king to have the Office of Ordnance purchase the 2,000 blades he had ready. He needed prompt payment because he was deeply in dept. He also needed a letter of protection so the bailiffs wouldn't arrest him for nonpayment of his debts.

Attorney General Bankes and Solicitor General Littleton confirmed his claims and that the 2,000 blades were ready. They indicated that Stone, had been, at great charge, making sword, rapier, "skein" (knife), and other blades for the king's stores and for the service of his subjects, which for the most part had, up to that time, been made in foreign countries. They indicated that Stone could produce 3,000 more blades by Michaelmas (September 29).

On July 9, 1636, Stone received a royal patent from Charles I for making swords and blades. The patent read as follows:

> *A special priviledge granted to Benjamin Stone, sword blade maker, and his assignees, for the term of 14 years next ensuing, (starting at Michaelmas) within England, Ireland, and Wales, to make and work all manner of sword blades, fauchions* (falchions) *skeynes* (skeines, or knives) *rapyer* (rapier) *blades and blasts* (blade poles) *serving for rests of muskets, of any fashion or kind whatsoever, according to a way or invention, by him devised, by the help of mill or mills and the same to sell at moderate rates, of diver* [diverse] *form & fashion, paying therefore XL's* [40 shillings] *yearly to the crown, amount during the said term, with the ordinary proviso for making this grant void in case it shall be found to be contrary to law and inconvient to the state.*

The sword blades mentioned in Stone's patent were mostly backswords (single-edged broadswords) with one wide fuller in the middle of the blade. The German bladesmiths engraved their names in the fuller, and many blades had the word "Hounslow" or one of its many variations (Hounslo, Hunslo, Hounsloe, Honslo, Hunsloe, Housslo, Houn) engraved in the fuller.

The "fauchions" mentioned were falchions—single-edged short swords—such as infantry hangers, naval cutlass', hunting swords, and pioneer swords, some with sawtooth blades. The "skeynes" were skenes (skeines), daggers, dirks, and knives. The "rapyer" (rapier) blades were long, narrow, straight blades with a wide fuller in the middle, mostly fitted with swept or cup hilts.

On December 12, 1636, Stone was made an official supplier of the Office of Ordnance. He now called himself "His majesties blade maker for the Office of Ordnance."

In early 1637, Stone petitioned the Privy Council to complain that the Office of Ordnance had just ordered 4,000 swords from the Cutlers Company. Stone said that their swords were old, of poor quality, and used foreign (German) blades. He asked that only English blades be accepted by the Office of Ordnance and indicated that he could turn out 500 blades a week and had large quantities ready to deliver. He also said that he had spent 6,000 pounds in perfecting the art of blade making.

On April 14, 1637, Stone delivered 500 new swords at 75 shillings, 6 deniers each to the Board of Ordanance for naval use. Later in 1637, Stone complained to the Privy Council that the swords supplied by the Cutlers Company to Capt. William Leggel, Master of the Tower Armoury, had not been proofed (inspected) by the Lieutenant of Ordnance before storing them in the tower.

At the end of 1637, Stone complained again to the Privy Council that the swords supplied by the Cutlers Company had still not been proofed and that he had 3,000 blades already mounted but had not received payment.

In 1638, Stone petitioned to the Council of War that he had spent 8,000 pounds on facilities to manufacture blades and was now delivering 1,000 good and serviceable swords to the Office of Ordnance monthly. He asked for a monopoly of sword supply to the Office of Ordnance over the Cutlers Company. Secretary of State Edward Nicholas did not agree, but he indicated that all good and serviceable swords would be taken from Stone.

Also in 1638, Stone again petitioned the Council of War about his 8,000 pound expense and the monopoly of sword supply he wanted for himself. He also wanted a warrant to prevent the cutlers of London from striking Spanish, German, and other foreign blade makers' marks on their blades.

In 1638, Stone delivered to the Office of Ordnance 5,000 sword and rapier blades. He subcontracted them to three other blade makers: the Nicholas Brothers, John Harvey, and John Hayes (probably from London). His subcontracted blades were to be delivered to Capt. William Legge, Master of the Tower Armoury, under Stone's name. Stone was contracted to hilt and make scabbards for 3,000 of the 5,000 sword blades. Robert South (royal cutler) and William Cave were each contracted to hilt and make scabbards for 1,000 of the 5,000 sword blades.

On July 6, 1639, Stone bought 200 swords at 75 shillings, 6 denier each from George Page, a London merchant, for resale to the Office of Ordnance.

In February 1640, the Cutlers Company of London informed its membership as follows:

> *There is to be provided for his majesties service, within one month, to the number of 3000 swords, ready trimmed* [assembled] *to be brought to the cutlers hall by several workman of this company. Every workman is to bring in those swords that he maketh (not trimmed with Irish* [basket] *hilts nor black blades* [iron] *blades) and every blade to be above nine handfull long.*

On March 14, 1640 Stone delivered 870 new swords at 75 shillings, 6 denier each to the Board of Ordnance. The debenture books of the Office of Ordnance showed that they had hired a London cutler named John Damm to proof Stone's blades delivered to the Tower Armoury. Payments were made to Damm on May 12, 1639, and August 10, 1640.

On June 30, 1642, Stone contracted with the Board of Ordnance to refurbish scabbards and make new scabbards for 400 swords for naval use. It was his last contract before the English Civil War of 1642–1651.

On May 19, 1643, during the English Civil War, Stone delivered some refurbished swords with scabbards to the royal stores at Oxford, the king's headquarters during the war.

In 1642, when parliamentary forces took over the blade and sword center at Hounslow, Stone must have followed King Charles I to his Oxford headquarters. He probably was the one who set up the king's sword mill at Wolvergate near Oxford. A blade mill was also set up at Glouster Hall in Oxford.

Stone was still in business in 1647 because on October 26, 1647, one of his apprentices, Thomas Hunt, became a freeman of the Cutlers Company of London. Since a term of at least seven years was required, Hunt's apprenticeship would have begun at Hounslow in 1640–1642 and continued at Oxford from 1642–1647.

When the parliamentary forces took over the Hounslow sword and blade center in 1642, they confiscated the mills of the German bladesmiths obedient to the king. The only bladesmiths to remain, obedient to the parliamentary forces, were the Germans Johann Kindt (Kinndt), Ceile Herder, and Johannes Dell (Bell) and Englishman Henry Risby. The other German bladesmiths followed King Charles I to his new headquarters at Oxford, where they would have worked at the blade mill at Glouster Hall, Oxford, or the sword mill at Wolvercote, near Oxford. They were Peter Munsten (English) the Younger, Caspar Fleisch, Clemens Horn the Younger, Johannes Hoppe (Hoppie) the Younger, Heinrich (Henry) Hoppe (Hoppie) the Elder, Johannes Meigen, Clemens (Clames) Meigen, and Caspar Karn (Carnis).

In 1643, Prince Rupert, a royalist commander, ordered 15,000 sword blades from Robert Porter of Birmingham. The parliamentary forces destroyed Porter's factory in retaliation.

In 1645, the wardens of the Cutlers Company of London provided 3,000 infantry swords with belts and 200 horseman's swords (all at 5 shillings each) to the parliamentary force's "New Model" army under Gen. Sir Thomas Fairfax.

In May 1649, some Hounslow bladesmiths petitioned the Council of State (Commonwealth under Oliver Cromwell) against unfair taxes on their residences and work houses (mills). Their tools and stock had been confiscated for nonpayment of taxes. The Council of State had the tools and stock returned and reduced their future taxes.

In 1650, Paul and Everard Ernions (Ernious) had a sword blade mill at Hounslow. In that year, an order was written to deliver to them ten trees out of Windsor Forest. The timber was to be used to repair their work houses at Hounslow.

On January 31, 1655, John Cooke of London petitioned the Council of State to encourage his manufacture of hollow ground (concave) small sword and rapier blades at Hounslow. In 1658, he supplied 600 Hounslow-style hangers to the Office of Ordnance for naval use (cutlasses), so he must have had a sword mill at Hounslow. He was located in London c. 1658 and used the cross and star mark in 1670.

William Walker had a sword mill at Hounslow. On April 1, 1653, Walker provided 1,000 new swords "with strange Irish [basket] hilts and large for ye hand, well joyned over ye shoulder of ye blade" (probably imported blades) to the Commonwealth under Cromwell.

On June 3, 1658, Walker supplied 200 Hounslow hangers with scabbards, 215 English hangers, and 171 Dutch (German) hangers, all at 45 shillings, 6 denier each, for sea service to the Commonwealth under Cromwell.

In 1659, Walker supplied 1,410 hangers for land service and 1,509 hangers for sea service (cutlasses) to the Commonwealth under Cromwell. In 1660, he supplied 1,000 hangers for sea service to the Board of Ordnance under King Charles II.

In 1660, John Gale (Hounslow 1636–1642, Oxford 1642–1659, Hounslow 1659–1670) petitioned King Charles II to be given the office of postmaster in the stage at Hounslow as compensation for the suffering he and his family endured for their loyalty to the king's father, Charles I. Gale described himself as the mill man to Charles I. He supposedly had worked at Hounslow and, when the parliamentary forces took over in 1642, was imprisoned, his house plundered, and his family turned out of doors. Eventually (c. 1645), he got himself to Oxford to work for King Charles I at his sword blade mill at Glouster Hall. Capt. William Legge, Master of the Tower Armoury in 1660, called Gale a sword blade maker who had worked at Oxford.

In 1672, Heinrich (Henry) Hoppe (Hoppie) the Elder and Peter Munster (English) the Younger, last survivors of the German bladesmiths at Hounslow, petitioned King Charles II for a charter to reestablish their craft and set up a factory (mill) in Hounslow to manufacture small swords with hollow ground (concave) blades. Hoppe and English were supposedly the last of the German swordsmiths who were brought to England by King Charles I in 1629. They were forced to follow the king to Oxford when parliamentary forces took over the Hounslow area with the start of the English Civil War in 1642. No charter was issued. Their 1672 petition reads as follows:

*Statement of Henry Hoppe & Peter English sword makers to the King* [King Charles II]; *that in 1629 they were brought over to England by William Heyden and the late King* [Charles I] *and set up their manufacturies at Hounslow; that in the wars they followed his majesty to Oxford, for which Cromwell took their mills* [at Hounslow Heath] *from them and converted them into powder mills; that they* [Hoppie and English] *only remain who know the Art* [making hollow ground blades] *and foreign* [Solingen, Prussia] *workmen are hard to obtain, as they are obliged to swear, on leaving the trade* [in Solingen, Prussia] *not to discover it* [show others the secret of making hollow ground blades] *on the pain of death; that his majesty ordered the late Colonel Legge* [Master of Armoury] *to see them* [Hoppe and English] *provided for, which he doubtless would have done had he lived; and that his majesty desire of setting up the said manufacture in England* [hollow ground blade mill] *may be performed by the instructions* [set up and operated by] *of the said Hoppie and English, if they receive his majesty's encouragement* [charter].

On March 19, 1674, Henry Hoppe the Elder and Peter

English talked to the masters of the Cutlers Company of London, again trying to establish a hollow ground sword blade factory in Hounslow. Nothing happened.

By 1675, production at Hounslow had slowed considerably. The John Ogilby map of 1675 (entitled *Itinerarium Anglias*) shows a sword mill on the Crane River at Hounslow located on the south side of the Staines Road, south of the Baber Bridge (not Benjamin Stone's old mill).

By 1685, there was very little sword production at Hounslow. The German swordsmiths had left for London or the new Shotley Bridge sword and blade making center, as follows:

- Peter Munsten (English) the Younger was in London, c. 1646–1675.
- Johann Konigs (Connyne) was in London, 1642–1661.
- Johnannes Bell (Dell) was in Shotley Bridge, 1685–1690.
- Heinrich Hoppe (Hoppie) the Elder was in London, 1646–1675.
- Heinrich Hoppe (Hoppie) the Younger was in London, 1646–1685.
- Joseph Hoppe (Hoppie) was in London, 1646–1685.
- Peter Henekels was at Shotley Bridge, 1685–1702.
(See Chapter 4 for more information on Shotley Bridge and the Hollow Sword Blade Company.)

## GERMAN BLADESMITHS WORKING AT HOUNSLOW

**Peter Munsten the Younger**
- At Hounslow 1629–1642.
- Immigrated from Solingen, Prussia, in 1629.
- Operated a sword blade mill.
- Signed his blades PETER MUNSTEN ME FECIT.
- Changed his name to Peter English.
- Followed King Charles I to Oxford in 1642 when parliamentary forces took over Hounslow.
- Moved to London 1649–c. 1675.
- His mill was converted to a powder mill by parliamentary forces.

**Johann Kindt (Kinndt)**
- At Hounslow 1629–1659.
- Immigrated from Solingen, Prussia, in 1629.
- Operated a sword blade mill.
- Signed his blades JOHAN KINNDT HOUNSLO 1630 and JONNES KINNDT FECIT HOSLO.
- Became a naturalized citizen.
- Changed his name to John Kennet.
- Stayed in Hounslow when parliamentary forces took over Hounslow in 1642.
- Because he provided swords to the parliamentary forces, he may have taken over Benjamin Stone's sword mill.
- In 1642, in a letter to the Parliamentary Office of State, Sir Walter Waller and Sir Arthur Hasselrig, parliamentary army commanders, asked for 200 swords from Kennet.

**Caspar Karn (Carnis)**
- At Hounslow 1629–1642.
- Immigrated to London from Solingen, Prussia, in 1621.
- Moved to Hounslow, c. 1629.
- Operated a sword blade mill.
- Signed his blades CASPAR CARNIS.
- Followed King Charles I to Oxford when parliamentary force took over Hounslow in 1642.
- His mill was converted to a powder mill by parlimentary forces.

**Johann Konigs (Connyne)**
- At Hounslow 1629–1642.
- Immigrated to London from Berg and Solingen, Prussia, in 1607.
- Worked in London, 1607–1629.
- Changed his name to John Connyne.
- Moved to Hounslow in 1629.
- Operated a sword blade mill.
- With the start of the English Civil War in 1642, moved back to London.

**Johannes Bell (Dell)**
- At Hounslow 1649–1688.
- Immigrated to London from Solingen, Prussia, in 1635.
- Moved to Hounslow in 1649.
- Operated a sword blade mill.
- Signed his blades JOHANNES DELL ME FECIT.
- Moved to Shotley Bridge in 1685.

**Caspar Fleisch**
- At Hounslow 1629–1642.
- Immigrated to Hounslow from Solingen, Prussia, in 1629.
- Operated a sword blade mill.
- Signed his blades CASPAR FLEISCH ME FECIT.
- Followed King Charles I to Oxford when parliamentary forces took over Hounslow in 1642.

**Clemens Horn the Younger**
- At Hounslow 1629–1642.
- Immigrated to Hounslow from Solingen, Prussia, in 1629.
- Followed King Charles I to Oxford when parliamentary forces took over Hounslow in 1642.

**Johannes Hoppe (Hoppie, Hoppy, Hopper)**
- At Hounslow 1634–1642.
- Immigrated to Greenwich from Solingen, Prussia, in 1629 and worked at the royal armouries there.
- Moved to Hounslow in 1634.
- Operated a sword blade mill.
- Signed his blades JONNES HOPPIE ME FECIT and ION HOPPIE ME FECIT HONSLO.
- Followed King Charles I to Oxford when parliamentary forces took over Hounslow in 1642.
- Moved to London in 1649 and died in 1662. (Listed in Chapter 9 as John Hoppy.)

**Heinrich Hoppe (Hoppie, Hoppy, Hopper) the Elder**
- At Hounslow 1629–1642.
- Immigrated to Hounslow from Solingen, Prussia, in 1629.
- Followed King Charles I to Oxford when parliamentary forces took over Hounslow in 1642.
- In c. 1649, moved to London with his sons Joseph and Henry the Younger and stayed until c. 1675.
- In London, listed as Henry Hoppe.

**Joseph Hoppe (Hoppie, Hoppy, Hopper)**
- At Hounslow 1629–1642.
- Son of Heinrich Hoppe the Elder.
- Born at Hounslow c. 1629.
- Followed King Charles I to Oxford when parliamentary forces took over Hounslow in 1642.
- In c. 1649, moved to London with his father and stayed until c. 1685.
- Master of the Cutlers Company in 1649.
- In London, listed as Joseph Hopper.

**Heinrich Hoppe (Hoppie, Hoppy, Hopper) the Younger**
- Son of Heinrich Hoppe the Elder.
- Born in Oxford c. 1642.
- In c. 1649, moved to London with his father and stayed until c. 1685.
- In 1685–1702, moved to Shotley Bridge.
- In Shotley Bridge, listed as Henry Hoppe.

**Ceile Herder**
- At Hounslow 1649–1659.
- Immigrated to Hounslow from Solingen, Prussia, in 1649.
- Worked at Hounslow during the parliamentary control period.
- Changed name to Cecil Furber.

**Johannes Meigen**
- At Hounslow 1629–1642.
- Immigrated to Hounslow from Solingen, Prussia, in 1629.
- Followed King Charles I to Oxford when parliamentary forces took over Hounslow in 1642.

**Clemens (Clemas) Meigen**
- At Hounslow 1629–1642.
- Immigrated to Hounslow from Solingen, Prussia, in 1629.
- Operated a sword blade mill.
- Signed his blades CLAMAS MEIGEN.
- Followed King Charles I to Oxford when parliamentary forces took over Hounslow in 1642.

**Peter Henckels (Henkell)**
- At Hounslow 1660–1685.
- Immigrated to Hounslow from Solingen, Prussia, c. 1660.
- In 1685, moved to Shotley Bridge as Peter Henkell as resided there until 1702.
- Called Peter Henkell at Shotley Bridge.

## ENGLISH SWORD AND BLADE MAKERS WORKING AT HOUNSLOW

**Joseph Jenks**
- Located in London, Parish of St. Anne Blackfriars, B1599–1625 (see London listing in Chapter 9).
- Son of John Jenks (a German).
- Moved to Hammersmith County of Middlesex (three miles from Blackfriars, London), 1625–1629.
- Moved to Isleworth (Hounslow Heath), 1629–1641.
- Operated a sword blade mill.
- Signed his blades JENCKES ME FECIT HOUNSLOE.
- Immigrated to America in 1641, settling in Lynn, Massachusets, in 1642.

**Richard (Recardes) Hopkins**
- At Hounslow 1655–1659.
- Operated a sword blade mill.
- Signed his blades RECARDUS HOPKINS FECIT HOUNSLO.
- Moved to London, 1659–c. 1675.
- Master of the Cutlers Company in 1671.

**Henry Risby**
- Located in London, c. 1612–1633.
- At Hounslow 1633–1649.
- Operated a sword blade mill.
- Signed his blades HENRY RISBY.

**Benjamin Stone**
- At Hounslow 1629–1642.
  (See detailed discussion above.)

**Paul and Everard Ernions (Ernious)**
- At Hounslow 1650.
- Operated a sword blade mill.

**John Cooke**
- At Hounslow 1655–1658.
- Operated a sword blade mill and sword mill.

**John Gale**
- At Hounslow 1630–1644.
- Imprisoned by parliamentary forces in 1642 when they took over Hounslow.
- His family was turned out of their home.
- Moved to Oxford c. 1645 and worked as mill man for King Charles I.

**William Walker**
- At Hounslow 1649–1660
- Operated a sword mill
- Supplied many swords to parliamentary forces and to the Board of Ordnance in 1660.

## SWORD MILL OWNERS AT HOUNSLOW

| | |
|---|---|
| Benjamin Store | 1629–1642 |
| Johamn Kindt | 1642–1659 (Stone's mill) |
| John Cooke | 1655–1659 |
| William Walker | 1649–1660 |

## SWORD BLADE MILL OWNERS AT HOUNSLOW

| | |
|---|---|
| Peter Munsten the Younger | 1629-1642 |
| Johann Kindt | 1629-1642 |
| Caspar Karn (Carnis) | 1629-1642 |
| Johamnes Dell (Bell) | 1649-1685 |
| Caspur Fleisch | 1629-1642 |
| Johamnes Hoppe | 1634-1642 |
| Clemens Meigen | 1629-1640 |
| Joseph Jenks | 1629-1641 |
| Richard Hopkins | 1655-1659 |
| Henry Risby | 1633-1649 |
| Paul and Everard Ernions | 1650 |

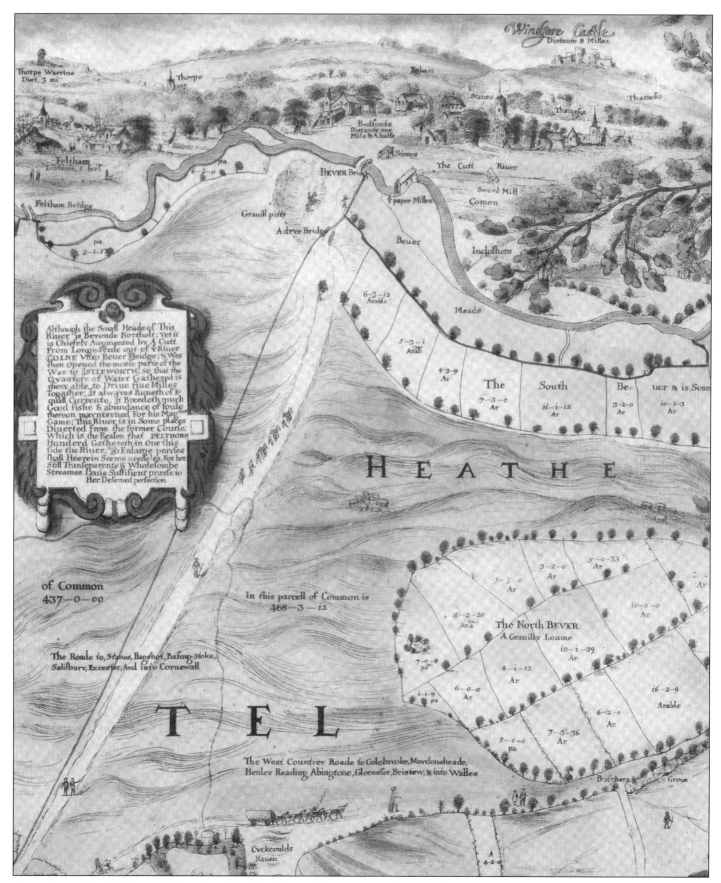

A portion of the 1635 map of Hounslow Heath by Moses Glover showing the Hounslow sword mill on the New Cutt River at top right. (Courtesy of Neil Robert Grigg)

# Chapter 4

## The Sword and Blade Making Center of Shotley Bridge (The Hollow Sword Blade Company)

round the mid 17th century, the small sword became very popular in Europe, replacing the rapier, a long, straight-bladed sword with a large hilt to balance the weapon. The small sword was a much shorter and lighter weapon, making it very deadly for dueling. It was carried by military officers, but it also became the most popular civilian weapon of the time. It had a simple hilt with a knuckle bow, two finger rings (*Pas d'ane*), and a disc-shaped counterguard. The most popular version featured a straight blade that was triangular in cross section. Each of the three sides were concave, i.e., hollowed out, creating a rigid but light blade. In England, the hollowing out of the blade was done by hand, a laborious job for the bladesmith. In Solingen, Prussia, however, a method was perfected to hollow out the three sides of the blade with small grinding wheels.

John Sanford (Sandford) of New Castle, John Parsons of London, Peter Justice of London, and John (Johannes) Bell (Dell), a Hounslow sword maker, noted the growing demand for small swords in England and the lack of English production. In 1685, they formed a syndicate to start a small sword manufactory in the Derwent River valley at Shotley Bridge that would use German production methods. With the help of King Charles II, the syndicate persuaded three German sword and blade makers from Hounslow—John (Johannes) Bell (Dell), Peter Henkels (Henckels-Henkell), and Heinrich (Henry) Hoppe (Hoppie) the Younger—to come to Shotley Bridge in 1685. (At that time, sword and blade production at Hounslow was coming to an end.)

Shotley Bridge was situated in the counties of Northumberland and Durham, near Consett, 2 miles southwest of Newcastle upon Tyne (River), England's second largest port (Newcastle was a large coal-producing area), and 12 1/2 miles northwest of Durham. The mills were located on the Derwent River, which separates the counties of Northumberland and Durham. Forges, mills, and workshops were built on the Northumberland side; workmen homes and stables were built on the Durham side. The rapidly moving river provided water power for the mill wheels, which ran the grindstones, buffing wheels, trip hammers, and other equipment. The Derwent River flowed into the Tyne and provided transportation of supplies from Newcastle upon Tyne to Shotley Bridge and finished swords and blades back to Newcastle upon Tyne. Sales warehouses were set up in London and later Newcastle upon Tyne.

Shotley Bridge had large deposits of iron ore, which the syndicate mined at Robinhood's Bay and transported to their forges. Later, several well-known iron founders and steel forgers located near Shotley Bridge provided the syndicate with raw materials for blade production. The Vinting family and later the Bertram family (John and William) built smelting mills and steel forges near Shotley Bridge. The Bertram family's operation, called Blackhall Mill, was located three miles downstream from the Shotley Bridge mills. There was plenty of wood in nearby forests to make charcoal for the syndicate's furnaces, and the Derwent River bed provided a rich source of grindstone grit for their grindstones.

In 1686, James II was raising an army to maintain himself as a Catholic monarch ruling over a Protestant England, and he needed a large supply of blades and swords. In early 1687, the syndicate, needing to increase production to meet this demand, brought nineteen blade makers and their families, led by Clemens Hoheman, to Shotley Bridge from Solingen, Prussia. Since their was a heavy duty on imported swords and blades to discourage excessive importation (only a limited number was allowed), the syndicate, by bringing in German sword and bladesmiths, increased English production and made larger profits. A six-year contract between the syndicate and the blade makers was signed and everything was built for the operation—homes for the smiths, hammer mills with trip hammers, blade forges with bellows, a sword blade mill with grinding and buffing wheels, workshops and warehouses, and tools and instruments. The Derwent River turned the waterwheels, which operated the grinding and polishing wheels and trip hammers. When the nineteen German families arrived in Shotley Bridge, sword blade forger Adam Ohlig and sword blade grinder and polisher Herman Mohll became the leaders of the group. A local sword cutler from New Castle, Thomas Carnforth, helped them set up their forges and mills. Mohll set up the sword blade mill where all blades were finished.

In late 1687 (possibly early 1688), the syndicate petitioned King James II for a patent granting them exclusive right to manufacture hollow sword blades in England. The petition stated that they had brought foreign workmen to England and proposed to make use of a mill unlike any other in the king's realm. It was a German-type mill that used small grinding wheels to hollow out triangular blades for small swords. The petition was referred by the Privy Council to the officers of the Board of Ordnance. Lord Dartmouth, Master of Ordnance, was shown a hollow sword blade made by an English bladesmith, cutler and gunsmith Francis Troulett of London, made in the old way (hollows made by hand) and one made by a German bladesmith in the new way (hollows made with grindstones). Lord Dartmouth referred the petition to John Hawgood, Master of the Cutlers Company, for his observations. Because of a claim for payment by Francis Troulett and growing unrest in the country (it was the eve of the Revolution of 1688), the grant of the patent was delayed. It was finally allowed, but the actual patent certificate was never issued because King James II didn't sign it—he was busy fighting to maintain his monarchy. Meanwhile, in early 1688, the syndicate signed a lease for Shotley Bridge land with William Johnson, owner of the Shotley Bridge Estate. (His son, Henry Johnson, later became a sword blade maker.)

On September 26, 1688, the court of Solingen, Prussia, issued a court order condemning Clemens Hoheman and the sword and bladesmiths who had left for England. Beginning around 1400, Solingen sword makers were required to swear an oath not to leave Solingen, set up business elsewhere, or divulge the highly prized techniques and equipment used by Solingen bladesmiths and *Schiefers* (grinders and polishers). The order read:

> We, Wilhelm Wassman, judge of the Court of Solingen, Mathias Wundes, Wilhelm Dinger, Wilhelm Vass [Voes] Johann Ganssland, Peter Voess [Voes], and the entire court of jurors of the town and parish of Solingen, have become aware of the fact that about a year ago [1687] Clemens Hoheman enticed away several craftsman, who had long been established and connected with this area, to the kingdom of England, and furthermore incited several more to depart, and as the infamy has become well known and as this merits the severest punishment, Clemens Hoheman is accused of being a seducer, deserving the severest punishment along with all the other people involved.

The court order went on to list the following individuals:

- Herman Mohll (Moll)
- Abraham Mohll (Moll)
- Johannes Clauberg, Clemen's son from Widdart
- Clemen Knechtgen (Knecht)
- Peter Theegarden (Tiergarden)
- (John) Johannes Voes (Voss)
- Johannes Vurkett (Vurkelt)
- Adolph Kratz
- Johannes Wupper of Feld
- Heinrich Wupper, Theis' son
- Johannes Wupper from Hesson
- Arnd (Arndt) Wupper from Hesson
- Heinrich Keuler (Keuller)(Koller)
- Adam Ohlig, Adam's son
- Johannes Hartkopt (Hartcop)
- Engel (Angell) Schimmelbusch
- Peter Kayser, Peter's son
- William Schaffe (not listed but did leave)
- Clemans Schaffe (not listed but did leave)

For some reason, Abraham and Heinrich Mohll were not listed as members of any sword or blade makers guild in Solingen.

All of the named men had summons delivered to their doors (four times) demanding that they present themselves before the court within six weeks or provide good reason in person or by a legal representative. Failure to present themselves meant the court would proceed against them in their absence, meaning seizure of all their property and assets or death.

The Hollow Sword Blade Company at Shotley Bridge began production right away, even though a royal patent had not been officially issued and signed by King James II. Between August 23 and August 28, 1690, the syndicate advertised in the *London Gazette*:

> *Where as great industry hath been used in erecting a manufacture for hollow sword blades at Newcastle* [Shotley Bridge] *by several able workmen brought from Germany, which now being brought to prefection, the undertakers thereof have thought fit to settle* [set up] *a warehouse at Mr Isaac Hadley's, at the* [sign of the] *Five Bells; New Street, near Shoe Lane* [in London] *whereas callers can be furnished with all sorts of sword blades at reasonable prices.*

On September 1, 1690, the syndicate (now with new members) petitioned the new king, William III, for a charter (not a patent for exclusive rights) to manufacture hollow sword blades in England. The members of the syndicate were Syndicate Governor Sir Stephen Evance (Evans), Syndicate Deputy Governor Robert Peter Reneau, First Assistant John Sanford (Sanford), First Assistant Peter Justice, Abraham Dashwood, and Thomas Evance (Evans). Sir Stephen Evance (knighted in 1690) was a London goldsmith, banker, merchant, and army contractor as well as commissioner of excise, with offices on Lombard Street. He also had a share in a privateer (ship) and interests in a diving invention and a mining project. Robert Peter Reneau was a London merchant who, it is believed, was the hands-on manager of the Shotley Bridge manufactory. The petition read:

> *They had been at very great charge* [expense] *and trouble in bringing from beyond the sea 19 or 20 families, in keeping them above* [housed] *these 2 years, building several mills and forges for making hollow sword blades in ye north of England. They pray for a charter of incorporation for the new mill during 14 years.*

The royal charter was granted on September 15, 1691, by King William III, and the syndicate officially became the Hollow Sword Blade Company. Issued to Sir Stephen Evance for making sword blades in England, its preamble read:

> *Our said subjects, at their great charge and management, have imported from foreign parts, divers* [diverse] *persons, who have exercised in their own country the said art of making hollow sword blades by the use of certain newly invented instruments, engines* [equipment] *and mills and by the contrivance of our said subjects, have prevailed upon them to expose themselves, to the hazard of their lives* [death penalty] *to impart to our said subjects their art and mystery* [hollow blade making techniques].

On March 28, 1691, Herman Mohll, Engel Schimmelbusch, Oliffe Groats, Adolph Kratz, and Johannes Voes rented a cottage with a courtyard close to the Shotley Bridge mills on the Derwent River and set up a grinding, polishing, and finishing mill. In the same year, Adam Ohlig built a cutler's hall at Shotley Bridge. In 1694, John Sanford, first assistant of the syndicate, leased a corn mill at Lintzford, paying a monetary rent plus one newly made and tempered sword blade.

The Hollow Sword Blade Company, although founded to produce small sword blades, in reality produced very few of them. The Revolution of 1688 and the next 20 years of war in Ireland, Scotland, Germany, and Spain resulted in a large demand for military swords. As a result, Shotley Bridge produced large quantities of standard blades for broadswords, rapiers, hangers, and bayonets. Many blades had "SHOTLEY" etched on one side and "BRIDG" on the other. Some had a bridge mark and others a crossed sword mark.

From 1690 to 1703, the syndicate advertised batches of sword blades for sale. A *London Gazette* advertisement running from July 10 to July 13, 1699, read:

> *The hollow sword blade company will put to the candle* [auction] *at Cutlers Hall, Clock Lane* [London] *what sword blades it has finished. They may be seen at the company's warehouse on New Street, near Fetter Lane* [in London] *3 days next before the sale.*

Shotley Bridge production seems to have dropped considerably after the war with France ended with the Treaty of Ryswick in 1697. Production may have stopped and the manufactory facilities closed in 1702. Several of the German sword and bladesmiths had died by that time, and there may have been a disagreement between them and the company over the price paid for their blades. Herman Mohll went back to Solingen with his family from December 1702 to December 1703.

The company made a new six-year agreement with the German bladesmiths (presumably with new prices for their blades) on April 27, 1703. The agreement was signed by Henry Woper (Wopper, Heinrich Wupper), John Woper (Wopper, Johannes Wupper), Peter Tiergarden (Theegarden), Adam Ohlig, and William Schafe (Schaffe), and production began again. The new agreement required the following:

- The German workmen were to make and finish good and sufficient sword blades for the Hollow Sword Blade Company.
- The sword blades were to be delivered by them to the company warehouse at Newcastle upon Tyne.
- The blades were to be in the size and type stated on the back of the agreement at the quoted price. Thirty-seven varieties of blades were called for, including rapier, cutlass, scimitar, hanger, and bayonet blades. The size of the blades and the number of hollows (fullers) on the blades was specified, and the top price was to be 1 pound, 10 shillings per dozen blades.
- The workmen were to preserve, maintain, and keep all of the company tools, machines, utensils, and instruments for making and finishing sword blades.

- The waterwheels, mills, and building were to be kept in repair.
- After six years, all of the above was to be returned to the company.
- Sword blades were to be sold only to the Hollow Sword Blade Company. If any were sold to others, all the workmen involved were to pay a 100 pound fine.
- All workmen were allowed to use the hammer mills owned by the company.

At this time, the Hollow Sword Blade Company began to buy blades from other sources besides the German bladesmiths at Shotley Bridge, including from Solingen bladesmiths. In December 1703, the company advertised in the *London Gazette*:

> *The Hollow Sword Blade company has lately received a considerable quantity of sword blades made at their mills at Shotley Bridge near Newcastle upon Tyne. They are now on sale at their warehouse in New St. near Fetter Lane.*

In December 1703, Herman Mohll returned from Solingen, arriving at Newcastle upon Tyne on the Dutch ship *St. Anne* from Rotterdam, Holland. Knowing that the Shotley Bridge production was low, Mohll brought with him some unfinished (not ground and polished) blades from Solingen to sell to the Hollow Sword Blade Company or anyone interested. He offered to sell some of these blades to Newcastle sword cutler Thomas Carnforth, who bought twenty dozen.

Mohll was determined not to pay the import duty prescribed by the government on foreign blades, so he tried to smuggle the blades into England in casks with false bottoms and casks filled with other duty-free commodities. After the *St. Anne* was anchored at the North Shields harbor, he hailed a group of watermen (transporters) and off-loaded 45 bundles of sword blades in their barges, instructing the men to take the bundles to a safe place until the next tide, when a member of the ship's company (Mohll) would move them up the Tyne to Gateshead near Shotley Bridge. The bundles were taken to the house of Thomas Davidson, one of the watermen. Some must have been off-loaded at the South Sheilds harbor because a bundle of hanger blades and a bundle of hollow blades were later found in the mud there.

A total of 64 dozen blades were smuggled into Newcastle by Mohll. Someone, however, informed the Custom House authorities, and when Mohll came to pick up the 45 bundles of blades at Davidson's house, he was arrested. Since he had no bail money, he was taken to Morpeth Gaol Prison and remained there for one month. He admitted that the sword blades were his, obtained in Solingen to sell in England, and they were to be taken to Robert Peter Reneau, deputy governor of the Hollow Sword Blade Company at Shotley Bridge. Mohll was accused of smuggling and, because there were many Scottish and Irish soldiers on the ship, he was also accused of attempting to sell sword blades to the Jacobites, a treasonable offense. Henry Woper (Heinrich Wupper), a sword blade maker at Shotley Bridge, and Thomas Carnforth, a sword cutler in New Castle, testified on Mohll's behalf. Mohll was released in January 1704 upon payment of fines by Robert Peter Reneau, and charges were withdrawn by direction of the English government. Mohll then set up his own sword blade mill, perhaps with craftsmen he brought back from Solingen.

In 1704, the Hollow Sword Blade Company decided to invest money in another enterprise. It spent 20,000 pounds (a huge amount in those days) on the purchase of Irish estates forfeited under attainders. In the same year, the company held a general court at its offices on Birchin Lane, London, for the purpose of raising new capital for the purchases. Everything was lost, however, when the Irish parliament, seeing how cheaply the estates were being sold and fearing the company would become too powerful in Ireland, refused to allow the Hollow Sword Blade Company to the purchase the estates.

Later in 1704, after its real estate speculation in Ireland failed, the company charter was obtained by a group of bankers headed by Sir George Caswell, a sheriff of the city of London, and Jacob Sawbridge. Both men were directors of the Southsea Company (another royal charter). The company name was changed to the Sword Blade Bank.

In the same year, William Cotesworth of Gateshead (near Shotley Bridge) purchased the sword blade making complex at Shotley Bridge from the Sword Blade Bank, including the forges, workshops, and hammer mills. The contract with the German bladesmiths now belonged to him. Cotesworth was a salt, dye, tallow, and corn merchant in Battle Bank, Gateshead. His partner was Henry Sutton. Cotesworth bought iron bar from Daniel (Den) Heyford of Roamley, Pontefract, to sell to the German bladesmiths, and he in turn sold sword blades to Heyford.

Between 1704 and 1709, Herman Mohll sold sword blades to London merchant and iron dealer Sir Ambrose Crawley, the owner of the Windaton Iron Works, located 15 miles from Shotley Bridge. He sold Crawley 37 varieties of sword blades varying in price from 7 to 14 shillings per dozen. Crawley had offices on Upper Thames Street, London.

A John Beardmore (probably the Shotley Bridge manager) wrote Cotesworth in January 1705 stating that swordsmith. Clemens Schaffe was very old and may not be able to carry out his work.

In February 1710, the Sword Blade Bank purchased sword blades from a John Saunthorp. Henry Benson, a company official in London, complained that the Saunthorp blades were soft and badly tempered.

In April 1710, after the six-year agreement ended, a three-year agreement was made between William Cotesworth and the German smiths. It was signed by Cotesworth and the 1703 agreement signees. The agreement called for sword blades to be purchased from the German smiths at 6 pence per dozen (lower than the earlier agreement).

Between November 30, 1710, and August 21, 1712, Coatesworth purchased 1,600 dozen sword blades from the German smiths (19,200 blades in 557 days, or 34 a day). The cost was 935 pounds, 13 shillings. Most of the blades were sent to the Hollow Sword Blade Company warehouses in London, which in turn sent some blades to Glascow for sale.

From 1711 to 1713, many of the German smiths were in financial trouble and going into debt, including Adam Oley (Ohlig), Henry Wopper (Wupper), John Wopper (Wupper) Zu Feld, John Wopper (Wupper) Zu Hesson, Peter Tiergarden (Theegarden), (Johannes) John Voose (Voes), (Johannes) John Hartcop (Hartcopt), William Voose (Voes), Abraham Mohll, Herman Mohll, John Mohll, Clemens Schaffe, and William Schaffe. Many owed money to their iron supplier, Daniel (Dan) Heyford. They complained often to William Cotesworth that he was slow in paying, which prevented them from paying for their iron supplies.

In 1715, Herman Mohll wrote William Cotesworth, asking him to prevent Daniel (Den) Heyford from taking the property of some of those who were indebted to him. Coatesworth was able to prevent this from happening.

Herman Mohll died on December 16, 1716, and left his sword blade mill to his son, William Mohll.

In 1721, sword grinder Johannes (John) Voes died. He was believed to be the manager of trade relations between Herman Mohll and the blade makers of his homeland in Solingen, Prussia. His will, signed by his wife, Jan Voes, and witnessed by William Buske and John Wopper (Wupper) Jr., stipulated that his estate in Germany, called *Auffenhewman*, in the county of Dusseldorf, was to be sold by his brother-in-law Johannes Smithart of Solingen.

In 1724, after the Sword Blade Bank went into liquidation, a new petition to renew the company charter was put forth before the House of Commons by Samuel Swinson, Henry Trollope, Thomas Beech, Loftus Brightwell, and Henry Symonds. The charter was not renewed.

On May 16, 1724, William Mohll the Elder (son of Herman Mohll) advertised his sword blade mill for sale in the *New Castle Courant*. Later that year, Robert Oley (Ohlig), son of Adam Oley (Ohlig), purchased William Mohll's sword blade mill and house. His brother Richard Oley (Ohlig) later became a partner. Robert and Richard Oley owned the mill until c. 1760.

William Cotesworth died in 1726, and the ownership of the Shotley Bridge works (forges, hammer mills, and workshops) was passed to his sons-in-law, Henry Carr and Harvey Elliss. In 1731, they sold the works to a London business for 200 pounds (probably to John Leaton).

Between c. 1726 and 1750, an Englishman named John Leaton, son of Thomas Leaton, had a sword mill at Shotley Bridge. He was a landowner and sheepherder. Swords stamped "JOHN LEATON" are known to exist.

From c. 1750 to 1764, an E. Leaton Blenkinson of Newcastle owned a sword mill at Shotley Bridge. He was a relation of John Leaton.

Between c. 1731 and 1750, Henry Johnson, son of William Johnson, had a sword mill at Shotley Bridge. William Johnson was the original owner of the Shotley Bridge Estate leased to the Hollow Sword Blade Company in 1688.

In 1754, Swedish engineer A.A. Angerstein visited Shotley Bridge while researching steel production techniques. He visited the sword mill of E. Leaton Blenkinson and found that there were only eight workmen at the mill and that the mill only bought four tons of steel from John Bertram's Blackhall steel mill that year.

Between c. 1760–1810, William Oley (Ohlig the Elder) and brother Nicholas Oley (Ohlig) the Elder operated a sword blade mill at Shotley Bridge. They were sons of Richard Oley (Ohlig). (Robert and Richard Oley had the mill from 1724 to 1760.) In 1787, they built a cutlers hall near their mill.

Sword etcher Thomas Bewich worked for brothers William Oley the Elder and Nicholas Oley the Elder. William Oley the Elder died in 1810; Nicholas Oley the Elder must have died too, since the estate included their sword blade mill as well as workshops, tenement houses, warehouses, a butcher shop, tools, and bellows. The sword blade mill was left to William Oley's sons—William Oley the Younger, Nicholas Oley the Younger, and Christopher Oley—who ran it from 1810 to c. 1830. From c. 1830 to 1845, a Joseph Oley ran the Oley sword mill at Shotley Bridge. A Nicholas Oley—called the last descendant of the Shotley Bridge Oley family—died in 1964.

Most of the Shotley Bridge sword mills had closed by 1840, and the German smiths had moved to Sheffield and Birmingham.

# GERMAN SWORDSMITHS AND BLADESMITHS (AND THEIR DESCENDANTS) WHO IMMIGRATED TO SHOTLEY BRIDGE IN 1687 FROM SOLINGEN, PRUSSIA

| Name | Known Dates in Shotley Bridge |
|---|---|
| **Johannes Clauberg (Cloberg)** | 1687 |
| • Son of Clemens Clauberg of Solingen, Prussia. | |
| **Johannes Hartkopt (Hartcop)** | 1687–1713 |
| • Later changed name to John Hartcop. | |
| **Peter Kayser (Keisser) the Younger** | 1687 |
| • Son of Peter Kayser the Elder of Solingen, Prussia. | |
| **Heinrich Keuller (Koller)** | 1687–1724 |
| • Later changed name to Henry Clewer. | |
| **Clemens Knetchen (Knecht) Zu Widdert the Younger** | 1687 |
| **Adolph Kratz** | 1687 |
| **Abraham Mohll (Moll)** | 1687–1711 |
| • Born c. 1640, in Solingen, Prussia. | |
| • Blade grinder (*Schliefer*). | |
| **Herman Mohll (Moll)** | 1687–D1716 |
| • Born 1662, in Solingen, Prussia. | |
| • Son of blade grinder Abraham Mohll. | |
| • Changed name to Herman Moll. | |
| • A leader of the German smiths and partner with Adam Ohlig. | |
| • Owned his own sword blade mill, c. 1704. | |
| **John Moll** | B1692–1718 |
| • Son of Herman Moll. | |
| **William Moll the Elder** | Bc1690–c. 1740 |
| • Son of Herman Moll. | |
| • Inherited Herman Moll's mill in 1716. | |
| • Sold his mill to Robert Ohlig in 1724. | |
| **John Moll the Elder** | Bc1710, Dc1770 |
| • Son of William Moll the Elder. | |
| **William Moll the Younger** | Bc1740, Dc1810 |
| • Son of John Moll the Elder. | |
| **John Moll the Younger** | Bc1780, Dc1846 |
| • Son of William Moll the Younger. | |
| • Moved to Birmingham in 1832 and eventually changed name to Mole. | |
| • Started the famous Mole sword making company. | |
| **Adam Ohlig (Oley) the Younger** | 1687–D1726 |
| • Born c. 1670, in Solingen, Prussia. | |
| • Blade forger. | |
| • Son of Adam Ohlig the Elder of Solingen, Prussia. | |
| • A leader of the German smiths and partner with Herman Mohll. | |
| **Adam Ohlig (Oley) III** | B1697–1715 |
| • Son of Adam Ohlig the Younger. | |
| **Nicholas Ohlig (Oley)** | B1698–1715 |
| • Son of Adam Ohlig the Younger. | |
| **Wilhelm (William) Ohlig (Oley)** | B1703–1715 |
| • Son of Adam Ohlig the Younger. | |
| **Richard Ohlig (Oley)** | B1716–1760 |
| • Son of Adam Ohlig the Younger. | |
| **Robert Ohlig (Oley) the Elder** | Bc1690–1760 |
| • Son of Adam Ohlig the Younger. | |
| • Bought the mill of William Mohll the Elder in 1724. | |

**Wilhelm (William) Ohlig (Oley) the Elder**  B1736–D1810.
- Son of Richard Ohlig.
- Inherited the sword mill of Robert Ohlig and Richard Ohlig c. 1760.
- Left his sword mill to sons William the Younger, Nicholas, and Christopher Ohlig.

**Nicholas Ohlig (Oley) the Elder**  Bc1730–1808
- Son of Richard Ohlig.
- Inherited sword mill of Robert Ohlig and Richard Ohlig c. 1760.

**William Ohlig (Oley)**  B1739–1808
- Son of Robert Ohlig (Oley) the Elder.

**Robert Ohlig (Oley) the Younger**  Bc1769, Dc1807
- Son of Nicholas Ohlig the Elder.
- Married in 1799.

**William Ohlig (Oley) the Younger**  Bc1756, 1810–1830
- Son of William Ohlig the Elder.
- Inherited his father's sword mill in 1810.

**Nicholas Ohlig (Oley) the Younger**  Bc1762, 1810 1830
- Son of William Ohlig the Elder.
- Inherited father's sword mill in 1810.

**Christopher Ohlig (Oley)**  Bc1762, 1810–1830
- Son of William Ohlig the Elder.
- Inherited his father's sword mill in 1810.

**Joseph Oley**  B1806–D1896
- Ran the Oley sword mill, c. 1830–1845.

**(Nicol) Nicholas Oley**  Bc1854–D1964
- Son of Joseph Oley.
- Lived to a very old age.

**Clemens Schaffe (Schafe)**  1687–c. 1711
**William Schaffe (Schafe)**  1690–c. 1711
- Son of Clemens Schaffe.

**Engel (Angell) Schimmelbusch**  1687–D1694
- Blade grinder.

**Peter Tiergarden (Theegarden)**  1687–D1714
**Johannes Voes (Voss-Vooz) the Elder**  1687–D1721
- Blade grinder.

**Johannes (John) Voes (Voss-Vooz) the Younger**  c. 1687–D1721
- Son of Johannes Voes the Elder.
- Later changed name to John Vaws (Faws).
- Married in 1700.

**William Voes (Vaws)**  1716–1730
- Son of Johnannes Voes the Younger.

**Johannes Vurkett (Vukett)**  1687–1713
**Johannes Wupper zu Feld**  1687–c. 1713
- Later changed name to John Wopper (Woper).

**Heinrich Wupper (Woppwe-Woper)**  1687–1725
- Son of Theis Wupper of Solingen, Prussia.
- Later changed name to Henry Wopper (Woper).

**Arndt (Arnt) Wupper zu Hesson**  1687–c. 1713
**Johannes Wupper zu Hesson the Younger**  1687–c1711
- Son of Johannes Wupper zu Hesson of Solingen, Prussia.

**Johannes (John) Wupper III**  B1692–1740
- Son of Johannes Wupper zu Hesson the Younger.

## OTHER GERMAN SWORDSMITHS AND BLADESMITHS WORKING AT SHOTLEY BRIDGE

**Johannes (John) Bell (Dell)**                                  1685–1810
- Came from Hounslow.

**Peter Henkels (Henkell, Henckels)**                            1685–c. 1720
- Came from Hounslow.

**William Henkels (Henkell, Henekels)**                          1707–1740
- Married Ann Voes in 1727.

**Heinrich Hoppie the Younger**                                  1685–c. 1720.
- Son of Heinrich Hoppie the Elder of Solingen, Prussia.
- Came from Hounslow.

**William Balfe**                                                1703–1711
- Worked for Johannes Wupper zu Hesson the Younger.

**John Himofan**                                                 1703–1711
- Worked for Johannes Wupper zu Hesson the Younger.

**Oliffe Groats**                                                1691–1715
- Blade grinder.

**Johannes Wolferts**                                            1704–1711
**William Palds**                                                1703–1709
**John Hindson**                                                 1711
**William Busche**                                               1710–1727
- Witness to John Voes the Elder's will in 1721.
- Possibly a descendant of A. Schimmelbusch.

**Thomas Beckwith**                                              1760–1780
- Engraver.
- Did work for William Ohlig the Elder and Nicholas Ohlig the Elder.

**John Bertram**                                                 1687–D1719
- Owner of Blackhall Mill (steel).
- Located on the Derwent River three miles downstream from Shotley Bridge.
- Made blister steel and later shear steel in 1714.
- Made steel for German bladesmiths at Shotley Bridge.

**William Bertram**                                              1719–1760
- Son of John Bertram.
- Made steel for German bladesmiths at Shotley Bridge.
- Had a blast furnace, smelt mill, and steel forge.
- Only four tons of steel made for the Shotley Bridge German bladesmiths in 1754; production declining.

**John Vinting (Vinting)**                                       1660–1690
- Had an early smelting mill in the area.

**William Vinting (Vintnig)**                                    B1685–1705–1735
- Son of John Vinting.
- Became an associate of John Bertram.

## ENGLISH SWORD MAKERS AT SHOTLEY BRIDGE WHO OPERATED SWORD MILLS

**John Leaton**                                                  c. 1724–1754
- A land owner and sheep herder.
- Son of Thomas Leaton.

**E. Leaton Blenkinsop**                                         1754
**Henry Johnson**                                                c. 1724–1755
- Son of William Johnson, owner of the Shotley Bridge Estate leased to the Shotley Bridge syndicate in 1688.

# Chapter 5

## German Edged Weapon Makers & Retailers Who Had Offices or Agents in London

| GERMAN FIRM | AGENT IN LONDON | ADDRESS & DATES IN BUSINESS |
| --- | --- | --- |
| J. (Johan) E. Bleckman<br>Solingen, Prussia, sword maker | A.G. Franklin & Co. 14 | South St., Finlay,<br>1858–1870 |
| Herman Boker & Co.<br>Solingen, Prussia, sword maker | Joseph Louis Merfield | 7 Savage Garden,<br>1869–1891 |
| Alexander Coppel & Co.<br>Solingen, Prussia, sword maker | S. Hecht & Co. | 19 Carter Lane, 1868–1870<br>20 Noble St., 1870–1877<br>14 Hamsell St., 1877–1878 |
| A. (August) & E. (Emil) Holler<br>Solingen, Prussia, sword makers | Lucas Platzholf & Co. | 3 Church Passage,<br>Guildhall, 1865–1868<br>30 Monkwell St.,<br>Cripplegate, 1868–1869 |
| F. (Frederick) W. (Wilhelm) Holler<br>Solingen, Prussia, sword maker | William Meyerstein & Co. | 6 Love Lane at Blood St.,<br>1870–1888<br>5 London Wall Ave.,<br>1888–1895 |
| S. (Samuel) Hoppe & Co. | Friedrich E.D. Hast | 34 Jewin St.,<br>Cripplegate, 1861–1862 |
| F. (Friedrich) Horster (the Younger)<br>Solingen, Prussia, sword maker | Schmitt & David | 102 Leaden Hill St.,<br>1862–1863 |

| | | |
|---|---|---|
| Carl Klonne & Co.<br>Suhl, Germany | S. Oppenheim & Son | 4 Bread St.,<br>Cheapside, 1859–1863 |
| H. (Herman) & W. (William) Lang<br>Solingen, Prussia, sword dealer | Rochussen & Co. | 13 St. Thomas Apostle,<br>1857–1858<br>75 Cannon St. (west end),<br>13 St. Thomas Apostle,<br>1858–1861<br>114 Fenchurch St., 1861–1863 |
| Heintzman & Rochussen | Second agent for Lang. | 9 Friday St., 1863–1867<br>23 Abchurch Lane, 1867–1869 |
| P. (Paul) D. Luneschloss<br>Solingen, Prussia, sword maker | H. Hartjen & Co. | 4 Falcon St., Aldergate,<br>1883–1897<br>1 & 2 Falcon St., Aldergate,<br>1897–1903<br>35 & 37 Noble St., 1903–1913<br>67 New Compton St., 1913–1914 |
| Schnitzer (August & Albert)<br>& Kirschbaum (Wilhelm B.S.)<br>Solingen, Prussia, sword maker | Own offices | 29 Walbrook St., 1858–1859<br>2 Walbrook Building, 1859–1864 |
| Rudolph Kirschbaum<br>Solingen, Prussia, sword maker | Joined Wilkinson & Son; name changed to Wilkinson Sword Co. Ltd. | 27 Pall Mall, 1890–1901<br>Oakley Works, 27 Pall Mall,<br>1901–1909<br>Oakley Works, 53 Pall Mall,<br>1909–1915 |
| Clemen & Jung<br>Solingen, Prussia, sword maker | Nathaniel Mills & Co.<br>Own offices | 25 Mary Ann St., 1865–1867<br>1867–1879 |
| A. (August) & A. (Albert) Schnitzler<br>Successors to Schnitzler<br>Kirschbaum | Frederick Blyth<br>Own offices | 72 Bath St., 1864–1870<br>1870–1873 |

# Chapter 6

## English Cutlers Who Were Appointed Sword Cutler to the Board of Ordnance

| Name | Dates of Appointment |
|---:|---|
| Stephen Archibald | 1644–1661 |
| Joseph Audley | 1662–1678 |
| Benjamin Banbury | 1681–1683 |
| Thomas Bigglestone | 1699–1716 |
| Richard Chapman | 1714–1733 |
| Samuel Chase | 1692–1712 |
| Thomas Cox | 1716–1730 |
| Peter English Jr. | 1689–1697 |
| John Hawgood | 1693–1707 |
| Thomas Hawgood | 1685–1715 |
| John Hill | 1688–1702 |
| Philip Mist | 1704–1711 |
| Michael Richardson | 1683–1690 |
| Guy Stone | 1715–1719 |
| John Woodcraft | 1693–1720 |
| Thomas Hollier | 1692–1720 |

# Chapter 7

## English Sword Blade Markings

**ROYAL CYPHERS**

George I, II, III
I     1714–1727
II    1727–1760
III   1760–1820

George IV
1820–1830

William IV
1830–1837

Victoria
1837–1901

Edward VII
1901–1910

George V
1910–1936

Edward VIII
1936

George VI
1936–1952

Elizabeth II
1952–2003

# ROYAL CROWNS

Georgian Pattern  Tudor  St. Edward's

## GOVERNMENT BLADE MARKS

| | |
|---|---|
| I.S.D. | Indian Stores Depot (London acceptance mark) |
| ♛/9 | Crown over a number (government inspector's mark) |
| ♛/S/9 | Crown over a letter and a number |
| |     Letter E – Enfield Armoury |
| |     Letter S – Solingen, Prussia |
| |     Letter W – Wilkinson Sword Co. |
| |     Letter B – Birmingham, England |
| 2/48 | English fraction    company or unit # / I.D. # of weapon |
| RA | Royal Artillery up to 1899 |
| RHA | Royal Horse Artillery after 1899 |
| RFA | Royal Field Artillery after 1899 |
| RGA | Royal Garrison Artillery after 1899 |
| R.N. | Royal Navy |
| H.M.S. | His or Her Majesty's ship |
| R.E. | Royal Engineers |
| D | Dragoons |
| L | Lancers |
| H | Hussars |
| LG | Life Guards |
| G.G. | Grenadier Guards |
| S.G. | Scots Guards |
| I.G. | Irish Guards |
| Y | Yeomanry |
| R.A.M.C. | Royal Army Medical Corps |
| W.G. | Welsh Guards |

## GOVERNMENT OWNERSHIP MARKS

| | |
|---|---|
| Crown over broad arrow | Board of Generals, 1750–1796 |
| BO under broad arrow | Board of Ordnance, 1796–1855 |
| WD under broad arrow | War Department, 1855–1998 |

# Chapter 8
## The Cutlers Company of London

The Cutlers Company of London was first chartered by King Henry V in 1416. The charter gave it the right to govern, discipline, and regulate its members and elect a master and two wardens for two-year tours of duty. (All masters are listed in Chapter 9.) Prior to this, there had been a Cutlers Society in 1154 and a House of Cutlers in 1285. In 1365, all sword and knife makers in London were required by law of King Edward III to put their mark on all of their goods. Each mark was registered by the House of Cutlers.

The Cutlers Company was made up of several crafts, or companies:

- Company of bladesmiths, which included makers of sword blades and knife blades.
- Company of hafters, which included hilt designers and makers (goldsmiths, silversmiths, and enamelers) and iron and brass hilt makers.
- Company of sheathers, which included makers of sword and knife leather scabbards.
- Company of cutlers, which assembled swords and knives after purchasing component parts (blades, hilts, grips, etc.).

A person wanting to join a company was required to serve as an apprentice to a member of the company for at least seven years. He then could be a journeymen for at least one year under company supervision before becoming a freeman (allowed to sell his wares in London).

In 1515, the company of bladesmiths who were part of the Cutlers Company of London became part of the Armourers Company of London.

In 1517, the bladesmiths move from the Cutlers to the Armourers Company was annulled by the crown. Individual bladesmiths were allowed to join either company.

In 1624, the crown contracted with the Cutlers Company of London to provide 4,000 swords a month. The blades were to be imported from Solingen, Prussia, and hilted and assembled by London cutlers.

In 1645, the Cutlers Company provided 3,000 infantry swords and belts and 200 horseman's swords and belts (at 5 schillings for each set) to the parlimentary forces (the "New Model" army) under Gen. Sir Thomas Fairfax. It used the dagger mark on foreign blades hilted by London cutlers along with the foreign cutlers mark. In 1664, only freeman cutlers of London could strike this mark. In 1682, there was a petition against Birmingham cutlers who were using the dagger mark.

# Chapter 9

# English Sword Makers, Cutlers, Dealers, and Craftsmen Who Mounted Swords

Early sword cutlers in England were actually sword assemblers. They bought component parts (blades, hilts, grips, sheaths, etc.), assembled complete swords, then sold the swords to individuals, military units, government ordnance departments, and merchants.

Although early English armourers and blacksmiths made blades, and military blades were made at Hounslow Heath and Shotley Bridge (as described in detail in Chapters 3 and 4), relatively few blades were made in England. Most were imported from foreign makers in Solingen, Prussia (the majority); Toledo, Spain; Milan, Italy; Paris, France; and Klingenthal, Alsase. Later, many blades were made in Birmingham.

Sword cutler/assemblers had to use the services of many special trades (they either employed them or contracted with them), including:

- Blade makers
- Hafters (hilt makers)
- Handle (grip) binders (wrapped handles with wire over leather or fish skin, or made a continuous wire wrap over wood handles)
- Handle (grip) makers
- Sheath (leather) makers
- Scabbard (metal) makers
- Hilt, blade, and scabbard decorators, including gold enlayers, fire gilders, enamelers, engravers, etchers, bluers, goldsmiths, silversmiths, and jewelers.

Of course, goldsmiths, silversmiths, and jewelers also assembled swords. Like the sword cutler/assembler, they bought component parts, then usually made the hilts; decorated the hilts, blades, and scabbards; and assembled the swords. Most artisans who did this were organized into combined companies, i.e, they were goldsmiths, silversmiths, and jewelers in one company.

By the middle of the 18th century, the term "cutler" also began to be applied to merchants and retailers who sold swords as a sideline. They bought swords from English assemblers and makers as well as swords mounted by goldsmiths and silversmiths, or they imported them. These companies included:

- Army and/or navy outfitters
- Arms warehouses

- Gold and silver lacemen (epaulettes, badges, etc.)
- Embroiderers (on uniforms)
- Uniform suppliers
- Tailors
- Clothiers
- Hatters (haberdashers, hat sellers)
- Hat, cap, and helmet makers
- Linen and woolen (cloth) drapers (makers)
- Linen, woolen (cloth), and silk mercers (dealers)
- Glovers
- Hosiers
- Accoutrement makers (belts, sashes, baldrics, etc.)
- Metal ornament makers
- Button makers (metal, pearl, etc.)
- Buckle makers

Sword makers and cutlers (both assemblers and retailers) sold swords to a variety of customers, including:

- Individual civilians
- Soldiers and sailors
- Military officers
- Military units (army or cavalry regiments; before 1788, regimental colonels purchased swords and equipment for their men)
- Naval ship captains and commanders
- Army and navy ordnance departments
- The king (for military needs)
- Royalty (for military units they commanded and for their own use)
- Royal chartered companies (i.e., English merchants who established commercial settlements in foreign countries to sell swords, guns, and military equipment to the people and royalty of those countries, such as the East India Company, which was given a royal charter in 1600 by Queen Elizabeth I and whose first settlement was established in Surat, India, in 1612.

These companies also purchased swords, guns, and military equipment to defend their settlements.)

Before 1788, the commanding officer (colonel) of English regiments purchased swords, equipment, and uniforms for his unit. The government gave him an allowance (the same for all regiments) for that purpose. In 1788, a board of ordnance was set up by King George III to establish the patterns for all English swords, including light and heavy cavalry swords. Cavalry officers were required to carry the same pattern as the troopers. The board also set up a proofing system to test swords before any contract could be issued to the sword maker. Many British cavalry swords had been failing in the field (broken blades, loose hilts, etc.), especially swords used by units sent to America during the Revolutionary War (1775–1783). The Treasury Department was required to levy import taxes on foreign swords to equalize prices so regiments could choose their own suppliers without worrying which ones had the cheapest prices.

• • • • •

The following listings, divided by city, cover all the individuals or companies that had anything to do with swords in England. All occupations and/or specialties are shown in the left margin. All of the masters of the Cutlers Company of London and all royal cutlers are indicated. Birth and death dates should be 100 percent accurate. Although I do not always use the term "circa," all other dates must be considered educated and well-researched approximations based on my summing up of all available information.

## LONDON

| | | | |
|---|---|---|---|
| Sword cutler | **James Abraham** | | 1707 |
| Hatter<br>Hosier<br>Sword cutler | **John Adams Sr.**<br>• Adams advertised as follows:<br>*John Adams*<br>*48 Fleet St.*<br>*Near Serjeants Inn*<br>*(At the sign of the hat and crossed daggers)* | 48 Fleet St. near Serjeants Inn | 1750–1758 |
| | **Adams (John) & Parr (Charles)** | 48 Fleet St. near Serjeants Inn | 1759–1766 |
| | **Adams (John) & Parr (Charles)** | 146 Fleet St.<br>at the corner of Wine Office Court | 1767 |
| | **John Adams**<br>(see Charles Parr) | 146 Fleet St.<br>at the corner of Wine Office Court | 1768–1780 |
| | **John M. Adams**<br>(see John Griffen) | | |
| Steel hilt maker | **Adams (Robert & James) Brothers**<br>• Robert Adams (hilt designer), B1728–D1792.<br>• James Adams, B1740–D1794.<br>• Made steel hilts with urn-shaped pommel caps and beaded knuckle bows (some mounted on small swords) | Durham Yard<br>Parish of St. James | 1760–1794 |
| Military tailor<br>Sword cutler | **W. (William) H. Adency & Son**<br>**Adency (William H.) & Boutroy (Henry)**<br>• Successor to Henry Boutroy.<br>(see Henry Boutroy) | 16 Sackville St.<br>16 Sackville St. | 1932–1939<br>1940–1965 |
| Bladesmith | **William Albon** | | 1376 |
| Cutler | **Thomas Alcroft**<br>• Master of the Cutlers Company, 1735. | | 1725–1740 |
| Goldsmith<br>Jeweler<br>Watchmaker<br>Hilt maker<br>Sword cutler | **John G. Alderhead**<br>• Had apprentices Thomas Ayres (1766–1773), William Ayres (1761–1768), John Ayres (1768–1775). | 114 Bishopsgate St.<br>near South Sea House<br>at the sign of the Ring and Pearl | 1742–D1795 |
| Silversmith<br>Goldsmith<br>Hilt maker<br>Sword cutler | **Charles Aldridge**<br>• Apprenticed to uncle Edward Aldridge Sr., 1758–1765.<br>**Charles Aldridge**<br>**Aldridge (Charles) & Green (Henry)**<br>• Plate workers.<br>**Aldridge (Charles) & Green (Henry)**<br>**C. (Charles) Aldridge**<br>• Mounted swords.<br>• Plate worker.<br>• Sold presentation swords that were mounted by Ray & Montaque.<br>(see Ray & Montaque)<br>**Charles Aldridge** | <br><br><br>Aldergate<br><br>62 St. Martins Le Grand<br>18 Aldergate St.<br><br><br><br><br>Falcon St., corner of Aldergate St. | B1743–1765<br><br>1766–1772<br>1773–1774<br><br>1775–1784<br>1785–1789<br><br><br><br><br>1790–1803 |

ENGLISH SWORD MAKERS, CUTLERS, DEALERS, AND CRAFTSMEN WHO MOUNTED SWORDS

| | | | |
|---|---|---|---|
| Goldsmith | Edward Aldridge (Sr.) | St. Leonards Court, Foster Lane | 1724–1738 |
| Silversmith | Edward Aldridge (Sr.) | Lilley Pot Lane, at the sign of the Golden Ewer | 1739–1746 |
| Hilt maker | | | |
| Sword Cutler | Edward Aldridge (Sr.) | Foster Lane, at the sign of the Golden Ewer | 1747–1752 |
| | Edward Aldridge (Sr.) & John Stamper | Foster Lane, at the sign of the Golden Ewer | 1753–1757 |
| | Edward Aldridge (Sr. & Jr.) | Foster Lane, at the sign of the Golden Ewer | 1758–1767 |
| | • Edward Aldridge Sr. died in 1767. | | |

| | | | |
|---|---|---|---|
| Goldsmith | Edward Aldridge (Jr.) | | Bc1737–1758 |
| Silversmith | • Apprenticed to Starling Wilford, 1751. | | |
| Hilt maker | • Apprenticed to Edward Aldridge (Sr.), 1751–1758 (see Edward Aldridge Sr.). | | |
| Sword cutler | • Nephew of Edward Aldridge (Sr.). | | |
| | Edward Aldridge (Sr. & Jr.) | Foster Lane, at the sign of the Golden Ewer | 1758–1767 |
| | • Nephew Charles Aldridge apprenticed to Edward Aldridge (Sr.), 1758–1765. | | |
| | • Edward Aldridge (Sr.) died in 1767. | | |
| | Edward Aldridge (Jr.) | George St., Parish of St. Martin's Le Grand | 1767–1774 |
| | Edward Aldridge (Jr.) & John Henry Vere | George St., Parish of St. Martin's Le Grand | 1775–1780 |
| | Edward Aldridge (Jr.) | Bishop Gate | 1781–D1801 |

| | | | |
|---|---|---|---|
| Silversmith | Thomas Allen | Charing Cross | 1630–D1708 |
| Hilt maker | • Supplied swords to the Board of Ordnance, 1698–1706. | A few doors from Mermaid Court Parish of St. Martin in the Fields, Westminster | |
| Sword cutler | • Sold silver-hilted small swords with hollow ground blades. | at the sign of the Angel & Crown | |

| | | | |
|---|---|---|---|
| Tailor | Allen (Thomas) & Collier (William) | 18 Bond St. | 1796–1799 |
| Woolen draper | Thomas Allen | 18 Bond St. | 1800–1814 |
| Sword cutler | Allen & Wilson | 18 Bond St. | 1815–1832 |
| | • Tailors to the Prince of Wales. | | |
| | • Succeeded by Wilson & Willman (see Wilson & Willman). | | |

| | | | |
|---|---|---|---|
| Sword cutler | Allen & Hanberry | | 1915–1920 |
| | • Sword cutlers to the army. | | |

| | | | |
|---|---|---|---|
| Sword cutler | John Allott | Parish of St. Mary Woolnoth | 1680–1700 |
| | • Supplied swords to Board of Ordnance, 1693–1695. | | |

| | | | |
|---|---|---|---|
| Army | George & William Almond | 1 Old Bond St. | 1843–1851 |
| accoutrement | George & William Almond | 14 St. James St. | 1852–1853 |
| maker | George & William Almond | 72 St. Martins Lane & 14 St. James St. | 1854–1855 |
| Sword cutler | George & William Almond | 1112 St. Martins Lane & 14 St. James St. | 1856 |
| | George & William Almond | 14 St. James St. & 37 Dean St, Soho | 1857–1858 |
| | George & William Almond | 14 St. James St. & 67 Willow Walk, Bermondsey & 112 St. Martins Lane | 1859–1862 |
| | George & William Almond | 14 St. James St. & 67 Willow Walk, Bermondsey | 1863–1874 |
| | George & William Almond | 67 Willow Walk, Bermondsey | 1875–1904 |
| | Almond (George & William) & Smith (John) | 67 Willow Walk, Bermondsey | 1905–1914 |

| | | | |
|---|---|---|---|
| Gold and silver | Amery (John), Pitter (John) & | corner of Bedford St., Covent Garden | 1777–1799 |

| Trade | Name / Notes | Address | Dates |
|---|---|---|---|
| lacemen<br>Sword cutler | **Slipper (Henry)**<br>• Successors to Brydges & Walker (see Brydges & Walker).<br>**Amery (John) & Pitter (John)**<br>• Succeeded by J. Pitter (see J. Pitter). | at the sign of the 3 Crowns<br><br>26 Bedford St., Covent Garden | <br><br>1800–1804 |
| Gun, pistol,<br>and sword dealer | **William Amies**<br>• Wholesale gun, pistol, and sword warehouse. | 5 Nags Head Ct.,<br>Gracechurch St. | 1811–1839 |
| Cutler | **William Anderson**<br>• Master of the Cutlers Company, 1866–1867. | | 1851–1872 |
| Sword cutler | **Anderson & Son** | St. James St. | 1910 |
| Cutler | **William John Andrews**<br>• Master of the Cutlers Company, 1749. | | 1739–1754 |
| Goldsmith<br>Jeweler<br>Watch maker<br>Hilt maker<br>Sword cutler | **Thomas Andrews**<br>• Located three doors from the Royal Exchange.<br>**Elizabeth Andrews**<br>• Widow of Thomas Andrews.<br>• Located three doors from the Royal Exchange. | 85 Cornhill St.<br><br>85 Cornhill St. | 1780–D1789<br><br>1789–1808 |
| Hatter<br>Army cap, hat,<br>and helmet maker<br>Sword cutler | **John Andrews**<br>**John Andrews**<br>**John Andrews**<br>• Sold M1796 light and heavy cavalry officers swords, life guard officer swords,<br>and M1821 light and heavy cavalry swords.<br>**William Andrews**<br>**William Andrews**<br>• Cutler by appointment of King William IV. | 4 Brewer St., Golden Square<br>18 Pall Mall<br>9 Pall Mall<br><br><br>9 Pall Mall<br>9 Pall Mall | 1810–1818<br>1818–1820<br>1821–1825<br><br><br>1826–1835<br>1836–1868 |
| Sword cutler | **Morris Angell & Sons** | 117 Saftesbury Ave. | 1900–1910 |
| Gold and silver<br>laceman<br>Sword cutler | **C. (Charles) Annakin** | 17 Blake St., York | 1860–1900 |
| Sword maker | **Stephen Archbald**<br>• Appointed sword cutler to the Board of Ordnance, 1644.<br>• Supplied swords to the Board of Ordnance, 1644–1661.<br>• Also repaired swords for the Board of Ordnance.<br>**Alice Archbald**<br>• Widow of Stephen Archbald.<br>• Supplied swords to the Board of Ordnance, 1661. | | 1640–D1661<br><br><br><br>1661–1670 |
| Lathe and<br>tool maker<br>Sword cutler | **Samuel Armfield**<br>**Armfield (Samuel) Ltd.** | Great St., Andrews St.<br>Great St., Andrews St. | 1820–1880<br>1881–1906 |
| Army and navy<br>outfitter<br>Sword cutler | **Army & Navy Cooperative Society**<br>**Army & Navy Cooperative Society**<br>**Army & Navy Stores Ltd.** | 117 Victoria St.<br>105 Victoria St.<br>105 Victoria St. | 1873–1881<br>1890–1934<br>1935–1964 |
| Cutler | **John Arnell**<br>• Master of the Cutlers Company, 1459–1460. | | 1449–1465 |

| | | | |
|---|---|---|---|
| Sword cutler | **John Arnold** | | 1607–D1635 |
| | • Was granted "fire tongs" mark in 1606. | | |
| Linen draper (maker) Sword cutler | **Arrow, Smith & Silver** | 3 Cornhill St. | 1795–1821 |
| | • Succeeded by Silver & Co. (see Silver & Co.). | | |
| Cutler | **Isaac Ash** | | 1636–1651 |
| | • Master of the Cutlers Company, 1646. | | |
| Cutler | **Jacob Ash** | | 1646–1661 |
| | • Master of the Cutlers Compant, 1656. | | |
| Goldsmith Hilt maker Buckle maker Sword cutler | **John Ash** | 18 Plumbtree, Bloomsbury | 1815–1825 |
| | • Successor to John Yardley (see John Yardley). | | |
| | **John Ash** | 19 Plumbtree, Bloomsbury | 1826–1828 |
| | **John Ash** | 42 Princess St., Leicester Square | 1829–1833 |
| | **John Ash** | 9 Hanover St., Longacre | 1834–1853 |
| | **John Ash** | 1 Green St., Leicester Square | 1854–1859 |
| Cutler | **Henry William Ashmole** | | 1887–1902 |
| | • Master of the Cutlers Company, 1897. | | |
| Cutler | **Richard Asser** | | 1423–1438 |
| | • Master of the Cutlers Company, 1433. | | |
| Cutler | **Thomas Assher** | | 1589–1604 |
| | • Master of the Cutlers Company, 1599. | | |
| Cutler | **Thomas Asshott** | | 1589–1603 |
| | • Master of the Cutlers Company, 1598. | | |
| Sword cutler | **William Aston** | 4 Yeate's Court, Lincoln's Inn Fields | 1802–1804 |
| Sword cutler | **Anthony Astwood** | Parish of St. Andrew, Holborn | 1603–1606 |
| Gold and silver laceman Sword cutler | **John & Matthew Atherly** | 14–15 Bridgewater Gardens | 1830–1832 |
| | **John Atherly** | 14–15 Bridgewater Gardens | 1833 |
| | **John Atherly** | 15 Bridgewater | 1834–1835 |
| | **Atherly (John) & Stillwell (Edward)** | 15 Bridgewater Gardens | 1836–1837 |
| | **Atherly (John) & Stillwell (Edward)** | 3 Bridgewater Square, Barbican | 1838–1851 |
| | • Next to Stillwell's father (Edward Swift Stillwell) at 4 Bridgewater Square Barbican. | | |
| | • Succeeded by Stillwell & Co. (see Stillwell & Co.). | | |
| Goldsmith Jeweler | **John Atherly** | 71–72 Burlington Arcade | 1848–1868 |
| | • Successor to Odell & Atherly (see Odell & Atherly). | | |
| Sword cutler | **John Lewis Atherly** | 71–72 Burlington Arcade | 1869–1919 |
| | • Succeeded by Emerson Ltd. (see Emerson Ltd.). | | |
| Cutler | **Richard Atkinson** | | 1566–1581 |
| | • Master of the Cutlers Company, 1576. | | |
| Sword maker | **Joseph Audley** | Royal Exchange, Parish of St. Bartholomew | 1661–1678 |

- Supplied plug bayonets and swords to the Board of Ordnance, 1661–1678.
- Repaired swords for the Board of Ordnance 1661–1678.
- Appointed sword cutler to the Board of Ordnance, May 1662.
- Cleaned 500 short swords/bayonets for the Board of Ordnance, 1663.
- Robert Stedman and Samuel Law also worked on the 500 swords.
- In 1672, Audley made 900 bayonets for dragoon matchlock muskets to be used by the Prince Rupert Regiment of Dragoons.

| | | | |
|---|---|---|---|
| Military tailor<br>Sword cutler | **Austin & Oaker**<br>• Successor to Thompson & Son (see Francis Thompson). | 11 Conduit St. | 1881–1890 |
| Gun maker<br>Bayonet maker | **T. (Thomas) Austin**<br>• Supplied plug bayonets to the Board of Ordnance, 1703–1704. | | 1688–1700 |
| Cutler | **John Ayland**<br>• Master of the Cutlers Company, 1541. | | 1531–1545 |
| Cutler | **John Aylett** | North side of Lombard St. | 1692–1700 |
| Goldsmith<br>Jeweler<br>Hilt maker | **Thomas Ayres**<br>• Apprentice to goldsmith John G. Alderhead, 1766–1773 (see John G. Alderhead).<br>• Freeman of the Goldsmiths Company, 1726. | | 1752–1775 |
| Buckle maker<br>Hardwaremen<br>Sword cutler | **Thomas Ayres**<br>• Partners and brothers: William and John Ayres.<br>• Both William and John also apprenticed with John G. Alderhead.<br>(William in 1751–1768, John in 1768–1775). | Bishopsgate | 1776–1785 |
| | **Thomas Ayres**<br>• Partners and brothers: William and John Ayres.<br>• Had apprentice and journeyman John Thomas Bennett (see John Thomas Bennett).<br>• Made presentation small sword for Admiral Lord Keith, British Fleet Commander, 1801. | 160 Feuchurch St.<br>Corner of Lime St., Parish of St. Benets | 1786–1811 |
| | **Ayres (Thomas, William & John)**<br>**& Bennett (John Thomas)**<br>• Became hardwaremen.<br>• Made a presentation small sword for Sir Phillip Vere Broke, captain of HMS *Shannon*,<br>for capturing the USS *Chesapeake* in 1813 during the War of 1812. | 160 Fenchurch St.<br>Corner of Lime St., Parish of St. Benets | 1812–1828 |
| | **Thomas Ayres**<br>• James Morisset and Ray & Montaque mounted some presentation swords sold by<br>Thomas Ayres and Ayres and Bennett (see James Morisset, Ray & Montaque). | 160 Fenchurch St.<br>Corner of Lime St., Parish of St. Benets | 1829–D1831 |
| Sword maker | **Thomas Backe**<br>• Accused of selling defective swords to the Board of Ordnance, 1632. | | 1630–1640 |
| Sword maker | **Daniel Badcocke** | Cripplegate | 1612–1627 |
| Silversmith<br>Hilt maker<br>Sword cutler | **William Badcocke**<br>• Made silver-hilted small swords. | | 1650–1680 |
| Cutler | **John Baldwin** | | 1757 |
| Cutler | **Adam Ball**<br>• Master of the Cutlers Company, 1682. | | 1672–1697 |

| | | | |
|---|---|---|---|
| Cutler | **Anthony Ball** | | 1697–1712 |
| | • Master of the Cutlers Company, 1707. | | |
| Gold and silver | **James Ball** | 4 Great New St. | 1814–1819 |
| laceman | **James Ball** | 3 New St. | 1820–1855 |
| Accoutrement | **James Ball** | 4 Great New St. | 1856–1903 |
| maker | **James Ball & Co.** | 8 Hatton Wall | 1904–1913 |
| Sword cutler | **James Ball & Co.** | 57 Hatton Garden | 1914–1921 |
| | **James Ball & Co. Ltd.** | 72-75 Turnmill St., Clerkenwell St. | 1922–1939 |
| Iron monger (dealer) | **Samuel Ballamy** | 78 Lower Thomas St. | 1759–1773 |
| | • Iron monger. | | |
| Sword cutler | **Samuel Ballamy** | 13 Red Lion Court, Watting St. | 1774–1775 |
| | • Became sword cutler. | | |
| Goldsmith | **William Ballantine** | | 1758–1777 |
| Hilt maker | **Mary Whitford & William Ballantine** | 6 Kings Head Court | 1778–1787 |
| Buckle maker | | Parish of St. Martin's Le Grand | |
| Sword cutler | **William Ballantine** | 6 Kings Head Court | 1788–1795 |
| | (see Samuel Whitford) | Parish of St. Martin's Le Grand | |
| Sword maker | **William Bals (Balls, Balser, Balse)** | | 1600–D1630 |
| | • Immigrated from Germany, c. 1621. | | |
| | • Brought the "leopard's head" mark with him from Germany. | | |
| | • Bought the "bunch of grapes" mark, 1621. | | |
| | • Had apprentice Benjamin Stone, 1605–1613. | | |
| | • Had apprentice Lambert Williams, 1621–1628. | | |
| | • Master of the Cutlers Company, 1627–1628. | | |
| | • Benjamin Stone obtained the "bunch of grapes" mark in 1631 after Bals died. | | |
| Gun and sword | **Benjamin Banbury** | | 1647–D1683 |
| engraver | • Appointed gun engraver and sword cutler to the Board of Ordnance. | | |
| Sword maker | • Also repaired swords for the Board of Ordnance, 1681–1683. | | |
| Sword cutler | **Miles Bancks (Banks)** | | 1610–1625 |
| | • Master of the Cutlers Company, 1617–1618. | | |
| | **William Bancks (Banks)** | | 1625–1640 |
| | • Son of Miles Bancks. | | |
| Armourer | **Edward Barker** | Cole Harbour | 1630–1655 |
| Sword maker | | Parish of All Saints the Less | |
| Gunsmith | • During the Civil War, supplied arms to the "New Model" army (parliamentary forces) under General Geneva. | | |
| | • Supplied 1,000 muskets and 1,000 bandoleers, July 5, 1645. | | |
| | • Supplied 200 potts (helmets) with three bars at 7 schillings each, July 10, 1645. | | |
| | • Supplied 1,000 matchlock muskets and 1,000 bandoleers, December 22, 1645. | | |
| | • Supplied swords to the Board of Ordnance, 1650 (to the monarchy after the Civil War). | | |
| Sword cutler | **Gregory Barker** | Parish of St. John Zachary | 1584–1599 |
| Gun maker | **Harman Barne** | | 1635–D1661 |
| Silversmith | • German gun maker to Prince Rupert. | | |
| Sword cutler | • Made silver-mounted pistols. | | |
| Gun and | **Frederick Barnes** | 109 Fenchurch St. | 1825–1847 |

| Trade | Name | Address | Dates |
|---|---|---|---|
| percussion cap wholesaler<br>Accoutrement maker<br>Sword cutler | F. (Frederick) Barnes & Co.<br>(See Birmingham, 1821–1903)<br>(See Sheffield, 1839–1904) | 3 Union Row, Tower Hill &<br>109 Fenchurch St. | 1848–1904 |
| Silversmith<br>Hilt maker<br>Sword cutler | Michael Barnett<br>Elizabeth Barnett<br>• Widow of Michael Barnett. | 36 Cock Lane, Smithfield<br>36 Cock Lane, Smithfield | 1781D–1823<br>1823–1830 |
| Cutler | Thomas Barnjum<br>• Master of the Cutlers Company, 1851. | | 1841–1856 |
| Gold and silver lacemen<br>Embroiderer<br>Gold and silver wire maker<br>Wire drawer<br>Gold and silver thread maker<br>Sword cutler | Bryant Barrett<br>Bryant Barrett & Son<br>Barrett (Bryant) & Corney (William)<br>Barrett (Bryant), Corney<br>(William) & Corney (Charles)<br>• Lacemen and embroiderers to the royal family, the Prince of Wales, and the Duke of York.<br>• Sold and embroidered gold lace decorations (four orders of knighthood)<br>  on the uniform of Lord Viscount (Admiral) Horatio Nelson before he left<br>  in his ship *Victory* in pursuit of the French fleet, which ended in the Battle of Trafalgar.<br>Barrett (Bryant), Corney<br>(William) & Corney (Charles)<br>• Succeeded by Charles Corney (see Charles Corney). | corner of Craven St., Strand<br>11 Strand<br>479 Strand<br>479 Strand<br><br>70 Little Britain, Aldersgate | 1774–1779<br>1780–1791<br>1792 1802<br>1803–1824<br><br>1825–1868 |
| Sword maker | Nathaniel Barrowey (Barrowe)<br>• Repaired swords for the Board of Ordnance, 1629–1636. | | 1625–1646 |
| Sword cutler | William Barrowey (Barrowe)<br>• Supplied swords to the Board of Ordnance, 1611. | | 1610–1620 |
| Sword cutler | Bartles & Co. | Hanover St. | 1897–1905 |
| Silversmith<br>Hilt maker<br>Sword cutler | Thomas Bass<br>• Apprenticed to cutler Joseph Reason, 1700–1707 (see Joseph Reason).<br>Thomas Bass<br>• Had apprentice Isaac Stuart, 1712–1720 (see Isaac Stuart). | <br><br>Dean St., Fetter Lane | Bc1686–1707<br><br>1708–1725 |
| Silversmith<br>Buckle maker<br>Hilt maker<br>Sword cutler | John Bassingwhite<br><br>John Bassingwhite<br>John Bassingwhite | Little Russell St., Drury Lane<br>at the French Bakers, 6 Dowlings Building<br>Purple Lane<br>Russell St. | 1760–1770<br><br>1771–1772<br>1773–1775 |
| Sword cutler | Richard Bates | | 1645–1650 |
| Cutler | Richard Batson<br>• Master of the Cutlers Company, 1654. | | 1644–1659 |
| Sword cutler | William Batson | | 1910–1933 |
| Gold and silver lacemen<br>Embroiderer<br>Sword cutler | Thomas Bayes | 86 St. Martins Lane,<br>Charing Cross | 1811–1815 |

| | | | |
|---|---|---|---|
| Goldsmith<br>Silversmith<br>Hilt maker<br>Sword cutler | **Richard Bayley (Bailey)**<br>• Apprenticed to Charles Overing, 1693–1703.<br>• Apprenticed to John Gibbons, 1703–1705.<br>**Richard Bayley**<br>• Had apprentice John Rowe, 1741–1748 (see John Rowe). | Foster Lane | Bc1685–1705<br><br><br>1706–1748 |
| Pike maker | **Thomas Bayman**<br>• In 1645, during the English Civil War, Bayman contracted for 700 long pikes (16 feetlong with steel heels) for the "New Model" army (parlimentary forces) under Gen. Sir Thomas Fairfax.<br>    June 9, 1645: 100 pikes at 4 shillings, 2d each<br>    July 5, 1645: 100 pikes at 3 shillings, 10d each<br>    Dec. 22, 1645: 500 pikes at 3 shillings, 10d each | Charter House | 1642–1649 |
| Cutler | **Joseph Bayzand**<br>• Master of the Cutlers Company, 1767. | | 1757–1772 |
| Cutler | **Henry Beale**<br>• Master of the Cutlers Company, 1653. | | 1643–1658 |
| | **Henry Bean**<br>(see William Newton) | | |
| Cutler | **James Bearan**<br>• Master of the Cutlers Company, 1785. | | 1775–1790 |
| Gun maker<br>Sword cutler | **James Beattie**<br>**J. (James) Beattie & Son**<br>**J. (James) Beattie & Co.** | 205 Regent St.<br>205 Regent St.<br>205 Regent St. | 1835–1864<br>1865–1879<br>1880–1894 |
| Cutler | **Charles George Beaumont**<br>• Master of the Cutlers Company, 1917. | | 1907–1922 |
| Cutler | **Edward Beaumont**<br>• Master of the Cutlers Company, 1900. | | 1890–1905 |
| Cutler | **James Beaumont**<br>• Master of the Cutlers Company, 1876. | | 1866–1880 |
| Cutler | **John Beaumont**<br>• Master of the Cutlers Company, 1768 and 1777. | | 1758–1782 |
| Cutler | **Mac Donald Beaumont**<br>• Master of the Cutlers Company, 1921. | | 1911–1926 |
| Cutler | **William Coppard Beaumont**<br>• Master of the Cutlers Company, 1916. | | 1906–1921 |
| Hilt maker | **John Beckett (Sr.)**<br>**John Beckett (Jr.)** | New St.<br>St. James St., Parish of St. James | 1690–1760<br>1761–1780 |
| Sword and knife maker | **Thomas Beckwith**<br>• Apprenticed to knife and sword maker Richard Matthew the Elder, 1584–1589 (see Richard Matthew the Elder). | Fleet Bridge | Bc1564–1589 |

| | Thomas Beckwith | Fleet Bridge | 1589–1610 |
|---|---|---|---|
| Steel worker | **Richard Beeks** | under the Royal Exchange, Cornhill | 1700–1710 |
| Hilt maker | • Made small swords with polished steel hilts. | | |
| Sword cutler | | | |
| Sword cutler | **Beddington & Co.** | | 1775–1795 |
| Goldsmith | **John Beedall** | Newman's Row, Lincohn's Inn Fields | 1770–1774 |
| Hilt maker | **John Beedall** | 1 Orange St., Red Lion Square | 1774–D1776 |
| Buckle maker | **Samuel Beedall** | 237 High Holborn | 1776–D1779 |
| Sword cutler | **Mary Beedall** | 237 High Holborn | 1779 |
| | • Widow of Samuel Beedall. | | |
| | **Mary Beedall & William Yardley** | 23 Thorney St., Bloomsbury | 1780–1785 |
| | • Succeeded by William Yardley. | | |
| | (see William Yardley) | | |
| Sword maker | **Johannes Bell (Dell)** | Solingen, Prussia | 1615–1635 |
| | • Immigrated to London, 1635 (brought "maiden head" mark with him). | | |
| | • Changed his name to John Bell. | | |
| | **John Bell** | London | 1635–1649 |
| | • In 1639, the Cutlers Company of London fined Bell 10 schillings for making brass sword hilts (old brass, made of copper by a cementation process with calamine, which contains zinc) and charcoal. It was considered a poor and defective metal. | | |
| | • The new method of making brass by alloying zinc and copper (discovered by L. Sarat in Paris, 1627) was also considered defective. | | |
| | • King Charles I had issued a proclamation prohibiting the use of brass in making girdles, belts, hangers, or military buckles (later sword hilts) in London. | | |
| | • Ironically, brass was used at Hounslow Heath for sword hilts. | | |
| | • Brass sword hilts were prohibited in London until 1683. | | |
| | • During the Civil War (1642–1649), Bell would have made swords for the parlimentary forces, since they controlled London. | | |
| | • In 1649, after Oliver Cromwell came to power, Bell moved to Hounslow Heath. | | |
| | **John Bell** | Hounslow Heath | 1649–1685 |
| | **John Bell** | Shotley Bridge | 1685–1690 |
| | • In 1688, Bell was part of a syndicate that petitioned King James II for a patent granting them exclusive right to manufacture hollow sword blades in England. It was not granted, but a new syndicate, without John Bell, was issued a charter on September 15, 1691, by King William III. | | |
| | (see Chapters 3 and 4) | | |
| Cutler | **John Bell** | | 1771–1786 |
| | • Master of the Cutlers Company, 1781. | | |
| Silversmith | **Joseph Bell** | | Bc1734–1755 |
| Hilt maker | • Apprenticed to Richard Hawkins, 1748–1755. | | |
| Sword cutler | **Joseph Bell** | Old St., near Iron Monger Row | 1756 |
| | **Joseph Bell** | Shoe Lane | 1757–1760 |
| Cutler | **Humphrey Bellamy** | | 1697–1712 |
| | • Master of the Cutlers Company, 1707. | | |
| Cutler | **John Benet (Bennett)** | Fleet St., Parish of St. Martin Ludgate | 1458 |
| Silversmith | **John Bennett (Sr.)** | | B1709–1730 |
| Goldsmith | • Apprenticed to John Carman Sr., 1723–1730 (see John Carman Sr.). | | |

| | | | |
|---|---|---|---|
| Jeweler<br>Sword and knife hilt maker<br>Sword and knife cutler<br>Gunsmith | John Bennett (Sr.)<br><br>John Bennett (Sr.) | New St., Shoe Lane<br>near the Three Turns<br>at the sign of the Crossed Daggers<br>Thread Needle St.<br>behind the Royal Exchange | 1731–1749<br>1717–1749<br><br>1750–D1773 |

- Had apprentice Thomas Dealtry the Elder, 1742–1748 (see Thomas Dealtry the Elder).
- Had apprentice John Bennett Jr. (born c. 1736), 1750–1760 (see John Bennett Jr.).
- Partner and son: John Bennett Jr., 1761–1773.
- Master of the Cutlers Company, 1750 and 1779.
- Advertised as follows:

    *John Bennett I*
    *Gun maker and sword cutler to his Royal Highness the Prince of Wales*
    *(at the sign of the Crossed Daggers)*
    *No. 67 Threadneedle St., opposite the northgate of the*
    *Royal Exchange, London*
    *Manufacturers of all sorts of arms*
    *Retail and exportation merchants and captains supplied with*
    *any quantity on shortest notice*

| | | | |
|---|---|---|---|
| | John Bennett (Jr.) | 67 Threadneedle St., Royal Exchange | 1773–1778 |
| | Bennett (John Jr.) & Ralph (Edward) | 67 Threadneedle St., Royal Exchange | 1779–1781 |
| | John Bennett (Jr.) | 67 Threadneedle St., Royal Exchange | 1782–1783 |

- Became gunsmith also.

| | | | |
|---|---|---|---|
| | John Bennett (Jr.) | 67 Threadneedle St., Royal Exchange<br>& 24 Tokenhouse Yard | 1784 |
| | John Bennett (Jr.) | 65 & 67 Threadneedle St.,<br>Royal Exchange | 1785–1790 |
| | John Bennett (Jr.) | 67 Threadneedle St., Royal Exchange | 1791–1795 |
| | John Bennett (Jr.) | 62 & 67 Threadneedle St.,<br>Royal Exchange | 1796 |
| | John Bennett (Jr.) | 67 Threadneedle St., Royal Exchange | 1797–D1802 |
| | Elizabeth Bennett | 67 Threadneedle St., Royal Exchange | 1803–1808 |

- Widow of John Bennett Jr.

| | | | |
|---|---|---|---|
| | Bennett (Elizabeth) & Lacy (J. D.) | 67 Threadneedle St., Royal Exchange | 1809 |
| | E. (Elizabeth) Bennett & Co | 67 Threadneedle St., Royal Exchange | 1810 |

- Partner J.D. Lacy.
- Succeeded by J.D. Lacy (see J.D. Lacy)

John Thomas Bennett
(see Thomas Ayres)

| | | | |
|---|---|---|---|
| Hatter<br>Hosier<br>Helmet maker<br>Sword cutler | R.S. Bennett | 55 Hounsditch | 1799–1800 |
| Sword cutler | Henry Benson | Parish of St. Dunston in the west | 1577–1592 |
| Sword cutler | Henry Benson | | 1742 |
| | J.W. Benson<br>(see Hunt & Roskell) | | |
| Sword cutler | Bent & Frazier | | 1875 |
| Gold and silver | George Benton | 29 Clerkenwell Close | 1853–1855 |

| | | | |
|---|---|---|---|
| laceman | • Successor to William Bullmore (see William Bullmore). | | |
| Embroiderer | George Benton | 28 Clerkenwell Close | 1856–1863 |
| Sword cutler | George Benton | 63 Kings Cross Road | 1864–1868 |
| Gold and silver thread maker Sword cutler | Benton & Johnson Ltd. | 63 Kings Cross Road | 1869–1965 |
| Sword cutler | Benton & Frazier | | 1870 |
| Sword cutler | Richard Berrey (Benney) | | 1606–1610 |
| Cutler | Tobie Berrey (Benney) • Master of the Cutlers Company, 1660. | | 1650–1665 |
| Sword cutler | Besch & Co. | 11 Hanover Square | 1790 |
| Blade etcher | Thomas Bewick • Etched blades for Nicholas and William Oley at Shotley Bridge, 1767–1800. (see Nicholas & William Oley) (see Chapter 4) | | |
| Silversmith Hilt maker Sword cutler | John Bibb • Signed an advertisement warning sword purchasers against counterfeit cast silver hilts. | | 1661 |
| Silversmith | Thomas Bibb | | Bc1667–1686 |
| Hilt maker Sword cutler | Thomas Bibb | Great Newport St. near St. Martins Lane | 1687–1757 |
| | • Had apprentice Henry Brunn, c. 1688–1695 (see Henry Brunn) • Master of the Cutlers Company, 1738. • Silversmith William Kersill made hilts for some Thomas Bibb small swords. | | |
| | Charles Bibb • Son of Thomas Bibb. | Great Newport St., near St. Martins Lane | 1758–1775 |
| | Charles Bibb & Co. | Great Newport St., near St. Martins Lane | 1776–1810 |
| | • Bibb advertised: *Charles Bibb, At the sign of the flaming sword in Great Newport St., Near St. Martins Lane* *Makes & sells all sorts of swords, and cuttoes* [hunting swords] *and belts (lace, silk and leather)* *Wholesale & Retail at reasonable rates* • Sold pierced and faceted steel-hilted and silver-hilted rapiers and small swords. | | |
| Silversmith Hilt maker Sword cutler | Edward Bickerstaff • Sold silver-hilted rapiers. | Russell St., Covent Garden at the sign of the Flaming Sword | 1680–1691 |
| | Thomas Bickerstaff | Prince St. Lincoln Inn Fields at the sign of the Halberd | 1692–1700 |
| Army hatter | Bicknell & Griffith | Old Bond St. | 1755–1758 |
| and hosier | Bicknell, James & Griffith | Old Bond St. at the sign of the Kings Arms | 1759–1773 |
| Glover Sword cutler | Bicknell & James | Old Bond St. | 1774–1778 |

| | | | |
|---|---|---|---|
| Sword cutler | Bicknell (William) & Son (George) | Old Bond St. | 1779–1789 |
| | William & George Bicknell | Old Bond St. | 1790–1793 |
| | William & George Bicknell | 57 Piccadilly | 1794–1802 |
| | William & George Bicknell | 1 Old Bond St. | 1803–1812 |
| | Bicknell (William & George) & Moove (W.) | 1 Old Bond St. | 1813–1837 |
| | • Hat, cap, and glove maker to the king. | | |
| | • Succeeded by J. Moore (see J. Moore). | | |
| Sword maker | Thomas Bigglestone | | 1695–1720 |
| Sword furnisher | • Supplied swords and plug bayonets to the Board of Ordnance, 1699–1711. | | |
| | • Also repaired swords for the Board of Ordnance. | | |
| | • Also a furbisher of edged weapons at the Tower of London. | | |
| | • Appointed sword cutler to the Board of Ordnance, March 15, 1716. | | |
| Cutler | Charles Biggs | | 1790–1813 |
| | • Master of the Cutlers Company, 1800 and 1807. | | |
| Sword cutler | Bilney & Ashdown | | 1887 |
| | Charles Bingham | | |
| | (see Newham & Bingham) | | |
| Dirk maker | W. Bingham & Co. | 29 Conduit St. | 1860–1881 |
| Sword cutler | Binnie & Mason | 31 Old Bond St. | 1850 |
| Gunsmith | Richard Birch | | Bc1591–1614 |
| Sword cutler | • Apprenticed to Edward Jones, 1605–1614. | | |
| | Richard Birch | Gunsmith St. | 1615–D1624 |
| | | Parish of St. Botolph Aldergate | |
| Cutler | Thomas Birdwhistle | | 1662–1677 |
| | • Master of the Cutlers Company, 1672. | | |
| Silversmith | James Birt | | 1820–1870 |
| Goldsmith | • Ray & Montaque mounted some of the presentation swords | | |
| Hilt maker | sold by Birt. | | |
| Sword cutler | | | |
| Armourer | James Bishop | 4 Long Lane Borough | 1805–1808 |
| Sword maker | | | |
| Silversmith | James Bishop | | 1750–1780 |
| Hilt maker | • Had apprentice Cornelius Bland, 1761–1768 (see Cornelius Bland). | | |
| Sword cutler | | | |
| Jeweler | William Bishop | 170 New Bond St. | 1826–1871 |
| Goldsmith | • Successor to Wicks & Bishop (see Wicks & Bishop). | | |
| Sword cutler | • Gun maker to Prince Albert. | | |
| Gun and pistol warehouse | • Did work for gun maker Joseph Manton. | | |
| Gold and silver lacemen | Thomas & Daniel Blachford | 67 Lombard St. | 1774–1779 |
| | Thomas & Daniel Blachford | 66 Lombard St. | 1780–1789 |
| Sword cutler | Daniel & Richard Blachford | 66 Lombard St. | 1790–1793 |

| | | | |
|---|---|---|---|
| | Daniel & Richard Blachford | 10 Change Alley, Cornhill | 1794–1798 |
| | Daniel & Richard Blachford | 9 Change Alley, Cornhill | 1799–1809 |
| | Daniel Blachford | 12 Leadenhall St. | 1809–1812 |
| Gold and silver lacemen | E. (Edward) M. Blachford | 67 Lombard St. | 1794–1810 |
| Belt maker | • Partner: Richard Blachford. | | |
| Sword cutler | Richard Blachford & Co. | 67 Lombard St. | 1811–1814 |
| | • Sword cutler and belt maker to the Honorable East India Company. | | |
| | Richard James & William Blachford | 67 Lombard St. | 1815–1821 |
| | Richard James Blachford | 67 Lombard St. | 1822–1823 |
| | Richard James Blachford | 49 Lombard St. | 1824 |
| Gun maker | George Blake | 168 Fenchurch St | 1784–1792 |
| | George Blake | 104 Gun Dock, Wapping | 1793–1805 |
| | George Blake | 95 Wapping, Old Stairs | 1806–D1807 |
| | Ann Blake | 95 Wapping, Old Stairs | 1807–1816 |
| | • Widow of George Blake. | | |
| Sword cutler | Blake (Ann) & Co. | 252 & 253 High St., Wapping | 1817–1823 |
| | John Alkin Blake | 252 & 253 High St., Wapping | 1824–1834 |
| | • Son of George & Ann Blake. | | |
| | John Alkin Blake & Co. | 252 & 253 High St., Wapping | 1835–1852 |
| | John Alkin Blake & Co. | 253 High St., Wapping | 1853–1854 |
| | John Alkin Blake & Co. | Upper East Smithfield St. | 1855–1864 |
| | • Succeeded by E. Yeomans. (see James Yeomans) | | |
| Gold and silver lacemen | V. & R. Blakemore | 46 Leadenhall St. | 1866–1874 |
| | V. & R. Blakemore | 8 Lime St. | 1875–1897 |
| Accoutrement maker | V. & R. Blakemore | 86 Leadenhall St. | 1898–1928 |
| | V. & R. Blakemore | 96–98 Leadenhall St. | 1829–1931 |
| Sword cutler | | | |
| Silver chaser | Cornelius Bland | | Bc1747–1771 |
| Silversmith | • Apprentised to James Bishop, 1761–1771 (see James Bishop). | | |
| Jeweler | Cornelius Bland | 62 Aldersgate | 1771–1772 |
| Hilt maker | Bland (Cornelius) & Natter (George Sigismund) | 185 Fleet St. | 1773–1781 |
| Sword cutler | • Had apprentice Thomas Young, 1773–1780 (see Thomas Young). (see George Sigismund) | | |
| | Cornelius Bland | Jewin St. | 1782–1787 |
| | • Had apprentice James Huell Bland (son), 1786–1787. | | |
| | Cornelius Bland | 116 Aldergate St. | 1788–1790 |
| | Cornelius Bland | 126 Bunhill Row | 1791–D1794 |
| | James Huell Bland & Elizabeth Bland | 126 Bunhill Row | 1794–1796 |
| | • Son and widow of Cornelius Bland. | | |
| | • James Huell Bland had apprenticed to Thomas Young, 1787–1793 (see Thomas Young). | | |
| Army and navy contractor | John Bland | 70 St. James St. | 1768–1787 |
| | Bland (John) & Foster (Robert) | 70 St. James St. | 1788–1791 |
| Accoutrement maker | • John Bland died in 1791. | | |
| | • Royal cutler to King George III and the Honourable East India Company. | | |
| Sword belt maker | • Sword cutlers and belt makers to King George III, the Prince of Wales, and the Duke of York | | |
| Sword cutler | • Succeeded by Robert Foster (see Robert Foster). | | |

| | | | |
|---|---|---|---|
| Silversmith<br>Goldsmith<br>Hilt maker<br>Sword cutler | **Thomas Bland & Son (James)**<br>**James Bland** | 106 Strand<br>106 Strand | 1740–1769<br>1770–1790 |
| Sword cutler | **Richard Blaney (Blayney)**<br>• Master of the Cutlers Company, 1679.<br>• Had apprentice John Widmore, 1678–1683 (see John Widmore).<br>**Mary Blaney**<br>• Widow of Richard Blaney.<br>• Partners and sons: Robert and Joseph Blaney.<br>**Joseph Blaney**<br>• Son of Richard and Mary Blaney.<br>**Thomas Blaney**<br>• Son of Joseph Blaney.<br>• Succeeded by John Tipping (see John Tipping). | Lombard St.<br>Parish of St. Edmund<br><br>Lombard St.<br>Parish of St. Edmund<br><br>Exchange Alley,<br>Parish of St. Mary Woolnoth<br>Exchange Alley,<br>Parish of St. Mary Woolnoth | 1678–D1691<br><br><br>1691–1695<br><br><br>1696–1698<br><br>1699–1704 |
| Silversmith<br>Sword cutler | **Robert Blaney**<br>• Son of Richard and Mary Blaney. | near the Royal Exchange | 1696–1710 |
| Sword cutler | **Henry Blathwaite** | | 1620 |
| Steel toy<br>(hardware) maker<br>Hilt maker<br>Sword cutler | **John Blounts**<br>• Imported sword blades from Berlin, Germany.<br>• Sold swords with polished steel hilts. | against the Meuse Gate, Charing Cross<br>at the sign of the Princes Head | 1700–1710 |
| Sword maker<br>Sales agent | **Frederick Blyth**<br>• Sales agent for A. & A. Schnitzler, a Solingen, Prussia, sword maker. | | 1870–1873 |
| Goldsmith<br>Silversmith<br>Hilt maker<br>Sword cutler | **Mark Bock**<br>**Mark Bock** | Kings Head Court, Shoe Lane<br>Cockspur St.<br>facing the Haymarket, Charing Cross | 1761–1773<br>1774–D1783 |
| Silversmith<br>Hilt maker<br>Sword cutler | **John Bockett (Buckett)** | St. James St., Westminster | 1760–1786 |
| Gold and silver<br>lacemen<br>Accoutrement<br>maker<br>Sword cutler | **Bodley & Etty**<br>• Successors to Edward Longdon.<br>**Bodley, Bodley & Etty**<br>**Bodley & Etty**<br>(succeeded by Thomas Wilson)<br>(see Thomas Wilson) | 31 Lombard St.<br><br>31 Lombard St.<br>31 Lombard St. | 1784–1814<br><br>1815–1839<br>1840–1851 |
| Haberdasher<br>Sword cutler | **Bolton (James) & Co.** | Ludgate Hill | 1723–1730 |
| Gun maker<br>Gun warehouse<br>Sword cutler | **Edward Bond**<br><br>**Edward Bond**<br>**Edward Bond**<br>• Had a gun and archery warehouse at 45 Cornhill St. | 59 Lombard St.<br>at the sign of the Golden Blunderbuss<br>59 Lombard St. & 31 Nicholas Ln.<br>59 Lombard St. | 1774–1779<br><br>1780–1786<br>1787–D1790 |

|  | | | |
|---|---|---|---|
| | Mary Bond | 59 Lombard St. & 45 Cornhill St. | 1790–1794 |
| | • Widow of Edward Bond. | | |
| | • Partner and son: Philip John Joseph Bond. | | |
| | Philip John Joseph Bond | 59 Lombard St. & 45 Cornhill St. | 1795–1799 |
| | Philip John Joseph Bond & William Thomas Bond (Sr.) | 59 Lombard St. & 45 Cornhill St. | 1800–1807 |
| | • In 1808, William Thomas Bond (Sr.) became propietor of the 59 Lombard St. shop. | | |
| | Philip John Joseph Bond | 45 Cornhill St. | 1808–D1816 |
| | Edward James Bond | 45 Cornhill St. | 1816–1825 |
| | • Son of Philip John Joseph Bond. | | |
| | • Began sword production. | | |
| | • Sword cutler and gun maker to the Honorable East India Company. | | |
| | Edward James & William Thomas Bond (Jr.) | 45 Cornhill St. | 1826–1834 |
| | • Sons of Philip John Joseph Bond. | | |
| | Edward Philip & William Thomas Bond (Jr.) | 45 Cornhill St., Hooper Square, Goodmans Field | 1835–1855 |
| | • Edward Philip was the son of Edward James Bond. | | |
| | Edward Philip & William Thomas Bond (Jr.) | 142 Leadenhall St., Hooper Square, Goodmans Field | 1856–1860 |
| | Edward Philip Bond | 142 Leadenhall St., Hooper Square, Goodmans Field | 1861–1869 |
| | Edward Philip Bond & William Thomas Bond III | 4 Northumberland Ave. | 1870–1879 |
| | Philip John Joseph Bond & William Thomas Bond (Sr.) | 59 Lombard St. & 45 Cornhill St. | 1880–1907 |
| Gun maker Sword cutler | William Thomas Bond (Sr.) | 59 Lombard St. | 1808–D1836 |
| | • William Thomas Bond died in 1836. | | |
| | • His widow, Elizabeth Bond, ran the company. | | |
| | Elizabeth Bond | 59 Lombard St. | 1836–1845 |
| | John Bonella (see C. Webb & Co.) | | |
| Cutler | Francis George Boot | | 1884–1899 |
| | • Master of the Cutlers Company, 1894. | | |
| Cutler | Henry Ross Boot | | 1891–1906 |
| | • Master of the Cutlers Company, 1901. | | |
| Cutler | Horace Boot | | 1892–1907 |
| | • Master of the Cutlers Company, 1902. | | |
| Cutler | Samuel Boot | | 1657–1672 |
| | • Master of the Cutlers Company, 1667. | | |
| Cutler | Samuel Poynton Boot | | 1874–1889 |
| | • Master of the Cutlers Company, 1884. | | |
| Cutler | Thomas Boot | | 1837–1864 |
| | • Master of the Cutlers Company, 1847 and 1859. | | |
| Gold and silver lacemen | Frederick Felix Boreman | 69 Sloane St. | 1843–1844 |
| | Frederick Felix Boreman | 69 Sloane St. & 432 Strand | 1845–1847 |

| | | | |
|---|---|---|---|
| Sword cutler | **Frederick Felix Boreman** | 432 Strand | 1848–1866 |
| Cutler | **Jarvis Boswell** | | 1627–1742 |

• Used "tongs" mark.
• Master of the Cutlers Company, 1737.

| | | | |
|---|---|---|---|
| Military tailor | **Henry Boutroy** | 24 Brewster St., Golden Square | 1853–1855 |
| Sword cutler | **Henry Boutroy** | 12 Sackville St., Piccadilly | 1856–1868 |
| | **Henry Boutroy & Son** | 12 Sackville St., Piccadilly | 1869–1939 |

• Succeeded by Aldency & Boutroy (see W.H. Adency).

| | | | |
|---|---|---|---|
| Sword maker | **C. Bowns** | Top Bank | 1799–1816 |
| Cutler | **Stephen Boyce** | | 1810–1825 |

• Master of the Cutlers Company, 1820.

| | | | |
|---|---|---|---|
| Sword cutler | **G. Boyton & Son** | Cherkenwell St. | 1855–1912 |
| Hilt maker | **Soloman Brabant (Braybant)** | | 1640–1645 |

• In 1645, the Cutlers Company of London seized 18 hilts illegally made by Brabant, a Frenchman, who was not a member of the Cutlers Company.

| | | | |
|---|---|---|---|
| Hatter | **Brabant (Braybant) & Godfrey** | Strand | 1765–1767 |
| Hosier<br>Sword cutler | **Brabant (Braybant) & Godfrey** | 2 Fleet St., Temple Bar | 1768–1784 |
| Cutler | **Thomas Bradford** | | 1650–D1684 |

• Used "crossed sceptres" mark.
• Master of the Cutlers Company, 1676.

| | | | |
|---|---|---|---|
| Pole arm maker | **Edmond Bradley** | | 1555–1560 |

• In 1557–1558, supplied pikes and halberd blades to the Drapers Company for the defense of their settlement in Calais, France, during the French War.

| | | | |
|---|---|---|---|
| Sword cutler<br>Draper<br>(cloth maker) | **Sir John Branch** | Cripplegate | 1557–1579 |

• On January 22, 1558, supplied swords and daggers to the Drapers Company for the defense of their settlement in Calais, France, during the French War.
• Master of the Drapers Company, 1572–1573, 1576–1580, 1583–1584.

| | | | |
|---|---|---|---|
| Gun maker | **William Brander** | Little Minories, Royal Exchange | 1758–1764 |
| Sword cutler | **William Brander** | 70 Minories | 1765–1789 |

• Son Martin Brander (B1760) apprenticed, 1774–1789.

| | | | |
|---|---|---|---|
| | **Martin Brander** | 70 Minories | 1790–1801 |
| | **Brander (Martin) & Potts (Thomas)** | 70 Minories & Goodmans Yard | 1802–1827 |

• Began making swords.
• Made M1796 heavy cavalry swords.
• Sword cutler and gun maker to the Board of Ordnance and the Honorable East India Co.
• Martin Brander died in 1827.
• Succeeded by Thomas Potts (see Thomas Potts).

| | | | |
|---|---|---|---|
| Goldsmith<br>Silversmith | **Hugh Brawne** | | Bc1682–1703 |

• Apprenticed to cutler Henry Panton, 1696–1703 (see Henry Panton).

| | | | |
|---|---|---|---|
| Hilt maker<br>Sword cutler | **Hugh Brawne** | Fleet St. | 1704–1720 |

| | | | |
|---|---|---|---|
| Sword maker | **George Brayfield (Brafield)** <br> Repaired swords for the Board of Ordnance, 1656. | | 1650–1660 |
| Silversmith <br> Knife and sword <br> hilt maker | **Moses Brent** <br> • Apprenticed to Dru Drury Sr., 1763–1767 (see Dru Drury Sr.). <br> • Worked for Dru Drury Jr., 1767–1774. | | Bc1758–1774 |
| Knife and <br> sword cutler | **Moses Brent** | Hind Court, Noble St., <br> Foster Lane | 1775–1781 |
| | **Moses Brent** <br> • Had apprentice Moses William Brent (son), 1794–1801. | Well Yard, Little Britain | 1782–1798 |
| | **Moses Brent** | 42 Little Britain | 1799 |
| | **Moses Brent** | 12 Kirby St., Hatton Garden | 1800–1812 |
| | **Moses Brent** | 22 Greville St., Leather Lane | 1813–1816 |
| | **Moses Brent** <br> • Did silverwork for Samuel Brunn (see Samuel Brunn). <br> • Made silver gun furniture for the gun trade. <br> • Made knife hilts. | 19 Leather Lane | 1817–1820 |
| Silversmith <br> Knife and sword <br> hilt maker | **Moses William Brent** <br> • Apprenticed to father Moses Brent, 1794–1801 (see Moses Brent). | | Bc1780–1801 |
| | **Moses William Brent** | 22 Greville St. | 1802–1816 |
| Knife and <br> sword cutler | **Brent (Moses William) & Peppin (S.W.)** | 22 Greville St. | 1817–1820 |
| Sword maker <br> Scabbard maker | **Bretts, Vandiest & Co.** <br> • Made naval cutlasses and steel scabbards. | | 1804–1810 |
| | **John Brewster** <br> (see George Gillott) | | |
| Sword cutler | **H. Brightwell & Co.** | 24 Cecil St. | 1897–1900 |
| Knife and <br> sword cutler | **Richard Briginshaw** <br> • On February 27, 1626, Briginshaw was asked to be an assistant in the yeomanry <br>   of the Cutlers Company. He refused and was fined. <br> • A sword in the London Musuem is marked on the blade "made in Hounsloe <br>   by Johannes Hoppie for Richard Briginshaw 1636." <br> • In 1658, he was nominated but not elected alderman. <br> • In the 1620s, Briginshaw bought unmarked knife blades and engraved them <br>   with the marks belonging to well-known blade makers of the Cutlers Company of London. <br> • In c. 1628, he had the master and warden of the Cutlers Company arrested. When his <br>   accusations were proved wrong, he was ordered to apologize to them by the Lord Mayor of London. | | 1620–1660 |
| Silversmith <br> Hilt maker | **John Brockus** <br> • Apprenticed to Robert Pilkington, 1741–1748. | | Bc1727–1748 |
| Sword cutler | **John Brockus** | Stanhope St., Chare Market | 1748–1759 |
| | **John Brockus** | Shoe Lane, Fleet St. | 1760–1780 |
| Silversmith <br> Hilt maker <br> Sword cutler | **Matthew (Matthieu) Brodier** <br> • Made silver-hilted hunting swords (cuttoes). | Newport Alley, <br> St. Ann's St., Westminster | 1750–1755 |
| Sword cutler | **Col. Lawrence Bromfield** | | 1650–1667 |
| Silversmith | **Albert George Brooker** | | 1945–1957 |

| | | | |
|---|---|---|---|
| Hilt maker<br>Sword cutler | • Supplied presentation swords to the Goldsmiths & Silversmiths Co.<br>(see Goldsmiths & Silversmiths Co.) | | |
| Gun maker<br>Sword cutler | Edward Brooks & Son<br>• Moved to Birmingham, 1847.<br>(see Birmingham listing) | 1 Fenchurch St. | 1840–1847 |
| Gun maker<br>Sword cutler | John Brown<br>Brown (John) & Co.<br>Brown (John) & Son<br>Brown, Son & Rowntree | 104 Strand<br>104 Strand<br>10 & 11 Princes St., Hanover Square<br>10 & 11 Princes St., Hanover Square | 1802–1809<br>1810–1819<br>1820–1849<br>1850–1860 |
| Sword dealer | Brown & Mannett<br>Brown & Mannett<br>• A sales agent for Birmingham and Sheffield sword makers.<br><br>James Brown<br>(see John Lambert) | 31 Broad St. Building<br>26 New City Chambers | 1861–1863<br>1864–1874 |
| Sword maker | Philip Brown<br>• Used the "cross and crown" mark. | | 1661–1681 |
| Tool maker<br>Gun and pistol<br>maker<br>Sword cutler | Richard Brown<br>Richard Brown & Co.<br>Richard Brown & Co.<br>Richard Brown | 24 Botolph Lane<br>38 Botolph Lane<br>23 Nicholas Lane at Lombard St.<br>28 Walbrook | 1815<br>1816–1819<br>1820–1822<br>1823 |
| Cutler | William Brown<br>• Master of the Cutlers Company, 1447. | | 1437–1452 |
| Cutler | Francis Browne<br>• Master of the Cutlers Company, 1686. | | 1676–1691 |
| Sword cutler | H. Browne<br><br>Henry Brownsmith<br>(see Meyer & Mortimer) | New Bond St. | 1850 |
| Silversmith<br>Hilt maker<br>Sword cutler | Charles Bruff<br>• Immigrated to Talbot County, Maryland, USA, 1668. | | 1650–1668 |
| Goldsmith<br>Silversmith<br>Hilt maker<br>Sword cutler | Anthony Brummidge | | 1650–1686 |
| Silversmith<br>Hilt maker<br>Sword cutler | Henry Brunn (Brun)<br>• Apprenticed to silversmith Thomas Bibb, c. 1688–1695.<br>Henry Brunn | New St. near Fetter Lane | Bc1673–1695<br><br>1696–1730 |
| Gun maker<br>Silversmith<br>Hilt maker<br>Sword cutler | Samuel Brunn (Brun)<br>John Knubley & Co.<br>• In 1795, Brunn bought the controlling interest in John Knubley & Co.<br>but kept the Knubley name. | 7 Charing Cross | B1770–1794<br>1795–1797 |

| | | | |
|---|---|---|---|
| | Samuel Brunn | 55 Charing Cross | 1798–1804 |
| | Samuel Brunn | 56 Charing Cross, opposite the Mewsgate | 1805–1820 |

- Sword cutler to the Patriotic Fund.
- Hilted imported German blades, bought from John Justus Runkel (Solingen and London).
- Gun maker and sword cutler to the Prince of Wales, 1800–1811.
- Gun maker and sword cutler to the Prince Regent, 1811–1820.
- Contractor to the Board of Ordnance, 1797–1809.
- Moses Brent did silver work for some of Brunn's guns.

| | | | |
|---|---|---|---|
| Sword cutler | John Bryan | Parish of St. Sepulcher | 1614–1629 |
| Currier Harness maker Japanner Button and ornament maker Sword cutler | Richard Bryan • Harness makers and japanners only. | 11 Great Chapel St., Westminster | 1810–1818 |
| | Robert Bryan (see Birmingham listing) | 11 Great Chapel St., Westminster | 1819–1821 |
| | Bryan (Robert) & Prince (John) • Became sword cutlers and button and ornament makers. | 11 Great Chapel St., Westminster | 1822–1823 |
| | Bryan (Robert) & Prince (John) | 11 Great Chapel St. & 33 Orchard St. | 1824–1825 |
| | Bryan (Robert) & Prince (John) | 9 Dacre St. | 1826–1827 |
| | Bryan (Robert) & Prince (John) | 8–9 Dacre St. | 1828–1836 |
| | Bryan (Robert) & Prince (John) & Co. | 8–9 Dacre St. | 1837–1857 |
| | John Bryan & Co. | 8–9 Dacre St. | 1858–1868 |
| | Bryan Bros. & Co. | 9 Dacre St. | 1869–1894 |
| | Bryan Bros. & Co. | 26 Grange Rd. | 1895–1896 |
| Hosier Sword cutler | Kempe Brydges | corner of Bedford St., Covent Garden | 1744–1766 |
| | Brydges (Kempe) & Walker (William) • Succeeded by Amery, Pitter & Slipper (see Amery, Pitter & Slipper). | corner of Bedford St., Covent Garden at the sign of the Three Crowns | 1767–1776 |
| Bladesmith | John Brykles (Brickles) | | 1376 |
| Cutler | William Brynkill • Master of the Cutler Company, 1453, 1454, 1461, 1462. | | 1493–1467 |
| Sword cutler | Thomas Buck • Had tenant Richard Hawes (see Richard Hawes). | Fleet St., Parish of St. Sepalchre without Newgate | 1550–D1566 |
| | John Buckett (see John Bockett) | | |
| Army and navy tailor and clothier Sword cutler | John Buckmaster | 22 Old Bond St. | 1816–1823 |
| | John Buckmaster | 17 Sackville St. | 1824–1825 |
| Army and navy tailor Clothier Sword and dirk cutler | William Buckmaster | 5 Old Bond St. | 1824–1825 |
| | William & Thomas Buckmaster | 5 Old Bond St. | 1829–1835 |
| | William & Thomas Buckmaster | 3 New Burlington St. | 1836–1840 |
| | William Buckmaster & Co. • Sold brass-mounted Black Watch dirks. | 3 New Burlington St. | 1841–1884 |
| Sword cutler | William Bull | Fleet St., Parish of St. Bride | 1621–1636 |

| | | | |
|---|---|---|---|
| Cutler | **John Bulley** | | 1714–1734 |
| | • Master of the Cutlers Company, 1724. | | |
| Silversmith<br>Hilt maker | **Thomas Bulley** | Shoe Lane | 1707–1723 |
| Sword cutler | **Thomas Bulley** | Hayden Yard Minories | 1724 |
| | • Also made silver furniture for pistols. | | |
| Gold and silver<br>lacemen<br>Embroiderer<br>Sword cutler | **Patrick Bullmore & Co.**<br>**William Bullmore**<br>• Succeeded by George Benton (see George Benton). | 17 Bedford St., Covent Garden<br>28 Clerkenwell Close | 1833–1850<br>1851–1855 |
| Antique dealer<br>Sword cutler | **William Bullock**<br>• Owned Bullock's London Musuem. | The Egyptian Temple,<br>Piccadilly | 1810–1820 |
| Silversmith<br>Goldsmith<br>Hilt maker<br>Sword cutler | **James Bult**<br>• Apprenticed to James Stamp, 1774–1780.<br>• Apprenticed to James Sutton, 1780–1781. | | Bc1760–1781 |
| | **James Bult & James Sutton**<br>**Samuel Goodbehere & Co.**<br>• Partner: James Bult. | 86 Cheapside<br>86 Cheapside | 1782–1783<br>1784–1785 |
| | **Goodbehere (Samuel) &**<br>**Wigan (Edward) & Co.**<br>• Partner: James Bult. | 86 Cheapside | 1786–1799 |
| | **Goodbehere (Samuel),**<br>**Wigan (Edward) & Bult (James)**<br>**(S. Goodbehere & Co.)** | 86 Cheapside | 1800–1818 |
| | **Goodbehere (Samuel) & Bult (James)**<br>**James Bult**<br>(see Samuel Goodbehere, Edward Wigan) | 86 Cheapside<br>86 Cheapside | 1818–1819<br>1819 |
| Sword cutler | **Wicker Bunt** | | 1585 |
| Sword cutler | **H. Burberry** | Haymarket | 1910 |
| Goldsmith<br>Medallion<br>maker<br>Medal maker<br>Engraver | **Edward Burch**<br>**Edward Burch**<br>• Designed and made sword hilts and scabbard medallions and decorations<br>   for Patriotic Fund swords made by Richard Teed (see Richard Teed). | Warwick St.<br>Charing Cross<br>24 Charlotte St. | 1771–1775<br>1776–1808 |
| | **William Burdan**<br>(see John Lambert) | | |
| Hardwaremen<br>(dealer)<br>Gun maker<br>Flint dealer<br>Sword cutler | **Burgon (John) & Jewson (Joseph)**<br>• Successor to Jewson & Burgon (see Joseph Jewson).<br>**John Burgon**<br>**John Burgon**<br>• Contractor to the Ordnance Dept., East India Co., and Hudsons Bay Co.<br>**John Burgon & Son** | 16 Fish Street Hill<br>16 Fish Street Hill<br>opposite the Monument<br>15 & 16 Fish Street Hill<br>opposite the Monument<br>15 & 16 Fish Street Hill<br>opposite the Monument | 1787–1789<br>1790–1791<br>1792–1811<br>1812–1818 |

|  |  |  |  |
|---|---|---|---|
|  | • Son Josiah Towry Burgon. |  |  |
|  | John Burgon & Son | 35 Bucklersbury | 1819–1842 |
|  | Josiah Towry Burgon | 35 Bucklersbury | 1843–1863 |
| Sword cutler | Joseph & Alfred Burnet & Co. | 150 Regent St | 1862–1864 |
| Cutler | Robert Burra | | 1844–1859 |
|  | • Master of the Cutlers Company, 1854. |  |  |
| Hatter<br>Sword cutler | Joshua Burton | 149 Fleet St. | 1771–1782 |
| Hatter | James Busain | 8 Gerrard St. | 1836–1750 |
| Accoutrement<br>maker | James Busain & Co. | 8 Gerrard St. | 1751–1858 |
|  | Busain (James) & Smith (Matthew) | 8 Gerrard St. | 1859–1877 |
| Sword cutler | Busain (James), Smith<br>(Matthew) & Co. | 8 Gerrard St. | 1878–1893 |
|  | Busain (James), Smith<br>(Matthew) & Co. | 39 Warwick St., Regent St. | 1894–1898 |
|  | • Succeeded by Hewson & Williams (see Hewson & Williams). |  |  |
| Sword maker | John Bushnell (Bushell) | | 1600–1665 |
|  | • Granted "cross keys" mark, 1806. |  |  |
| Sword and<br>knife maker | Christopher Butler | | 1606–1607 |
| Sword and<br>bayonet maker | John Butler | | 1659–1700 |
|  | • Made plug bayonets. |  |  |
|  | • Used the "cross daggers" mark. |  |  |
| Sword cutler | Thomas Byewater | Willow Walk Bermondsey | 1630–1670 |
|  | • Master of the Cutlers Company, 1639 and 1663. |  |  |
| Sword cutler | C. Caffin | 13 Railway St., Chatham | 1910 |
| Sword cutler | Campbell, Whitaker & Co. | Conduit St. | 1890 |
| Military<br>outfitter | S. (Samuel) Campbell & Co | 71 Jermyn St. | 1751–1860 |
|  | • In 1861, Saul Isaac & Co. (partners J. Hart and B.F. Hart), a New York City, USA,<br>military goods dealer, bought S. Campbell & Co. |  |  |
| Army and navy<br>contractor | Isaac (Saul), Campbell (Samuel) & Co. | 71 Jermyn St. | 1861–1864 |
| Sword cutler | • Exported M1853 (variant) cavalry sabers and M1822 (variant) foot officers swords<br>to the Confederate states during the American Civil War (1861–1865). |  |  |
|  | • Swords marked "Isaac Campbell & Co." or "Isaac & Co." |  |  |
| Cutler | George Frederick Carden | | 1852–1867 |
|  | • Master of the Cutlers Company, 1862. |  |  |
| Sword cutler | Henry Carnegie Carden | | 1876 |
| Cutler | James Carden | | 1807–1822 |
|  | • Master of the Cutlers Company, 1817. |  |  |
| Sword cutler | Sir Robert Walter Carden | | 1843–1881 |

| | | | |
|---|---|---|---|
| | • Master of the Cutlers Company, 1853, 1865, and 1876. | | |
| Cutler | **Peter Cargill** | | 1763–1788 |
| | • Master of the Cutlers Company, 1773. | | |
| Goldsmith | **John Carmen (Sr.)** | | Bc1694–1716 |
| Silversmith | • Apprenticed to Thomas Vicaridge, c. 1708–1715 (see Thomas Vicaridge). | | |
| Hilt maker | • Succeeded Thomas Vicaridge, 1716. | | |
| Sword cutler | **John Carmen (Sr.)** | New St., Chancery Lane | 1716–D1743 |
| | • Had apprentice John Bennett, 1723–1736 (see John Bennett). | | |
| | • Had apprentice William Strange, 1716–1724 (see William Strange). | | |
| | **John Carmen (Jr.)** | New St., Chancery Lane | 1743–1751 |
| | • Son of John Carmen Sr. | | |
| | **John Carmen (Jr.)** | near Hatton Garden, Holborn | 1752–D1764 |
| | • Master of the Cutlers Company, 1761. | | |
| | • His trade card read: | | |
| | *Working goldsmith & sword cutler at the sign of the Ewer & Swords near Barletts Buildings, Holborn. Makes & Sells all sorts of Gold & Silver work at ye lowest prices, like wise choice of swords & cuttoes in silver, steel & brass at reasonable rates.* | | |
| | • Mounted small swords. | | |
| | **Mary Carmen** | near Hatton Garden, Holborn | 1864 |
| | • Widow of John Carmen (Jr.). | | |
| | **Edward Carmen** | near Hatton Garden, Holborn | 1865–1880 |
| | • Son of John Carmen (Jr.). | | |
| Sword maker | **Moses Carr** | | 1680–1696 |
| | • Granted the "sceptre and cross" mark, 1686. | | |
| Sword cutler | **Carr & Son** | 52 Conduit St. | 1950 |
| | • Sold swords made by Wilkinson Sword Company. | | |
| Cutler | **Robert Carrington** | | 1649–1664 |
| | • Master of the Cutlers Company, 1659. | | |
| Silversmith | **John Carter** | | 1661 |
| Hilt maker | • Signed an advertisement warning sword purchasers against counterfeit cast silver hilts. | | |
| Sword cutler | | | |
| Silversmith | **John Carter** | Westmoreland Building | 1768–1769 |
| Goldsmith | • Sold some silver work for Parker & Wakelin (see Edward Wakelin). | | |
| Hilt maker | **John Carter** | Bartholomew Close | 1770–1777 |
| Sword cutler | **Robert Makpeace & Richard Carter** | Bartholomew Close | 1777–1778 |
| | **Carter (Richard) & Smith** | 14 Westmoreland Building | 1778–1779 |
| | **(Daniel) & Sharp (Robert)** | Aldergate St. | |
| | • Succeeded by Smith & Sharp. | | |
| | (see Robert Makepeace, Daniel Smith & Robert Sharp) | | |
| Steel toy (hardware) maker Hilt maker | **Peter Carter** | | 1625–1630 |
| | • Made steel hilts and pommels. | | |
| Silversmith | **Richard Carter** | | Bc1757–1776 |
| Goldsmith | **Robert Makepeace & Richard Carter** | Bartholomew Close | 1777–1778 |
| Hilt maker | **Carter (Richard), Smith (Daniel)** | 14 Westmoreland Building | 1778–1779 |

| Trade | Name | Address | Dates |
|---|---|---|---|
| Sword cutler | & Sharp (Robert)<br>(see Daniel Smith & Robert Sharp) | Aldergate St. | |
| Armourer<br>Sword maker<br>Blade maker | Richard Carter<br>• Worked at the Greenwich Royal Armouries.<br>Richard Carter<br>• Master of the Cutlers Company, 1548.<br>• In 1557 and 1558, supplied swords, daggers, and sword parts to the Drapers Company for the defense of their settlement in Calais, France. | Greenwich | 1528–1540<br><br>1540–1570 |
| Sword maker<br>Sword blade maker | William Carter<br>• Worked at the Greenwich Royal Armouries, 1528–1541. | | 1528–1541 |
| Cutler | William Carter<br>• Master of the Cutlers Company, 1662. | | 1652–1667 |
| Sword cutler | William Carter | | 1768 |
| Military tailor<br>Accoutrement maker<br>Sword cutler | W. (William) S. Carver<br>W. (William) S. Carver<br>W. (William) S. Carver<br>• Succeeded Gardner & Carver (see Gardner & Carver). | 70 Jermyn St.<br>16 King St., Parish of St. James<br>35 Bury St., Parish of St. James | 1891–1899<br>1900–1910<br>1911–1916 |
| | Sir George Caswell<br>(see the Sword Bank) | | |
| Hatter<br>Army cap, helmet and accoutrement maker<br>Military outfitter<br>Sword cutler | John Alexander Cater<br>John Alexander Cater<br>John Alexander Cater & Sons<br>Francis John & John Alexander Cater<br>F. (Francis) J. (John) Cater<br>• Became sword cutlers.<br>F. (Francis) J. (John) Cater<br>F. (Francis) J. (John) Cater<br>Cater (Francis John) & Co.<br>• Hatter to the royal family, the king, and the Duke of York.<br>• Francis John Cater died in 1844.<br>Mary Ann & William Charles Cater<br>• Widow and son of Francis John Cater.<br>Cater (William Charles) & Co.<br>W. (William) C. (Charles) Cater<br>• Sold swords made by the Wilkinson Company.<br>• William Charles Cater died in 1899.<br>William Cater & Co.<br>William Cater & Co. Ltd.<br>William Cater & Co. Ltd. | 55 Pall Mall<br>67 Pall Mall<br>67 Pall Mall<br>67 Pall Mall<br>67 Pall Mall<br><br>60 Pall Mall<br>56 Pall Mall<br>56 Pall Mall<br><br><br>56 Pall Mall<br><br>56 Pall Mall<br>56 Pall Mall<br><br><br>56 Pall Mall<br>56 Pall Mall<br>62 Pall Mall | 1776–1781<br>1782–1799<br>1800–1805<br>1806–1811<br>1812–1820<br><br>1821–1831<br>1832–1839<br>1840–D1844<br><br><br>1844–1850<br><br>1851–1860<br>1861–D1899<br><br><br>1899–1911<br>1912–1921<br>1922–1923 |
| Hatter<br>Sword cutler | Joseph Cator | 58 Bishopsgate without | 1778–1808 |
| Cutler | John Catour<br>• Master of the Cutlers Company, 1457, 1458, and 1465. | | 1447–1470 |
| Sword cutler | William Cave | | 1615–1648 |

- Supplied swords to the Board of Ordnance, 1621–1639.
- Benjamin Stone (see Chapter 3) delivered 5,000 sword and rapier blades to the Board of Ordnance, 1638. Stone hilted and made scabbards for 3,000, while the Board of Ordnance contracted with Cave to hilt and make scabbards for 1,000. Robert South (royal cutler) hilted and made scabbards for 1,000.
- Master of the Cutlers Company, 1640.

| | | | |
|---|---|---|---|
| Cutler | **John Chaddle** | | 1407–1423 |
| | • Master of the Cutlers Company, 1417. | | |
| Sword cutler | **James & John Chalk** | 1 Strand | 1809–1910 |
| | **James Chalk** | 1 Strand | 1811–1815 |
| Gun maker<br>Sword cutler | **Charles Chambers** | 36 Rosemary Lane | 1776–1800 |
| Bladesmith | **John Chambers (Chambre)** | | 1430–1461 |
| | • Master of the Bladesmiths Company, 1437. | | |
| Cutler | **John Chambers** | | 1749–1764 |
| | • Master of the Bladesmiths Company, 1759. | | |
| Cutler | **William Chambers** | | 1731–1746 |
| | • Master of the Cutlers Company, 1741. | | |
| Military tailor<br>Sword cutler | **William Chambers** | 8 Maddox St., Bond St. | 1830–1852 |
| | • Merged with William Newton, 1853 (see William Newton). | | |
| Silversmith<br>Hilt maker | **Richard Chapman** | Lombard St. | 1676–1722 |
| | • Appointed sword cutler to the Board of Ordnance, December 24, 1714. | | |
| Sword cutler | **Richard Chapman** | Adams Court, Broad St. | 1723–1740 |
| | • Master of the Cutlers Company, 1727 and 1733. | | |
| Gold and silver<br>lacemen<br>Sword cutler | **John Charlton**<br>**Thomas Charlton**<br>**Thomas Charlton** | 53 Strand<br>53 Strand<br>60 Quadrant St., Regent St. | 1800–1810<br>1811–1825<br>1826 |
| Cutler | **Yeeling Charlwood** | | 1777–1792 |
| | • Master of the Cutler Company, 1787. | | |
| Bladesmith | **John Charyet** | | 1465–1480 |
| | • Became freeman of Bladesmiths Company, 1465. | | |
| Sword maker | **Anthony Chase** | Parish of St. Mary Woolnoth<br>Langbourne Ward, North Precinct | 1690–D1691 |
| | **Samuel Chase** | Parish of St. Mary Woolnoth<br>Langbourne Ward, North Precinct | 1691–D1723 |
| | • Son of Anthony Chase.<br>• Supplied plug bayonets to the Board of Ordnance, 1708–1711.<br>• Supplied swords to the Board of Ordnance, 1692–1712.<br>• Also repaired swords for the Board of Ordnance. | | |
| Silversmith<br>Hilt maker | **Thomas Chawner** | | B1734–1758 |
| | • Apprenticed to Ebenezer Coker, 1754–1758. | | |
| Sword cutler | **Thomas & William Chawner** | 60 Pater Noster Row | 1759–1766 |
| | • Thomas and William were brothers. | | |

| | Thomas Chawner | 60 Pater Noster Row at Red Lion Street, Clerkenwell | 1767–1782 |
|---|---|---|---|
| | Thomas Chawner | 9 Ave Maria Lane | 1783–1785 |
| | • Made urn pommel silver-hilted small swords. | | D1802 |

Cutler **Richard James Cheeswright** 1880–1896
- Master of the Cutlers Company, 1890 and 1891.

Cutler **Frederick Richard Cheeswright** 1910–1925
- Master of the Cutlers Company, 1920.

Sword cutler **Thomas Chesshire (Cheshire)** Parish of St. Dunston in the west 1595–1639
- Freeman of the Cutlers Company, 1603.
- Master of the Cutlers Company, 1615 and 1616.
- Supplied swords to the Board of Ordnance, 1624–1627.
- Believed to be an associate of royal cutler Robert South (South also became a freeman of the Cutlers Company in 1603).
- Chesshire, although he was not the king's royal cutler, did make some richly decorated swords for the king's wardrobe and therefore could have worked with Robert South, who was the royal cutler from 1603–1642.
- On June 13, 1621, by order of the Court of the Cutlers Company, master John Porter, past master Thomas Chesshire, and Robert South were to inspect a sword blade mill (believed to be the mill of Thomas Murray) and report their findings to the court.

Sword cutler **Ralph Chesshire** London 1600–1630

| | | | |
|---|---|---|---|
| Jeweler Goldsmith | Thomas Chesson | near Queen St., Cheapside at the sign of the Unicorn and Pearl | 1732–1748 |
| Silversmith Engraver | Thomas Chesson | 32 Ludgate Hill at the sign of the Golden Salmon and Pearl | 1749–1757 |
| Hilt maker Sword cutler | • Chesson advertised on his trade card: | | |

*Makes & sells all sorts of jewelers work, large & small plate,
particulary to deal in all kinds of second plate & watches of which
he has great variety with many other curious things at reasonable rates
Coat of Arms, crests & ciphers engraved in stone, steel or silver*
- Succeeded by Henry Hurt (see Henry Hurt).

Cutler **William Chiffinch** 1675–1690
- Master of the Cutlers Company, 1685.

Gold and silver lacemen
Sword cutler **John Childe** 186 Strand 1829–1831
- Succeesor to Wilding & Childe.
(see Samuel Wilding)

Gun maker
Sword cutler **William Childe** 280 Strand 1829–1850
- Succeeded by Joseph Wilbraham (see Joseph Wilbraham).

**James Christopher**
(see C. Flight)

| Hatter | Christy & Co. | 1 Old Bond St. | 1852–1869 |
|---|---|---|---|
| Army | • Successor to William Moore (see William Moore). | | |
| Contractor | J.E. & W. Christy | 35 Gracechurch St. | 1870–1881 |
| Sword cutler | Christy & Co. Ltd. | 35 Gracechurch St. | 1882–1920 |

|  |  |  |  |
|---|---|---|---|
|  | Christy & Co. Ltd. | 175 Bermondsey St. | 1921–1954 |
|  | Christy & Co. Ltd. | 8 & 10 Lower James St. & 175 Bermondsey St. | 1955–1959 |
|  | Christy & Co. Ltd. | 175 Bermondsey St. | 1960–1963 |
|  | Christy & Co. Ltd. | 1 Old Bond St. & 175 Bermondsey St. | 1963–1978 |
|  | • Succeeded Scott & Co. at 1 Old Bond St. (see Scott & Co.). |  |  |
|  | Christy & Co. Ltd. | Higher Hillgate, Stockport | 1979–2003 |
| Sword cutler | Jasper Churchill |  | 1634 |
| Cutler | Anthony Clapham • Master of the Cutlers Company, 1689. |  | 1679–1694 |
| Silversmith Goldsmith | Joseph Clare (Sr.) • Apprenticed to Nathaniel Locke, 1702–1712. |  | Bc1698–1712 |
| Hilt maker | Joseph Clare (Sr.) | Wood St. by Love Lane | 1713–1720 |
| Sword cutler | Joseph Clare (Sr.) | Lombard St. at the sign of the Blackmoor's Head | 1721–D1728 |
| Silversmith Goldsmith | Joseph Clare (Jr.) • Son of Joseph Clare Sr. |  | Bc1718–1739 |
| Hilt maker | • Apprenticed to Jeremiah Marlow, 1732–1739 (see Jeremiah Marlow). |  |  |
| Sword cutler | Joseph Clare Jr. | Deanes Court Parish of St. Martins Le Grand | 1739–1768 |
|  | John Clark (see Meyer & Mortimer) |  |  |
|  | Henry Clarke (see John Field) |  |  |
| Steel toy (hardware) maker | Clarke (Richard) & Green (Samuel) • Steel toy (hardware) makers and gun dealers. | 102 Cheapside | 1784 |
| Gun dealer | Clarke (Richard) & Green (Samuel) • Became silversmiths, goldsmiths, and jewelers. | 62 Cheapside | 1785–1786 |
| Silversmith Goldsmith | Richard Clarke • Began mounting swords. | 58 Featherstone St. | 1787–1788 |
| Jeweler | Richard Clarke | 126 Bunhill St. | 1789 |
| Sword cutler | Richard Clarke | 67 Wheler St., Spital Field | 1790–1795 |
|  | Richard Clarke & Son | 15 Ratcliff Row, City Road | 1796–1799 |
|  | Richard Clarke & Son | 4 Ship Court, Old Bailey | 1800–1829 |
| Sword cutler | Alexius Clayton |  | 1715 |
| Military tailor Silk mercer (dealer) Sword cutler | William Clementson • Succeeded by Francis Thompson (see Francis Thompson). | 344 Oxford St. | 1800–1819 |
| Cutler | Francis Cob • Master of the Cutlers Company, 1634. |  | 1624–1639 |
| Sword cutler | Cobb & Co. | Baker St. | 1850 |

| | | | |
|---|---|---|---|
| Sword maker | **Thomas Cole** | | 1620–1630 |
| | • Repaired swords for the Board of Ordnance, 1625. | | |
| Sword cutler | **E. Coles & Co.** | 20 Old Burlington St. | 1860–1865 |
| Sword cutler | **Lewis Coles & Co.** | 1 New Burlington St. | 1866–1870 |
| Sword maker | **Henry Collett** | | 1640–1660 |
| | • Master of the Cutlers Company, 1645. | | |
| | • Supplied swords to the Board of Ordnance, 1651–1656. | | |
| | • Also repaired swords for the Board of Ordnance, 1651–1656. | | |
| Bladesmith | **William Collett (Colet)** | | 1469–1480 |
| | • Became freeman of the Bladesmiths Company, 1469. | | |
| Sword cutler | **Richard Colley** | | 1821 |
| Beaver hat maker<br>Haberdasher of hats<br>Sword cutler | **Thomas Collier (Collyer)** | Exchange Alley, Cornhill<br>at the sign of the Kings Arms and Beaver Hat | 1700–1730 |
| | **William Collier**<br>(see Allen & Collier) | | |
| Cutler | **William Collins** | | 1741–1756 |
| | • Master of the Cutlers Company, 1751. | | |
| Sword maker | **Abraham Colmer** | | 1625–1635 |
| | • Repaired swords for the Board of Ordnance, 1630. | | |
| | **John Connyne**<br>(see Johann Konigs) | | |
| Jeweler<br>Goldsmith<br>Hilt maker<br>Sword cutler<br>Hardwaremen<br>(dealer) | **William Constable & Co.** | 45 New Bond St. | 1783–1784 |
| | • Partner: Thomas Tookey. | | |
| | • Successors to Thomas Tookey (see Thomas Tookey). | | |
| | **Constable (William) & Tookey (Thomas)** | 45 New Bond St. | 1785–1791 |
| | • Thomas Tookey died in 1791. | | |
| | • His widow, Ann Tookey succeeded him (see Thomas Tookey). | | |
| | **William Constable** | 45 New Bond St. | 1791–1793 |
| | • Succeeded by Gray & Constable (see Thomas Gray). | | |
| | **John Cooke**<br>(see Chapter 3) | | |
| | **Thomas Cooke**<br>(see Richard Gurney) | | |
| | **Richard Cooling**<br>(see James Poole) | | |
| Sword cutler | **John Cooper** | | 1615–1630 |
| | • Supplied swords to the Board of Ordnance, 1621. | | |
| Sword cutler | **Henry Coote** | Conduit St. | 1850 |

| | | | |
|---|---|---|---|
| Bladesmith | **John Cope** | | 1480–1500 |
| | • Master of the Bladesmiths Company, 1488. | | |
| Cutler | **James Copous** | | 1831–1846 |
| | • Master of the Cutlers Company, 1841. | | |
| Sword cutler | **John Copous** | | 1781 |
| Cutler | **Thomas Copous** | | 1766–1791 |
| | • Master of the Cutlers Company, 1776 and 1786. | | |
| Goldsmith<br>Hilt maker<br>Sword cutler | **Benjamin Corbett** | | Bc1704–1725 |
| | • Apprenticed to cutler Charles Jackson, 1718 (see Charles Jackson). | | |
| | • Apprenticed to goldsmith John Welles, 1718–c. 1722 (see John Welles). | | |
| | • Apprenticed to goldsmith Thomas Moulden, c. 1722–1725 (see Thomas Moulden). | | |
| | **Benjamin Corbett** | Gutter Lane | 1726–1740 |
| Sword cutler | **William Cordwell** | | 1640–1655 |
| | • Supplied swords to the Board of Ordnance, 1650. | | |
| Hatter<br>Hosier<br>Sword cutler | **James Corneck** | 8 Cheapside<br>near St. Pauls Churchyard<br>at the sign of the Leg and Beaver Hat | 1767–1786 |
| Gold and silver<br>lacemen<br>Sword cutler | **Charles Corney** | 70 Little Britain, Aldersgate | 1869–1870 |
| | • Successor to Barrett, Corney & Corney<br>   (see Bryant Barrett). | | |
| | **William Corney**<br>(see Bryant Barrett) | | |
| Cutler | **Charles Cotton** | | 1726–1753 |
| | • Master of the Cutlers Company, 1736 and 1748. | | |
| Cutler | **Benjamin Cox** | | 1734–1749 |
| | • Master of the Cutlers Company, 1744. | | |
| Goldsmith<br>Enameler<br>Jeweler<br>Steel toy<br>(hardware)<br>maker<br>Sword cutler | **James Cox** | Shoe Lane | 1757–1788 |
| | • Made hilts inlayed with jewels and enamel. | | |
| Sword maker | **N.F. Cox** | Great Peter St., Westminster | 1837–1860 |
| | • Exhibited fencing foils at the London International Exposition, 1851. | | |
| Sword cutler | **Thomas Cox** | | 1696–1738 |
| | • Master of the Cutlers Company, 1728. | | |
| | • Appointed sword cutler to the Board of Ordnance, June 10, 1716. | | |
| Silversmith<br>Hilt maker | **John Craig**<br>**John Craig** | <br>corner of Norris St.<br>and the Haymarket, | Bc1682–1713<br>1714–1729 |

|  |  |  |  |
|---|---|---|---|
|  | George Wickes & John Craig | Parish of St. James corner of Norris St. and the Haymarket, Parish of St. James | 1730–1735 |
|  | • John Craig's son David Craig apprenticed to George Wickes, 1731–1735. <br> • John Craig became ill in 1735 and died in 1736. | | |
|  | Ann Craig <br> • Widow of John Craig. | corner of Norris St. and the Haymarket, Parish of St. James | 1735–1737 |
|  | Ann Craig & John Neville <br> • Ann Craig died in 1745. <br> • Succeeded by John Neville (see John Neville). | corner of Norris St. and the Haymarket, Parish of St. James | 1738–1745 |
| Iron works operator <br> Iron monger (dealer) <br> Merchant <br> Steel toy (hardware) maker | Ambrose Crawley | Greenwich | 1670–1682 |
| | Ambrose Crawley | Sunderland | 1682–1690 |
| | Ambrose Crawley | Winlaton Iron Works & Swalwell Iron Works | 1690–1715 |
| | • Winlaton Iron Works was located 15 miles from Shotley Bridge. <br> • Between 1704 and 1709, Crawley bought 37 varieties of sword blades from Herman Mohll, owner of a Shotley Bridge grinding and finishing sword blade mill. He paid from 7 to 14 shillings a dozen. <br> • At Winlaton, he made files, knives, saws, chisels, and hammers. <br> • At Swalwell, he made chains, pumps, cannon carriages, and ships anchors. | | |
| Cutler | John Craythorne <br> • Master of the Cutlers Company, 1560. | | 1550–D1569 |
| Sword cutler | Crellin & Co. | Parish of St. James | 1820–1825 |
| Silversmith <br> Goldsmith <br> Hilt maker <br> Sword cutler | Thomas Cressy <br> • Mounted small swords. | Pall Mall | 1710–D1720 |
| Gun maker <br> Bayonet maker <br> Furbisher | Henry Cripps (Crips) <br> • Master furbisher at the Tower of London, 1703–1704. <br> • Supplied plug bayonets to the Board of Ordnance. | | 1645–1705 |
| Goldsmith <br> Silversmith <br> Jeweler <br> Hilt maker <br> Sword cutler | John Cripps (Crips) | 21 Great St., Parish of St. Thomas Apostle | 1760–1767 |
| | John Cripps (Crips) | 43 Friday St. | 1768–1780 |
| | Cripps (John) & Francillan (William) | 43 Friday St. | 1781–1789 |
| | Cripps (John) & Francillan (William) | 24 Norfolk St., Strand | 1790–1816 |
| Belt maker <br> Sword cutler | John David Cripps (Crips) | 23 Cursitor St., Chancery Lane | 1829–1834 |
| | John David Cripps & Co. | 23 Cursitor St., Chancery Lane | 1835–1845 |
| | • Succeeded by Thomas Warren (see Thomas Warren). | | |
| Goldsmith <br> Silversmith <br> Hilt maker <br> Sword cutler | William Cripps (Crips) <br> • Apprenticed to David Willaume, 1731–1737. | | B1717–1742 |
| | William Cripps | Compton St. at the sign of the Crown & Golden Ball | 1743–1745 |
| | William Cripps | St. James St. at the sign of the Golden Ball | 1746–D1767 |
| | Mark Cripps | St. James St. | 1767–D1776 |

| | | | |
|---|---|---|---|
| | • Son of William Cripps. | at the sign of the Golden Ball | |
| Cutler | **Thomas Cross**<br>• Master of the Cutlers Company, 1750. | | 1740–1755 |
| Sword maker<br>Bayonet maker | **Nicholas Croucher (Crutcher)** | **North end, Long Lane<br>Aldersgate without** | 1692–1700 |
| | • Sword cutler to Samuel Pepys of the Naval Ordnance Department.<br>• Made plug bayonets, 1692.<br>• Had apprentice John Hopkins, 1692–1700 (see John Hopkins). | | |
| | **Nicholas Croucher** | **St. Pauls Churchyard, Cheapside** | 1701–1720 |
| | • His advertisement read as follows:<br>   *Nicholas Croucher, sword cutler*<br>   *(At the sign of ye Flaming Sword)*<br>   *St. Pauls churchyard ye corner of ye Booksellers Row*<br>   *– Cheapside*<br>   *Maketh & Selleth all sorts of swords, hangers, bayonets, corselets*<br>   *(body armour, chest and back plates) with all manner of belts*<br>   *(for swords) and also forbisheth old swords and hangers, at reasonable rates* | | |
| Cutler | **Thomas Croucher**<br>• Master of the Cutlers Company, 1754. | | 1744–1759 |
| Goldsmith<br>Hilt maker<br>Buckle maker<br>Sword cutler | **William Crowder (Sr.) (Crow)**<br>**William Crowder (Sr.)**<br>**Henry Crowder**<br>• Son of William Crowder (Sr.). | **2 Cox Court, Aldergate St.**<br>**126 Bunhill Row**<br>**1 Artillery Row<br>West Bunhill Row** | 1779–1786<br>1787–D1791<br>1792–1840 |
| Goldsmith<br>Hilt maker<br>Buckle maker<br>Sword cutler | **William Crowder (Jr.) (Crow)**<br>• Son of William Crowder Sr.<br>• Apprenticed to Gilder David Moorewater. 1792–1797.<br>**William Crowder (Jr.)** | **16 Curitor St., Charing Cross** | Bc1778–1797<br><br><br>1798–1810 |
| Army cap<br>and<br>accoutrement<br>maker<br>Sword belt<br>maker<br>Sword cutler | **James Cullum**<br>**James Cullum**<br>• Master of the Cutlers Company, 1755.<br>• James Cullum retired in 1785, died in 1786.<br>• His son Thomas Cullum (born 1752), who had worked in his father's shop, took over the business.<br>**Thomas Cullum**<br>• Appointed sword cutler and belt maker<br>  to His Majesty the King (George III), the Duke<br>  of Gloucester, and the Duke of Cumberland.<br>• Sold high-quality swords with silver hilts.<br>• Exported swords to the USA.<br>**Thomas Cullum**<br>• Thomas Cullum died in 1790.<br>• His son Matthew Cullum and Henry<br>  Spaulding took over the 12 Charing Cross shop.<br>  (see Henry Spaulding)<br>**Mary Cullum** | <br>**12 Charing Cross<br>Parish of St. Martin in the Field,<br>Westminster**<br><br>**12 Charing Cross<br>Parish of St. Martin in the Field, Westminster**<br><br><br><br><br>**9 Charing Cross<br>Parish of St. Martin in the Field,<br>Westminster**<br><br><br><br>**9 Charing Cross<br>Parish of St. Martin in the Field,<br>Westminster** | B1709–1748<br>1749–1785<br><br><br><br>1785–1789<br><br><br><br><br><br>D1790<br><br><br><br><br>1790–1795 |
| | • Mary Cullum, in poor health, hired John Prosser to manage the shops in April 1795.<br>• She sold the shop to John Prosser in June 1795 and died later that year.<br>  (see John Prosser) | | |

| Trade | Name | Address | Dates |
|---|---|---|---|
| Surgical instrument maker / Hilt maker / Sword cutler | **Cullum (Matthew) & Spalding (Henry)**<br>• Matthew was the son of Thomas and Mary Cullum.<br>• Same address as his father, Thomas Cullum.<br>**Matthew Cullum** | 12 Charing Cross<br>Parish of St. Martin in the Field, Westminster<br>12 Charing Cross<br>Parish of St. Martin in the Field, Westminster | 1790–1792<br><br><br>June 1792–1805 |
| Gold and silver lacemen / Hatter / Sword cutler | **Donald Currie (Curry)** | 6 Haymarket | 1811–1820 |
| Sword maker | **John Damm**<br>• Repaired swords for the Board of Ordnance, 1638–1642.<br>• From 1638–1642, proofed (inspected) sword blades delivered to the Tower by Benjamin Stone of Hounslow. | | 1630–1645 |
| Sword cutler | **J. (John) Daniels & Co.** | | 1820–1830 |
| Military tailor / Sword cutler | **Davidson, Dunnell & Bannister**<br>**Davidson, Bannister & Newham**<br>**Davidson & Newham**<br>**Davidson & Co.**<br>• Succeeded by John Le Gassick (see John Le Gassick). | 12 Cork St., Bond<br>12 Cork St., Bond<br>12 Cork St., Bond<br>12 Cork St., Bond | 1809–1822<br>1823–1832<br>1833–1837<br>1838–1844 |
| Sword cutler | **David Davies**<br>• Sword cutler and belt maker to the king.<br>• Exported swords, including eaglehead pommel infantry and artillery officers swords, to the USA.<br>**Davies (David) & Strong (John)** | 10 St. James<br><br><br><br>78 St. James St. | 1796–1811<br><br><br><br>1811–1815 |
| Sword cutler | **Edward Davies**<br>**Edward Davies**<br>**Edward Davies** | Covent Garden<br>17 Castle St., Holborn<br>4 Water St. Blackfriars | 1800–1801<br>1802–1805<br>1806–1810 |
| Sword cutler | **J. (John) Davies** | 340 Oxford St. | 1805–1827 |
| Sword cutler | **John Davies** | Fleet St., Parish of St. Bride | 1588–1603 |
| Gold and silver lacemen / Sword cutler | **Davies (Matthew) & Son**<br>• Lacemen to the army.<br>**Matthew Davies & Co.** | Hanover St.<br><br>103 St. Martins Lane | 1850–1859<br><br>1860–1889 |
| Cutler | **George Frederick Davis**<br>• Master of the Cutlers Company, 1855. | | 1845–1860 |
| Cutler | **William Davis**<br>• Master of the Cutlers Company, 1825 and 1826. | | 1815–1831 |
| Gun maker / Sword cutler | **George Henry Daw**<br>• Successor to Witton & Daw (see John Witton).<br>**Daw (George Henry) & Co.**<br>**Daw (George Henry) Gun Company** | 57 Threadneedle St.<br><br>57 Threadneedle St.<br>19 Great Winchester St. | 1861–1876<br><br>1877–1889<br>1890–1892 |
| Gun maker / Sword cutler | **David William Daw** | 21 Great St., Bishopgate | 1854–1856 |

| | | | |
|---|---|---|---|
| Sword cutler | **William Dawe** | | 1589–1604 |
| Gun maker | **Samuel Deacon** | 44 Goodge St. | 1832–1835 |
| Sword cutler | **Samuel Deacon** | 17 Frederick St., Hampstead Rd. | 1836 |
| Goldsmith<br>Silversmith | **Thomas Dealtry the Elder**<br>• Apprenticed to John Bennett Sr., 1742–1748 (see John Bennett Sr.). | | Bc1727–1762 |
| Hilt maker<br>Sword cutler<br>General cutler<br>Gun dealer | **Thomas Dealtry the Elder**<br>• His name appears on silver-mounted pistols.<br>• Master of the Cutlers Company, 1771.<br>• A 1780 trade card reads:<br>    *Thomas Dealtry*<br>    *Swords and all other cutlery ware*<br>    *At sign of the Flaming Sword*<br>    *Sweetings Alley, Royal Exchange*<br>    *Canes neatly fitted* | Sweeting Alley<br>Royal Exchange, Cornhill | 1763–1780 |
| | **Thomas Dealtry the Elder**<br>• Thomas Dealtry died in 1783.<br>• His widow Mary continued the business. | 85 Cornhill | 1781–D1783 |
| | **Mary Dealtry**<br>• Her son Thomas took over in 1785. | 85 Cornhill | 1783–1785 |
| | **Thomas Dealtry the Younger**<br>• A 1786 trade card reads:<br>    *Dealtry No. 85 Cornhill*<br>    *Makes & sells sword, arms & accoutrements*<br>• Mounted small swords. | 85 Cornhill | 1786–D1812 |
| Steel toy<br>(hardware) maker<br>Hilt maker | **William Deards (Sr.)** | Fleet St.<br>opposite St. Dunstan's Church | 1724–D1761 |
| Sword cutler | **William Deards (Jr.)** | Strand<br>West end of Craven St. | 1761–1764 |
| | **William (Jr.) & Mary Deards**<br><br>• Son and widow of William Deards (Sr.). | End of Pall Mall<br>near the Parish of St. James,<br>Haymarket, at the sign of the Star | 1765–1776 |
| | **William Deards (Jr.)** | Dover St., Piccadilly | 1777–1787 |
| Iron monger<br>(dealer) | **George & John Dean**<br>• Wholesale iron monger (iron dealer). | 41 Fish St. Hill,<br>corner of Arthur St. (at the monument) | 1830–1837 |
| Gun and pistol<br>warehouse | **George & John Dean**<br>• Gun and pistol warehouse. | 46 King William St. | 1838–1845 |
| Sword cutler<br>Jeweler | **George & John Dean** | 30 King William St. | 1846–1851 |
| Gold and silver<br>lacemen | **Joseph Dean** | 4 Strand | 1791–1803 |
| | **Joseph Dean** | 2 Craven St., Strand | 1804–1806 |
| Sword cutler | **Joseph Dean** | 9 Strand | 1807–D1808 |
| | **Prisella Dean**<br>• Widow of Joseph Dean. | 9 Strand | 1809–1822 |
| Sword blade<br>maker | **William Dean** | | 1687 |

| | | | |
|---|---|---|---|
| Sword cutler | **Deane & Son** | **London Bridge** | **1840–1880** |
| | • Sold M1821 light cavalry sabers. | | |
| Sword cutler | **J. Dece & Sons** | **13 Condust St.** | **1950–1960** |
| | • Sold swords made by the Wilkinson Sword Company. | | |
| Silversmith<br>Hilt maker<br>Sword cutler | **William, Henry & Louis Dee**<br>• Made silver sword hilts and decorated some presentation swords for<br>• Widdowson & Veal (see Widdowson & Veal). | | **1850–1875** |
| Jeweler | **Delafons (Joseph Sr.) & Sons** | **12 Broadway, Blackfriars** | **1790–1793** |
| Goldsmith | **Joseph Delafons (Jr.)** | **12 Broadway, Blackfriars** | **1794–1799** |
| Hilt maker | **Joseph Delafons (Jr.)** | **12 Broadway, Blackfriars** | **1800–1816** |
| Sword cutler | **Delafons (Joseph (Jr.) & Sons** | **41 & 42 Sackville St. &**<br>**17 Great Ryder St.** | **1817–1827** |
| | • At the Sackville Street address, successors to Gray & Constable (see Thomas Gray). | | |
| | **Delafons (Joseph (Jr.) & Sons** | **41 & 42 Sackville St.** | **1828–1830** |
| | **James Delafons** | **41 & 42 Sackville St.** | **1831–1833** |
| Cutler | **Thomas Dermer** | | **1664–1679** |
| | • Master of the Cutlers Company, 1674. | | |
| Hafter<br>(sword hilt<br>maker) | **John DeWare** | **Parish of St. Mildred poultry** | **1280–1315** |
| Cutler | **John Deye (Deyer, Defer)** | | **1457–1480** |
| | • Master of the Cutlers Company, 1467–1475. | | |
| Sword maker | **John Deyer (Defer)** | **Parish of the St. Clemen Danes** | **1660–1670** |
| | • The Cutlers Company of London denied Deyer employment with sword maker and armourer Edward Yonger the Younger (a German) in 1661 because he had not served an apprenticeship. He probably had immigrated to London from Solingen, Prussia. | | |
| Sword blade<br>maker<br>Damascener | **William Dickenson**<br>• Made Damascus sword blades, 1586–1587. | | **1580–1590** |
| Sword cutler | **David Dixon** | **Fenchurch St.** | **1827** |
| Bladesmith | **John Dober** | | **1480–1488** |
| Jeweler | **Phillip George Dodd** | **11 Camomile St., Bishopsgate** | **1827–1835** |
| Diamond | **Phillip George Dodd** | **25 Leadenhall St.** | **1836–1841** |
| merchant | **Phillip George Dodd** | **79 Cornhill** | **1842–1854** |
| Sword cutler | **Phillip George Dodd** | **45 & 79 Cornhill** | **1855–1856** |
| | **P. Phillip) G. (George) Dodd & Son** | **45 Cornhill** | **1857–1864** |
| | **P. (Phillip) G. (George) Dodd & Son** | **42 Cornhill** | **1865** |
| | **P. (Phillip) G. (George) Dodd & Son** | **42 Cornhill & 146 Leadenhall St.** | **1866–1923** |
| | **P. (Phillip) G. (George) Dodd & Son** | **42 Cornhill** | **1924–1965** |
| Sword cutler | **Thomas Dodson (Dowdon)** | **Fleet St., Parish of St. Bride** | **1582–D1589** |
| | **John Dodson (Dowdon)** | **Fleet St., Parish of St. Bride** | **1590–1595** |

| | | | |
|---|---|---|---|
| Gun maker<br>Sword cutler | **Thomas Doo** | 15 Maddox St. | 1880–1889 |
| Sword cutler | **D. Dote** | Conduit St. | 1875 |
| | **Benjamin Doughty**<br>(see Phipson, Doughty & Co.) | | |
| Cutler | **Edward Dowling**<br>• Master of the Cutlers Company, 1794. | | 1784–1799 |
| Cutler | **John Dowling**<br>• Master of the Cutlers Company, 1806. | | 1796–1811 |
| Goldsmith<br>Silversmith<br>Jeweler | **Dru Drury (Sr.)**<br>• Apprenticed to William Gardiner, 1703–1710 (see William Gardiner).<br>• Became freeman of the Goldsmiths Company, 1711. | | B1688–1710 |
| Sword and knife<br>hilt maker<br>Sword and<br>knife cutler | **Dru Drury (Sr.)**<br>• Had apprentice Thomas Moulden, 1713–1717.<br>• Had apprentice and journeyman John Pont, 1713–1731.<br>• Cutler to the king. | Noble St., near Goldsmiths Hall | 1711–1716 |
| | **Dru Drury (Sr.)**<br>• Had apprentice Samuel Mansell, 1754–1761.<br>• Had apprentice Moses Brent, 1763–1767.<br>• Made some M1757 basket hilt broadswords. | Lad Lane, Wood St. | 1717–D1767 |
| | **Dru Drury (Jr.)**<br>• Apprenticed to his father Dru Drury Sr., 1739–1745.<br>• Freeman of the Goldsmiths Company, 1746.<br>• Worked in his father's shop, 1746–1766.<br>• Succeeded his father, 1767.<br>• Entered his mark with the Goldsmiths Company, 1767. | | B1725–1766 |
| | **Dru Drury (Jr.)**<br>• Bought knife blades from Samuel Trickett (Sr.) and James Woolhouse of Sheffield. | Lad Lane, Wood St. | 1767–1769 |
| | **Jeffrey's (Nathaniel) & Drury (Dru Jr.)**<br>• Nathaniel Jeffrey's died, 1779. | 32 Strand, corner of Villiers St.,<br>Westminster | 1770–1779 |
| | **Dru Drury (Jr.) & Son (William)**<br>• Appointed cutler to His Majesty the King.<br>• Subcontracted with many goldsmiths and silversmiths to decorate his hilts.<br>• Son and partner William Drury (born 1752) apprenticed and worked for his father, 1766–1795.<br>• Supplied Scottish pistols to the Ordnance Department for the 74th and 75th Regiment of Foot.<br>• Dru Drury Jr. retired, 1795.<br>• Dru Drury Jr. died, 1804.<br>• William succeeded Dru Drury Jr., 1796 | 32 Strand, corner of Villiers St.,<br>Westminster | 1780–1796 |
| Jeweler<br>Goldsmith<br>Silversmith | **William Drury**<br>• William Drury died in 1826, succeeded by Dru Drury III (born 1767). | 32 Strand, corner of Villiers St.,<br>Westminster | 1796–D1826 |
| Sword and knife<br>hilt maker | **D. (Dru) Drury III & Son** | 32 Strand, corner of Villiers St.,<br>Westminster | 1827–1831 |
| Sword and<br>knife cutler | **George Drury**<br>• Son of Dru Drury III.<br>• The Drurys sold many iron basket hilt swords, military swords, small swords<br>  knives, and dirks.<br>• The Drurys supplied basket hilt swords to English Highland regiments. | 32 Strand, corner of Villiers St.,<br>Westminster | 1832–1851 |

- The Drurys bought blades from Birmingham sword and blade makers Samuel Harvey I, II, and III.

| | | | |
|---|---|---|---|
| Cutler | **James Dunnage** | | 1799–1814 |
| | • Master of the Cutlers Company 1809. | | |
| Cutler | **John Dunnage** | | 1781–1796 |
| | • Master of the Cutlers Company, 1791. | | |
| Cutler | **Thomas Dunnage** | | 1853–1868 |
| | • Master of the Cutlers Company, 1863. | | |
| Cutler | **Thomas Alfred Dunnage** | | 1893–1903 |
| | • Master of the Cutlers Company, 1903. | | |
| Silversmith Goldsmith Hilt maker Cutler | **Leslie G. Durbin** | | 1940–1960 |
| | • Designed and supplied presentation swords to the Goldsmiths and Silversmiths Co. Ltd. | | |
| Goldsmith Silversmith Cord wainer Scabbard mount maker Silver hilt maker | **John Dutton** | | Bc1752–1773 |
| | • Apprenticed to Thomas Burch, goldsmith and cord winder (maker) leather worker, 1766–1773. | | |
| | • Made silver mounts for leather-covered goods (leather sword scabbards). | | |
| | • Made silver hilts. | | |
| | **John Dutton** | 6 Dartsmouth Row, Westminster | 1773–1805 |
| | • Partner and son: Joseph Dutton, c. 1786. | | |
| | **Woolley (James), Deakin (Thomas) & Co.** | 74 Edmund St. | 1805–1809 |
| | • Partners: John and Joseph Dutton. | | |
| | **Woolley (James), Deakin (Thomas), Dutton (John & Joseph), & Johnston (Richard)** | 74 Edmund St. | 1809–1811 |
| | (see Thomas Deakin, Birmingham) | | |
| | (see Richard Johnston) | | |
| | **Woolley (James), Deakin (Thomas) & Dutton (John & Joseph)** | 74 Edmund St. | 1811–1814 |
| | • Succeeded by Woolley & Sargant. | | |
| | (see William Sargant) | | |
| | (see James Woolley, Birmingham) | | |
| Sword cutler | **Ede & Son** | 93 & 94 Chancery Lane | 1840–1900 |
| | • Sold Wilkinson Sword Company swords. | | |
| | **Ede & Ravenscroft** | 93 & 94 Chancery Lane | 1900 |
| Army and navy outfitter Sword cutler | **Benjamin Edginton Ltd.** | | 1870–1875 |
| | • Joined S.W. Silver & Co., 1876. | | |
| | (see Silver & Co.) | | |
| Goldsmith Hilt maker Sword cutler | **John Edington (Egington)** | Handsworth near Soho | 1775–1778 |
| | **John Edington (Egington)** | 10 Portland St., Soho Square | 1799–1818 |
| | • Sold M1796 infantry officers swords. | | |
| | • John Edington died in 1811. | | |
| | • His wife took over the business, 1811–1818. | | |
| | **John James Edington** | 10 Portland St., Soho Square | 1819–1840 |
| | • Son of John Edington. | | |
| | • Apprenticed to Stephen Gaubert when his father died in 1811 (1811–1818). | | |

| | | | |
|---|---|---|---|
| Sword cutler | **John Edkins** | | 1871–1886 |

• Master of the Cutlers Company, 1881.

| | | | |
|---|---|---|---|
| Sword cutler | **John Edmed** | Parish of St. Magnus | 1621–1628 |

| | | | |
|---|---|---|---|
| Pike maker | **John Edwards** | Philip Lane | 1642–1649 |

- Between April 3, 1645, and September 22, 1645, during the English Civil War, made 1,600 long pikes at 4 shillings, 2d each for the "New Model" army (parliamentary forces) under Gen. Sir Thomas Fairfax.
- Infantry long pikes were 16 feet long with steel heads.
- Contract dates and amounts:

    April 3, 1645     600 pikes
    May 15, 1645     200 pikes
    July 10, 1645     200 pikes
    September 22, 1645     600 pikes
    December 22, 1645     500 pikes

- The pikes were to be delivered to the Tower 125 at a time by the last day of December 1645 and in January, February, and March of 1646.
- In December 1645, had three more contracts for 250 long pikes, totaling 750 at 3 shillings, 10d each.

| | | | |
|---|---|---|---|
| Military tailor | **Joel Edwards** | 9 Greville St., Hatton Garden | 1802–1828 |
| Sword cutler | **Joel Edwards** | 2 Hanover Square | 1829–1831 |
| | **Joel Edwards** | 9 Hanover Square | 1832–1837 |
| | **Joel Edwards & Sons** | 9 Hanover Square | 1838–1929 |
| | **Joel Edwards & Sons** | 13 Braton St. | 1930–1941 |
| | **Joel Edwards & Sons** | 5 Cork St. | 1942–1943 |

(see William Frisbee)

| | | | |
|---|---|---|---|
| Sword maker | **John Edwards** | Philip Lane, near Cripplegate | 1629–1650 |

• Made swords and pikes.

| | | | |
|---|---|---|---|
| Pole arm maker | **William Edwards** | Philip Lane, near Cripplegate | 1650–1675 |

- Bandoleer maker to the Board of Ordnance, 1656.
- Advertised in the *Mercurius Publicus*, 1661:

    *Pole arms*
    *Leading staves for captains*
    *Partizons for lieutenants*
    *Ensign staves*
    *Halberts of all sorts*
    *Pikes*
    *Coronet staves for horse*
    *Javelins for Sheriffs officers to ride curcuits*
    *Divers* [diverse] *other instruments*

| | | | |
|---|---|---|---|
| Military tailor | **Thomas Edwards** | 4 Charing Cross | 1846–1860 |
| Sword cutler | | | |

• Succeeded William Vernon at that address (see William Vernon).

| | | | |
|---|---|---|---|
| Silversmith | **Thomas Edwards** | | 1661 |
| Hilt maker | | | |
| Sword cutler | | | |

• Signed an advertisement warning sword purchasers about counterfeit cast silver hilts.

| | | | |
|---|---|---|---|
| Gun maker | **Urs Christian Egg** | Oberbuchsiten Switzerland | B1748–1772 |
| Sword cutler | | | |

- Immigrated to London, 1772.
- Changed his name to Durs Egg.
- Worked for gun maker John Twigg at 132 Strand, 1772–1778.

|  |  |  |  |
|---|---|---|---|
|  | Durs Egg | 24 Princes St., Leicester Fields | 1778–1787 |

• Appointed gunsmith to His Majesty the King and the Prince of Wales.
• Had a gun factory at 35 Mansell St., 1788–1815.

|  |  |  |
|---|---|---|
| Durs Egg | 38 Haymarket & 1 Coventry St. | 1788–1804 |
| Durs Egg | 132 Strand, near Somerset House | 1805–1815 |
| Durs Egg | 1 Opera Colonnade, Pall Mall | 1816–1838 |
| Durs Egg | 10 Opera Colonnade, Pall Mall | 1839–1841 |
| Durs Egg | 4 Opera Colonnade, Pall Mall | 1842–1866 |

• Also made bayonets for the Board of Ordnance.
• Made M1788 light cavalry sabers.
• Bought blades from J.J. Runkel (see J. Runkel).
• Durs Egg died in 1831, but his business continued under his name.
• His son, John, and grandsons continued the business, even though they had their own separate business's (see below).

| | | | |
|---|---|---|---|
| Gun maker | John Egg | | Bc1772–1826 |

• Son of Durs Egg.
• Apprenticed under his father, Durs Egg, c. 1787–1792.
• Worked at his father's shop, 1792–1826.

| | | |
|---|---|---|
| John Egg | 26 Princes St., Leicester Fields | 1826–1837 |
| John Egg | 20 Haymarket | D1838 |

• Ran his father's business from his father's death in 1831 until he died in 1838.
• John's nephews, Charles and Henry Egg (sons of his brother, Joseph), continued to run Durs Egg's business, 1838–1868.

| | | | |
|---|---|---|---|
| Gun maker<br>Sword cutler | Joseph Egg | | Bc1783–1802 |

• Son of Durs Egg.

| | | |
|---|---|---|
| Tatham (Henry) & Egg (Joseph) | 37 Charing Cross | 1803–1816 |
| Joseph Egg | 1 Piccadilly | 1817–1831 |
| Joseph Egg | 28 Tichbourne St., Halborn & 1 Piccadilly | 1832–1834 |
| Joseph Egg & Son (Charles) | 28 Tichbourne St., Halborn & 1 Piccadilly | 1835–1840 |

• Joseph Egg died in 1837.

| | | |
|---|---|---|
| Charles & Henry Egg | 28 Tichbourne St., Halborn & 1 Piccadilly | 1841–1850 |

• Sons of Joseph Egg.

| | | |
|---|---|---|
| Henry Egg | 28 Tichbourne St., Halborn & 1 Piccadilly | 1851–D1868 |

• Henry Egg died in 1868.

| | | |
|---|---|---|
| Henry William Egg | 28 Tichbourne St., Halborn & 1 Piccadilly | 1869–1886 |

• Son of Henry Egg.

| | | | |
|---|---|---|---|
| Goldsmith<br>Silversmith | George Richard Elkington | 74 Hatton Garden | 1840–1843 |
| | • Succeeded John George Fearn (see John George Fearn). | | |
| Gold and silver ornamentor | George Richard Elkington | 22 Regent St. | 1844 |
| | Elkington (George Richard) & Co. | 22 Regent St. | 1844–1886 |
| Electroplater | Elkington Co. Ltd. | 22 Regent St. | 1887–1935 |
| Gilder | • By appointment, sword cutler and goldsmith to the Prince of Wales. | | |
| Hilt maker | Elkington Co. Ltd. | 136 Regent St. | 1936–1955 |
| Sword cutler | Elkington Co. Ltd. | Terminal House, Grosvenor Gardens | 1956–1964 |
| | Elkington Sales Ltd. | Aldwych House, Aldwych | 1965–1968 |
| Cutler | George Ellis | | 1594–1610 |
| | • Master of the Cutlers Company, 1604 and 1605. | | |
| Bladesmith | Richard Elyot (Eliot) | Parish of St. Sepulchre | 1410–1420 |
| | • Master of the Bladesmiths Company, 1417. | | |

# THE PETER ENGLISH FAMILY

| | | | |
|---|---|---|---|
| Sword maker | **Peter Munsten the Younger** | Solingen, Prussia | c1609–1629 |

- Immigrated to the sword and blade center at Hounslow Heath outside of London, 1629.
- Set up a sword and blade factory (mill).
- Changed his name to Peter English.

| | | | |
|---|---|---|---|
| | **Peter English (Sr.)** | Hounslow Heath | 1629–1642 |

- At the onset of the Civil War in 1642, parliamentary forces seized his factory and converted it to a powder mill.
- English then moved to Oxford to work for King Charles I.

| | | | |
|---|---|---|---|
| | **Peter English (Sr.)** | Oxford | 1642–c1649 |

- At the end of the Civil War, moved to London.

| | | | |
|---|---|---|---|
| | **Peter English (Sr.)** | London | c1649–1675 |

- In 1672 Henry Hoppe (Hoppie) and Peter Munsten (English) petitioned King Charles II for a charter to set up a factory (mill) to make hollow ground small sword blades (see Chapter 3). No charter was issued.
- A factory (mill) was chartered by King William III in 1691 for the manufacture of hollow ground sword blades to a new syndicate (see Chapter 4).

| | | | |
|---|---|---|---|
| Sword maker | **Peter English (Jr.)** | London | 1675–1697 |

- Son of Peter English (Sr.) of Hounslow Heath, Oxford, and London.
- Appointed as sword cutler to the Board of Ordnance, March 25, 1689.
- Also an arms furbisher at the Tower of London, 1686–1697, receiving a quarterly retaining fee.
- On December 15, 1687, he was paid 143 pounds, 7 shilling, 4 denier for sword work (described as *"new ground, glazed, oyled, scabbered and ye hilts blackt"*).

**Paul Ernions**
(see Chapter 3)

**Everard Ernions**
(see Chapter 3)

| | | | |
|---|---|---|---|
| Military tailor Sword cutler | **Errington & Harrison** | 26 Great Tower | 1925–1926 |

- Succeeded by Glen, Errington, & Harrison (see Glen, Errington, & Harrison).

| | | | |
|---|---|---|---|
| Army accoutrement maker | Sir James Esdaile (Sr.) | 110 Bunhill Row | 1767–1790 |
| | James (Jr.), John & Joseph Esdaile | 110 Bunhill Row | 1791–1821 |
| | James Esdaile (Jr.) | 110 Bunhill Row | 1822–1833 |
| Sword cutler | Evan Evans | 8 George Yard, Lombard St. | 1771–1776 |
| Army outfitter | Evan Evans | 134 Leadenhill St. | 1777–1778 |
| Woolen draper (cloth maker) | Evans (Evan), Passman (John) & Co. | 134 Leadenhill St. | 1779–1790 |
| Sword cutler | Evans (Evan) & Welch (James) | 134 Leadenhill St. | 1791–1794 |

- Succeeded by Welch & Stalker (see Welch & Stalker).

| | | | |
|---|---|---|---|
| Goldsmith Hilt maker | John Evans | | Bc1789–1810 |

- Apprenticed to Joseph Clarke, 1803–1810.

| | | | |
|---|---|---|---|
| Sword cutler | John Evans & Samuel Wheatley | 3 Old Change St., Goswell St. | 1810–1829 |
| | John Evans | 3 Od Change St., Goswell St. | 1830–1832 |

**Stephen & Thomas Evans**
(see Chapter 4)

| | | | |
|---|---|---|---|
| Cutler | John Llewellyn Evans | | 1864–1879 |

- Master of the Cutlers Company, 1874.

| | | | |
|---|---|---|---|
| Cutler | **William Falkener**<br>• Master of the Cutlers Company, 1739. | | 1729–1744 |
| | **Thomas Fattorini**<br>(see Birmingham listing) | | |
| Cutler | **Edward Faulkingham**<br>• Master of the Cutlers Company, 1683. | | 1673–1788 |
| Hatter<br>Gold and silver<br>laceman<br>Sword cutler | **Edward Fayle**<br>• Advertised in 1768:<br>    *Hatter & Sword Cutler*<br>    *(At the sign of the Hat & Cross daggers)*<br>    *Near Serjeants Inn in Fleet St.*<br>    *(#48) sells all sorts of hats*<br>    *Gold & Silver laces, swords and hangers etc.*<br>    *Wholesale & Retail at the lowest prices* | 48 Fleet St., near Serjeants Inn | 1766–1775 |
| Silversmith<br>Hilt maker<br>Sword cutler | **George Fayle (Sr.)** | Gillots Court,<br>Little Old Bailey | May 1787–July 1787 |
| | **George Fayle (Sr.)**<br>• Sons and partners: George (Jr.) and John. | Dogwell Court, White Friars | July 1767–D1771 |
| | **Sarah Fayle**<br>• Widow of George Fayle Sr.<br>• Sons and partners: George (Jr.) and John. | Kings Head Court<br>Helen Lane | 1771 |
| | **George Fayle (Jr.)**<br>• Son of George (Sr.) and Sarah Fayle.<br>• Also a buckle maker. | Dogwell Court<br>White Friars | 1772–1782 |
| | **John Fayle**<br>• Son of George (Sr.) & Sarah Fayle.<br>• Made silver hilts for Edward Loxham. | 31 Wilderuess Lane<br>Salisbury Ct., Fleet St. | 1772–1785 |
| Goldsmith | **John George Fearn** | 114 Strand | 1802–1803 |
| Jeweler | **John George Fearn** | 73 Strand | 1804–1820 |
| Hilt maker | **John George Fearn** | 73 Strand & 18 Cornhill | 1821–1823 |
| Sword cutler | **John George Fearn** | 18 Cornhill | 1824–1834 |
| | **John George Fearn** | 18 Cornhill & 22 Regent St. | 1835–1841 |
| | **John George Fearn** | 22 Regent St. | 1842–1843 |
| | • Succeeded by Elkington & Co. (see Elkington & Co.). | | |
| Goldsmith | **William Fearn** | 43 Lombard St. | 1772–1779 |
| Jeweler | **Samuel Fearn** | 43 Lombard St. | 1780–1796 |
| Hilt maker | **John Fearn** | 43 Lombard St. | 1797–1798 |
| Sword cutler | **Joseph Fearn** | 43 Lombard St. | 1799–1805 |
| | **Joseph Fearn** | 10 Cornhill | 1805–1813 |
| | • Goldsmith and jeweler to the Prince of Wales (later Prince Regent). | | |
| Silversmith<br>Hilt maker<br>Sword cutler | **Matthew Feesey** | Pall Mall | 1725–1758 |
| | **John Feesey** | Pall Mall | 1759–1795 |
| | • Appointed sword cutler to His Majesty the King.<br>• Made silver-hilted small swords.<br>• Made "state" sword of U.S. president George Washington (a silver-hilted small sword). | | |
| Sword cutler | **John Fell** | | 1644 |

| | | | |
|---|---|---|---|
| Sword cutler | **Roger Fentham** | Westminster | 1597–1621 |
| Sword cutler | **Henry Fentham**<br>• Son of Roger Fentham. | Westminster | 1617–1632 |
| | **Fenton Bros.**<br>(see Sheffield listing) | | |
| Goldsmith<br>Hilt maker<br>Sword cutler | **John Field** | Fleet St.,<br>Parish of St. Bride | 1594–1604 |
| Silversmith<br>Gun maker<br>Hilt maker | **John Field (Sr.)** | 233 High Holborn | Bc1760<br>1783–1788 |
| | **Field (John Sr.) & Clarke (Henry)** | 233 High Holborn | 1789 |
| | **John Field (Jr.)**<br>• John Field Sr. died in 1791.<br>• John Field Jr. (born c. 1780) took over. | 233 High Holborn | 1790–D1791 |
| | **Field (John Jr.) & Co.**<br>• Partner: William Parker.<br>• Started to mount swords and make guns.<br>• William Parker's daughter married John Field Jr. | 233 High Holborn | 1791–1793 |
| | **Field (John Jr.) & Parker (William)** | 233 High Holborn | 1794–1796 |
| | **William Parker & Co.**<br>• Partner: John Field Jr.<br>• Appointed gun makers to His Royal Highness the Duke of Kent. | 233 High Holborn | 1797–1839 |
| | **William Parker & John Field (Jr.)**<br>• Appointed gun makers to His Majesty the King and, later, Her Majesty the Queen, the Board of Ordnance, and the East India Company.<br>• William Parker died in 1840.<br>• John Field Jr. took over the company but kept William Parker as part of the official company name. The sons of John Field now joined the company. | 233 High Holborn | 1840–1841 |
| | **William Parker & Field (John Jr.) & Sons**<br>• Partner and son: John William Parker Field.<br>• Partner and son: William Shakespeare Parker Field. | 233 High Holburn | 1842–1849 |
| | **William Parker & Field (John Jr.) & Sons** | 58 Mansell St. & 233 High Holburn | 1850–1870 |
| | **William Parker & Field (John Jr.) & Sons** | 62 Tenters St. & 233 High Holburn | 1871–1875 |
| | **William Parker & Field (John Jr.) & Sons** | 233 High Holburn | 1876 |
| | **William Parker & Field (John Jr.) & Sons** | 59 Leman St. | 1877–1879 |
| | **William Parker & Field (John Jr.) & Sons** | 122 Leman St. | 1880–1882 |
| | **William Parker & Field (John Jr.) & Sons** | 22 Tavistock St., Covent Garden | 1883–1886 |
| Sword cutler | **Charles Finch** | | 1607 |
| Metal button maker<br>Silversmith<br>Sword cutler | **Thomas Firming**<br>• Immigrated from Paris, France. | 3 Kings Court<br>Lombard St. | 1677–1690 |
| Badge maker<br>Metal button maker<br>Silversmith<br>Sword cutler | **Samuel Firming (Sr.)**<br>• Immigrated from Paris, France.<br>• Changed his name to Firmin. | Opposite the New Church, Strand | 1763–1768 |
| | **Samuel Firmin (Sr.)**<br>• Appointed button maker to His Majesty the King and His Royal Highness the Prince of Wales. | near Somerset House, Strand | 1769–1782 |

|  |  |  |  |
|---|---|---|---|
|  | Firmin (Samuel Sr.) & Son (Phillip Sr.) | 153 Strand | 1783–1796 |
|  | Firmin (Samuel (Sr.) & Westall (Henry) | 153 Strand | 1797–1811 |
|  | Phillip Firmin (Sr.) | 153 Strand | 1812–1814 |
|  | Firmin (Phillip Sr.) & Langsdale (John) | 153 Strand & 10 Clare St., Drury Lane | 1815–1821 |
|  | Firmin (Phillip Venner) & Sons (Richard and Samuel Jr.) | 153 Strand & 10 Clare St., Drury Lane | 1822–1828 |
|  | P. (Phillip Venner) & R. (Richard) Firmin<br>• Partner and son: Samuel Firmin Jr. | 153 Strand & 10 Clare St. Drury Lane & Whitehorse Yard | 1829–1834 |
|  | P. (Phillip Venner), R. (Richard) & S. (Samuel Jr.) Firmin | 153 Strand | 1835–1837 |
|  | Samuel Firmin (Jr.) & Sons | 153 Strand | 1838 |
|  | Firmin (Samuel Jr.) & King (Henry) | 153 Strand & 13 Conduit St., Bond St. | 1839 |
|  | S. (Samuel (Jr.) & P. (Phillip Jr.) Firmin<br>• Philip Jr. was the son of Samuel Jr.<br>• Became sword cutlers in 1840. | 153 Strand & 13 Conduit St., Bond St. | 1840–1849 |
|  | Philip (Jr.) Firmin & Son (Philip Victor)<br>• Exhibited army and navy swords at the London International Exposition, 1851. | 153 Strand & 13 Conduit St., Bond St. | 1850–1854 |
|  | P.V. (Philip Victor) Firmin & Sons | 153 Strand & 13 Conduit St., Bond St. | 1855 |
|  | Firmin (Philip Victor) & Sons Ltd. | 153 Strand & 13 Conduit St., Bond St. | 1856–1860 |
|  | Firmin (Philip Victor) & Sons Ltd. | 153 & 154 Strand & 13 Conduit St. | 1861–1863 |
|  | Firmin (Philip Victor) & Sons Ltd. | 153 to 155 Strand & 13 Conduit St. | 1864–1879 |
|  | Firmin & Sons Ltd. | 153 to 155 Strand & 47 Warwick St. | 1880–1894 |
|  | Firmin & Sons Ltd. | 108–109 St. Martins Lane & 47 Warwick St | 1895–1904 |
|  | Firmin & Sons Ltd. | 108–109 St. Martins Lane & 6 Warwick St | 1905–1915 |
|  | Firmin & Sons Ltd.<br>• Button makers only.<br>• Offices in Birmingham, c. 1980–1998.<br>• Offices in Paris, France, 1861–1894.<br>• Offices in Dublin, Ireland, 1861–1894. | 8 Cork St. | 1916–2003 |
| Gun maker | Charles Fisher | 8 Princes St., Leicester Square, Soho | 1831–1877 |
| Sword cutler | Charles Fisher<br>• Gun and archery warehouse. | 16 Wardour St. | 1878–1881 |
| Cutler | George Fisher<br>• Master of the Cutlers Company, 1747. |  | 1737–1752 |
| Silversmith<br>Hilt maker<br>Sword cutler | James Fisher | 7 Gun Powder Alley, Shoe Lane | 1779–1800 |
| Cutler | John Fisher<br>• Master of the Cutlers Company, 1730 and 1731. |  | 1720–1736 |
| Cutler | Joseph Fisher (Sr.)<br>• Master of the Cutlers Company, 1772. |  | 1762–1777 |
| Cutler | Joseph Fisher (Jr.)<br>• Master of the Cutlers Company, 1860. |  | 1850–1865 |

| | | | |
|---|---|---|---|
| Cutler | **John Flampstead** | | 1657–1672 |
| | • Master of the Cutlers Company, 1667. | | |
| Sword cutler | Isaah Fleureau | Lower end of St. James St. at the sign of the Kings Arms | 1743–1765 |
| | Isaah Fleureau | Haymarket | 1766–1783 |
| | Isaah Fleureau | 5 Newport St. | 1784–1790 |
| Military tailor | C. Charles) Flight | 17 Golden Square | 1824–1832 |
| Breeches | C. (Charles) Flight | 14 St. James St. | 1833–1836 |
| (pant) maker | Flight (Charles) & Son | 12 Park Place, Parish of St. James | 1837–1840 |
| Sword cutler | Flight (Charles) & Christopher (James) | 12 Park Place, Parish of St. James | 1841 |
| | Charles Flight & Co. Ltd. | 32 Pall Mall | 1842–1849 |
| | Charles Flight & Co. Ltd. | 24 Sackville St. | 1850–1851 |
| | Charles Flight & Co. Ltd. | 9 Opera Arcade | 1852–1854 |
| | Charles Flight & Co. Ltd. | 24 Jermyn St. | 1855 |
| | Charles Flight & Co. Ltd. | 41 Great Pulteney St., Golden Square | 1856–1857 |
| Military tailor | F. (Frederick) W. (William) Flight & Son | 5 New Burlington St. | 1900–1919 |
| Breeches | Flights (Frederick & William) Ltd. | 4 New Burlington St. | 1920–1950 |
| (pant) maker | Flights (Frederick & William) Ltd. | 97 Bond St. | 1951–1956 |
| Sword cutler | • Had branches in Winchester, Camberly, Catterick Camp, Salisbury, and Woolwich. | | |
| | • Sold swords made by the Wilkinson Sword Company. | | |
| Sword cutler | **Abraham Flint (Flynt)** | | 1600–1620 |
| Silversmith | Nicholas Flint | 1 Clements Lane, Temple Bar | 1776–1777 |
| Hilt maker | Nicholas Flint | 29 Greenhills Rents, Smithfield | 1778–1788 |
| Sword cutler | | | |
| Sword maker | **William Floyd** | | 1607 |
| Silversmith | Andrew Fogelberg | St. Annes Court, Dean St., Soho | 1767–1772 |
| Goldsmith | | | |
| Hilt maker | Andrew Fogelberg | 29 Church St. Parish of St. Anne, Soho | 1773–1779 |
| Sword cutler | | | |
| | Fogelberg (Andrew) & Gilbert (Stephen) | 29 Church St. Parish of St. Anne, Soho | 1780–1793 |
| | Andrew Fogelberg | 29 Church St. Parish of St. Anne, Soho | 1794–D1815 |
| | • Mounted some Patriotic Fund swords. | | |
| Sword cutler | **B. Folkand** | Jermyn St., Parish of St. James | 1821–1825 |
| Sword maker | **William Foote** | | 1683–1711 |
| | • Used the "bunch of grapes with long stem" mark. | | |
| Military tailor | James Forrest | 20 Cork St., Bond St. | 1839–1843 |
| Sword cutler | Forrest (James) & Kerry (Richard) | 20 Cork St., Bond St. | 1844–1846 |
| | Richard Kerry | 20 Cord St., Bond St. | 1847 |
| | Mary Anne Forrest & Son (George) | 20 Cord St., Bond St. | 1848–1853 |
| | Forrest (Mary Anne) & Son (George) | 20 Cord St., Bond St. | 1854–1867 |
| | George Forrest | 20 Cord St., Bond St. | 1868–1881 |

| | | | |
|---|---|---|---|
| Sword cutler | **William Forster**<br>• Supplied swords to the Board of Ordnance, 1611. | | 1600–1615 |
| Silversmith | **John Foster**<br>• Apprenticed to John Bishop, 1753–1763. | | Bc1738–1764 |
| Hilt maker<br>Sword cutler | **John & George Foster**<br>• John and George were brothers.<br>**John Foster** | Carter Lane<br><br>Little Carter Lane | 1764<br><br>1764 |
| Goldsmith<br>Hilt maker<br>Sword cutler | **John Foster**<br>**E. (Edward) Foster** | 1 Bartletts Passage, Holborn<br>32 Bartletts Passage, Holborn | 1797–1808<br>1809–1816 |
| Army and navy<br>accoutrement<br>maker<br><br>Sword belt<br>maker<br>Sword cutler | **Robert Foster**<br>• Successor to Bland & Foster (see Bland & Foster).<br>• Appointed sword cutler to His Majesty the King, His Royal Highness the Prince of Wales, and the Duke of York.<br>**Robert Foster & Richard Johnston**<br>• Robert Foster died in 1798.<br>• Succeeded by Richard Johnston at this address (see Richard Johnston). | 68 St. James, Pall Mall<br><br><br><br>68 St. James, Pall Mall | 1791–1797<br><br><br><br>1798 |
| Silversmith<br>Jeweler<br>Hilt maker<br>Sword cutler | **Thomas Foster (Sr.)**<br>• Apprenticed to Josiah Daniel, 1754–1768.<br>**Thomas Foster (Sr.)**<br><br>**John Foster**<br><br>• Son and successor to Thomas Foster (Sr.).<br>**John Foster**<br>**Hannah Foster**<br>• Widow of John Foster.<br>**Richard Foster**<br>• Son of John Foster and Hannah Foster.<br><br>**Thomas Foster (Jr.)**<br>• Son of John Foster Sr.<br>**Thomas Foster (Jr.)** | <br><br>16 Kings Head Court,<br>Fetter Lane<br>16 Kings Head Court,<br>Fetter Lane<br><br>65 Fetter Lane, Fleet St.<br>65 Fetter Lane, Fleet St.<br><br>65 Fetter Lane, Fleet St.<br><br><br>66 Fetter Lane, Fleet St.<br><br>18 Castle St., Holborn | Bc1740–1768<br><br>1769–1777<br><br>1778–1789<br><br><br>1789–D1795<br>1795–1798<br><br>1799–1805<br><br><br>1806–1808<br><br>1808–1810 |
| Cutler | **John Foulds**<br>• Master of the Cutlers Company, 1801. | | 1791–1806 |
| Cutler | **Robert Fowlzer**<br>• Master of the Cutlers Company, 1644. | | 1634–1649 |
| Gold and silver<br>lacemen<br>Embroiderer<br>Sword cutler | **Benjamin Fox**<br>• Successor to Pitter & Fox (see Pitter & Fox).<br>• Succeeded by C. & T. Lonsdale (see C. & T. Lonsdale). | 28 Bedford St., Covent Garden | 1829–1832 |
| Bladesmith<br>Sword maker | **John Foxtone** | | 1382 |
| Silversmith | **Anthony Francia** | Wood St. | 1729–1740 |

| | | | |
|---|---|---|---|
| Hilt maker<br>Sword cutler | **William Francillan**<br>(see John Cripps) | | |
| | **Joseph Franklin**<br>(see Nicoll & Franklin) | | |
| Sales agent | **A.G. Franklin & Co.** | 14 South St., Finlay | 1858–1870 |
| | • Sales agent for J. (Johan) E. Bleckman, sword maker of Solingen, Prussia, 1858–1870. | | |
| Sheather (leather scabbard) maker | **Daniel Frazer** | | 1800–1812 |
| Sword cutler | **Freeman & Roth** | Old Bond St. | 1825–1880 |
| | • Sold M1831 general officers swords. | | |
| Sword cutler | **Thomas Freeman** | Fenchurch St. | 1630–1660 |
| | • On July 5, 1645, during the English Civil War, contracted for 2,000 swords (infantry) with belts at 4 schillings, 6d each for the "New Model" army (parliamentary forces) under Gen. Sir Thomas Fairfax. The swords were to have "Dutch" (Solingen made) blades.<br>• Freeman could have been acting for the Cutlers Company since the contract calls him "wardon."<br>• Master of the Cutlers Company, 1651. | | |
| Goldsmith<br>Silversmith | **William Frisbee**<br>• Apprenticed to John Crouch, 1774–1779. | | Bc1759–1779 |
| Hilt maker | **William Frisbee** | | 1779–1790 |
| Sword cutler | **William Frisbee & John Edwards** | 48 Jewin St. | 1791 |
| | **Frisbee (William) & Storr (Paul)**<br>(see Paul Storr) | 5 Cocklane, Snowhill | 1792 |
| | **William Frisbee** | 5 Cocklane, Snowhill | 1793–1800 |
| | **William Frisbee** | Inner Court, Bridewell Hospital | 1801–1810 |
| | **William & John Frisbee** | Inner Court, Bridewell Hospital | 1811–1820 |
| | • William Frisbee died in 1820. | | |
| Cutler | **Capt. John Frith** | | 1693–1708 |
| | • Master of the Cutlers Company, 1703. | | |
| Goldsmith<br>Silversmith | **Ralph Frith (Sr.)**<br>• Apprenticed to James Smith, 1721–1727. | | Bc1705–1727 |
| Hilt maker<br>Sword cutler | **Ralph Frith (Sr.)** | Shoreditch<br>at the sign of the Golden Cup | 1728–1780 |
| Goldsmith<br>Silversmith<br>Hilt maker<br>Sword cutler | **Ralph Frith (Jr.)**<br>• Hilted small swords. | Charing Cross | 1756–1780 |
| Goldsmith<br>Ornament maker | **Frost (William) & Son (John)** | 11 Air St., Piccadilly | 1814–1817 |
| Sword cutler | **William Frost & William Thompson**<br>(see William Thompson) | 11 Air St., Piccadilly | 1818–1826 |
| Hatter | **Daniel Fry** | Ludgate St. | 1760–1770 |

| | | | |
|---|---|---|---|
| Sword cutler | • Successor to Thomas Holden (see Thomas Holden). | at the sign of Two Eagles | |
| Gun maker<br>Sword cutler | George Fuller | 2 Dean St., Soho | B1801, 1832–1834 |
| | George Fuller | Caroline St., Parish of St. Pancras | 1835–1841 |
| | George Fuller | 104 Wardour St. | 1842–1845 |
| | George Fuller | 30 Southampton St. | 1846–1855 |
| | George Fuller | 280 Strand | 1856–1871 |
| | • Successor to George Wilbraham at 280 Strand. | | |
| | George Fuller | 12 Wych St., Strand | 1872–1873 |
| | George Fuller | 6 Newcastle St., Strand | 1874–1877 |
| | George Fuller | 3 Waterloo Rd. | 1878–1879 |
| | George Fuller | 5 Waterloo Rd. | 1880 |
| Hatter | John Fuller | 2 Charing Cross | 1790–1808 |
| Sword cutler | Joseph Fuller | 2 Charing Cross | 1809–1820 |
| Silversmith<br>Hilt maker<br>Sword cutler | Thomas Fuller | | 1661 |
| | • Signed an advertisement warning sword purchasers about counterfeit cast silver hilts. | | |
| Knife and<br>sword cutler | Heinrich Vollenweidner | Solingen, Prussia | Bc1546–1566 |
| | • Immigrated to England, 1566.<br>• Changed his name to Henry Fulwater. | | |
| | Henry Fulwater | Bride Lane, Farrington without Parish of St. Bride | 1566–1573 |
| | • Apprenticed to Ralph Cole, 1566–1573. | | |
| | Henry Fulwater | Parish of St. Anne, Brackfriars | 1573–D1603 |
| | • Son Jacob born, 1583.<br>• Daughter Sarah Fulwater born, 1573; married sword and knife maker John Jenks (Jenckes) in 1595 (see John Jenks).<br>• Had apprentices:<br>    Jengken Phellepp, 1575–1582<br>    James Long, 1577–1585<br>    Thomas Marson, 1583–1590<br>    Robert Hewlet, 1583–1590<br>    Richard Jacob, 1583–1590<br>    Thomas Witherspoon, 1583–1590 | | |
| Knife and<br>sword cutler | Jacob Fulwater | | B1583–Dc1635 |
| | • Son of Henry Fulwater.<br>• Made freeman of the Cutlers Company, 1608.<br>• Sold 30 swords to London cutler Benjamin Stone, 1621. | | |
| Cutler | Francis Fulwell | | 1623–1638 |
| | • Master of the Cutlers Company, 1633. | | |
| Cutler | William Fulwell | | 1651–1666 |
| | • Master of the Cutlers Company, 1661. | | |
| Gun maker<br>Sword cutler | Nicholas Furlong | 124 Cock Hill, Ratcliff | 1835–1838 |
| | Nicholas Furlong | 112 High St., Shadwell | 1839–1840 |
| | Nicholas Furlong | Church Row, Stepney | 1841–1854 |
| | Nicholas Furlong | 26 Silver St., Stepney | 1855–1857 |
| Swordsmith | William Fyniel of Windsor | London Bridge | 1310 |

- Called a Gladiarius (swordsmith).
- Admitted as a freeman to the Cutlers Company, 1310.

**John Gale**
(see Chapter 3)

| | | | |
|---|---|---|---|
| Gun maker<br>Sword cutler | **Gameson (Henry) & William (Joseph)**<br>• Successors to Lacy & Witton (see J.D. Lacy).<br>**Gameson (Henry) & Co.**<br>• Partner: Joseph Williams.<br>• Succeeded by Joseph Williams (see Joseph Williams). | 67 Threadneedle St.<br><br>67 Threadneedle St. | 1827–1830<br><br>1831–1833 |
| Army<br>accoutrement<br>maker<br>Saddler<br>Sword cutler | **John Garden**<br>**Garden (John) & Stratton (Henry)**<br>**Garden (John) & Stratton (Henry)**<br>**Hugh Garden**<br>**Hugh Garden & Son**<br>(Robert Spring Garden)<br>**Garden (Hugh) & Son**<br>(Robert Spring Garden)<br>**Robert S. (Spring) Garden** | 203 Piccadilly<br>203 Piccadilly<br>200 Piccadilly<br>200 Piccadilly<br>200 Piccadilly<br><br>200 Piccadilly<br><br>29 Piccadilly | 1793–1819<br>1820–1823<br>1824–1826<br>1827–1851<br>1852<br><br>1853–1861<br><br>1862–1877 |
| Cutler | **John Gardiner**<br>• Master of the Cutlers Company, 1596 and 1597. | | 1586–1602 |
| Goldsmith<br>Silversmith<br>Sword cutler | **William Gardiner**<br>• Apprenticed to Richard Jordon, 1681–1688.<br>• Used the "cross keys" mark.<br>**William Gardiner**<br>• Had apprenticee Dru Drury Sr., 1703–1710.<br>(see Dru Drury Sr.) | <br><br><br>Ely Court, Holborn<br>near Hatton Garden | Bc1667–1688<br><br><br>1689–1720 |
| Military tailor<br>Accoutrement<br>maker<br>Sword cutler | **Gardner (Henry) & Carver (William S.)**<br>**W. (William) S. Carver**<br>**W. (William) S. Carver**<br>**W. (William) S. Carver** | 70 Jermyn St.<br>70 Jermyn St.<br>16 King St., Parish of St. James<br>35 Bury St. | 1886–1890<br>1891–1899<br>1900–1910<br>1911–1916 |
| Sword cutler | **Humphrey Gardner** | Parish of St. Giles in the Field,<br>Middlesex | 1613–1623 |
| Sword cutler | **Samuel Gardner & Co.**<br>• Sold swords made by the Wilkinson Sword Company. | 1 Clifford St., Savile Row | 1950 |
| Goldsmith<br>Jeweler<br>Hilt maker<br>Sword cutler | **Robert Garrand (Sr.)**<br>• Apprenticed to Stephen Unwin, 1773–1780.<br>**John Wakelin & William Taylor**<br>• Employee Robert Garrard Sr.<br><br>**John Wakelin & Robert Garrard (Sr.)**<br>• Successor to Wakelin & Taylor (see Edward Wakelin).<br>**Robert Garrard (Sr.)**<br>**Robert Garrard (Jr.) & Brothers**<br>**R. (Robert Jr.) J. (James) &**<br>**S. (Sebastian) Garrard**<br>**Robert (Jr.) & Sebastian Garrard** | <br><br>Panton St.<br>two doors from the Haymarket,<br>Parish of St. James<br>31 Panton St.<br><br>31 Panton St.<br>31 Panton St.<br>31 Panton St.<br><br>29 to 31 Panton St.<br>& 25 Haymarket | B1758–1780<br><br>1780–1791<br><br><br>1792–1801<br><br>1802–D1818<br>1818–1820<br>1821–1835<br><br>1836–1842 |

|  |  |  |  |
|---|---|---|---|
|  | Robert (Jr.) & Sebastian Garrard & Co. | 29 to 31 Panton St. & 25 Haymarket | 1843–1859 |
|  | Robert (Jr.) & Sebastian Garrard & Co. | 29 to 32 Panton St. 25 Haymarket | 1860–1863 |
|  | Robert (Jr.) & Sebastian Garrard | 29 to 32 Panton St. & 25 Haymarket | 1864–1884 |
|  | R. (Robert Jr.) & S. (Sebastian) Garrard & Co. | 29 to 32 Panton St. & 25 Haymarket | 1885–1909 |
|  | Garrard & Co. Ltd. | 39 to 42 Panton St. & 25 Haymarket | 1910–1911 |
|  | Garrard & Co. Ltd. | 17 Grafton St. | 1912–1952 |
| Goldsmith Hilt maker Sword cutler | **William Garrard** • Apprenticed to Samuel Laundry, 1729–1732. • Apprenticed to Jeffery Griffith, 1732. • Apprenticed to Robert Maidman, 1732–1734. |  | Bc1715–1734 |
|  | William Garrard | Staining Lane | 1734–1738 |
|  | William Garrard | Noble St. | 1739–1748 |
|  | William Garrard | Short's Building, Clerkenwell | 1749–1754 |
|  | William Garrard | Noble St. | 1755–1775 |
|  | **Marion Garret** • Immigrated to London, c. 1495. | Normandy France | Bc1477–c1495 |
| Sword and knife maker Sword cutler | **Marion Garret** • Became a member of the Cutlers Company, c. 1497. • Became a naturalized citizen, 1514. • Royal cutler to King Henry VIII; paid 20 shillings a year for keeping the king's swords. • Granted pierced star mark, 1517. • Delivered several swords and knives to the Kings Wardrobe of Robes, October 3, 1538. • Several officials maintained the king's wardrobe, including the Yoeman of Our Robes, Groom of Our Robes, Page of Our Robes, and other assistants. | London | c1495–1547 |
| Sword maker Sword belt maker Button and badge maker Military ornament maker | **J.R. Gaunt & Son Ltd.** • Successors to Edward Thurkle. • Had a branch in Canada at 594 Trans Canada Place, Lonqueuil, Quebec, 1908–1916, and at 63 & 65 Beaver Hall Hill, Montreal, 1917–1969. • Had a branch in Birmingham. | 5 Denmark St., Charing Cross Rd. | 1900–1905 |
|  | J.R. Gaunt & Son Ltd. | 53 Conduit St., Bond St. | 1906–1918 |
|  | J.R. Gaunt & Son Ltd. | 60 Conduit St., Bond St. | 1919–1924 |
|  | J.R. Gaunt & Son Ltd. | 60 Conduit St., Bond St., 2 New Burlington Place | 1925–1937 |
|  | J.R. Gaunt & Son Ltd. | 60 Conduit St., Bond St. | 1938–1939 |
|  | J.R. Gaunt & Son Ltd. | 5 Warwick St. | 1940–1966 |
|  | J.R. Gaunt & Son Ltd. | 1 to 8 Bateman Building | 1967–1969 |
|  | • A 1901 advertisement read: *J.R. Gaunt & Son Ltd. (Edward Thurkle's successors) Presentation swords (From 10 to 300 Guineas) Proved swords (can be obtained from any Naval or Military outfitter) Manufacturers of swords, buttons and badges Military, Naval & Volunteer ornaments & buttons* • A 1917 advertisement read: *J.R. Gaunt & Son* |  |  |

*Military Equiptment Manufacturers*
*Badges*
*Buttons*
*Swords*
*Helmets*
*Caps*
*Belts*
*Gold Lace &*
*Embroidery*
*Enamelled badges & jewelry*

| | | | |
|---|---|---|---|
| Cutler | **Thomas Geary** | | 1736–1757 |
| | • Master of the Cutlers Company, 1746. | | |
| Cutler | **Richard Geast** | | 1689–1704 |
| | • Master of the Cutlers Company, 1699. | | |
| Cutler | **John Gent** | | 1793–1808 |
| | • Master of the Cutlers Company, 1803. | | |
| Cutler | **Walter Gibbons** | | 1647–1652 |
| | • Master of the Cutlers Company, 1657. | | |
| Cutler | **George Gibson** | | 1750–1763 |
| | • Master of the Cutlers Company, 1760. | | |
| Cutler | **Gibson, Thomson & Craig** | | 1798–1803 |
| Army saddler | Gibson (Henry) & Son (Robert) | Little Mills | 1790–1794 |
| Army and navy contractor | Gibson (Henry), Son (Robert) & Peat (John) | 9 Coventry St. | 1795–1799 |
| Sword cutler | Gibson (Henry) & Peat (John) | 29 Coventry St. | 1800–1809 |
| | Gibson (Henry) & Peat (John) | 1 Whitcomb St. | 1810–1813 |
| | Gibson (Henry) & Peat (John) | Coventry St. | 1814–1819 |
| | Gibson (Henry) & Peat (John) | 25 Princes St., Coventry St. | 1820–1831 |
| | Gibson (Henry) & Peat (John) | 25 Princes St., Coventry St. | 1832–1839 |
| | Robert Gibson & Co. | 10 Princes St., Leicester Square | 1840–1845 |
| | Robert Gibson & Co. | 1 New Coventry St. | 1846–1847 |
| | Robert Gibson & Co. | 6 New Coventry St. | 1848–1857 |
| | Gibson & Co. | 6 New Coventry St. | 1858 |
| | Gibson & Co. | 6 & 7 New Coventry St. | 1859–1891 |
| | Gibson & Co. | 8 & 9 New Coventry St. | 1892–1895 |
| Silversmith Goldsmith Hilt maker Sword cutler | **William Gibson & John Langman** | | 1890–1910 |
| Hosier Tailor Army and navy Outfitter Sword cutler | J. (James) Gieve & Sons Ltd. | 21 George St., Hanover Square | 1900–1906 |
| | Gieve (James), Matthews (Henry G.) Seagrove (Edwin Augustus) Ltd. | 21 George St., Hanover Square | 1907–1911 |
| | Gieves Ltd. | 65 So. Motton St. | 1912–1919 |
| | Gieves Ltd. | 21 Old Bond St. | 1920–1941 |
| | Gieves Ltd. | 80 Piccadilly | 1942–1946 |
| | Gieves Ltd. | 27 Old Bond St. | 1947–1960 |
| | Gieves Ltd. | 142 Fenchurch St. & 27 Old Bond St. | 1961–1969 |

| | | | |
|---|---|---|---|
| | • Sold swords made by the Wilkinson Sword Company. (see Devenport & Portsmouth) | | |
| Goldsmith Jeweler Hilt maker Sword cutler | **Gilbert (Philip) & Jefferys (Thomas)** | 20 Cockspur St. | 1802–1805 |
| | • Successors to Jeffreys and Gilbert (see Thomas Jeffreys). | | |
| | **Philip Gilbert** | 20 & 36 Cockspur St. | 1806–1808 |
| | **Philip Gilbert** | 20 Cockspur St. | 1809–1828 |
| | **Philip Gilbert** | 5 St. James Square | 1829–1831 |
| Sword cutler | **Robert Gilbert** | | 1642–1649 |
| | • On September 18, 1645, during the English Civil War, Gilbert contracted for 500 (infantry) swords with belts at 4 shillings, 6d each with "Dutch" (Solingen, Prussia) blades for the "New Model" army (parliamentary forces) under Gen. Sir Thomas Fairfax. | | |
| Silversmith Hilt maker Sword cutler | **Stephen Gilbert** | Panton St. | 1770–1779 |
| | **Fogelberg (Andrew) & Gilbert (Stephen)** | 29 Church St. Parish of St. Anne, Soho | 1780–1793 |
| Sword cutler | **Charles Henry Gilks** | 37 Minories & 3 Union Row, Minories | 1855–1857 |
| | **Charles Henry Gilks** | 67 Minories & 3 Union Row, Minories | 1858–1862 |
| | **Wilson (John) & Gilks (Charles Henry)** | 3 Union Row, Minories | 1863–1868 |
| | **C. (Charles) H. (Henry) Gilks & Co.** | 3 Union Row, Minories | 1869–1880 |
| Gun maker Saw maker File maker Steel toy (hardware) maker Sword blade maker Sword maker | **Thomas Gill (Sr.) & Co.** | 11 Charing Cross | 1783–1798 |
| | • Began sword and blade production in London and Birmingham, 1783 (see Birmingham listing) | | |
| | **Thomas Gill (Sr.)** | 22 Norfolk St., Strand | 1799–D1801 |
| | • Thomas Gill Sr. died in 1801. | | |
| | **T. (Thomas Jr.) & J. (James Jr.) Gill** | 83 St. James St. | 1801–1808 |
| | • Sons of Thomas Gill Sr. | | |
| | • James Gill (Jr.) died in 1808. | | |
| | **Thomas Gill (Jr.)** | 83 St. James St. | 1809–1816 |
| | **Thomas Gill (Jr.) & Co.** | 6 Princes St., Leicester Square | 1817–1818 |
| | **T. (Thomas) Gill (Jr.)** | 6 Princes St., Leicester Square | 1819–D1826 |
| | • Many swords with "Gill" or "G" marked on the blades were exported to the USA. | | |
| | • Thomas Gill Jr. wrote an article called "Fine & Delicate Steel Works" in which he describes the use of cut steel beads and studs, which he set into some of his sword hilts. | | |
| Army and navy tailor and outfitter Sword and knife cutler | **Henry Gillot** | Strand | 1787–1827 |
| | **George Gillott** | 36 Strand | 1828–1840 |
| | **George & William Henry Gillott** | 36 Strand | 1841–1852 |
| | **Gillott Brothers (George & William Henry) & Hassell (John)** | 36 Strand | 1853–1867 |
| | • Made midshipmen's dirks. | | |
| | • George Gillott left in 1867. | | |
| | (see Portsea listing) | | |
| | **Gillott (William Henry) & Hassell (John)** | 2 New Burlington St. | 1868–1940 |
| | **Gillott (William Henry) & Hassell (John)** | 169 Grafton St. | 1941–1954 |
| Navy tailor Draper (cloth maker) Sword cutler | **George Gillott** | 18 Princes St., Hanover Square | 1847–D1868 |
| | • George Gillott died in 1868. | | |
| | **George Henry Gillott** | 18 Princes St., Hanover Square | 1868–1871 |
| | • Son of George Gillott. | | |

|  |  |  |  |
|---|---|---|---|
|  | George Henry Gillott | 2 Princes St., Hanover Square | 1872–1878 |
|  | Gillott (George Henry) & Brewster (John) | 2 Princes St., Hanover Square | 1879–1881 |
|  | George Henry Gillott | 2 & 6 Princes St., Hanover Square | 1882–1884 |
|  | George Henry Gillott | 2 Princes St., Hanover Square | 1885–1889 |
| Silversmith | Joseph Girdler | 13 Rose St., Long Acre | 1770–1792 |
| Buckle maker | Joseph Girdler | 13 Covent Garden | 1793–1802 |
| Fine steel worker | John Girdler | 33 Bridge Road, Lambert | 1803–1804 |
| Silver and steel hilt maker | John Girdler | 7 Crescent Place, St. George's Fields | 1805–1810 |
| Sword cutler |  |  |  |
| Sword maker |  |  |  |
| Sword maker | Robert Girdler | 20 Queen St. at the sign of the Seven Dials | 1811–1815 |
| Sword maker | John Glascocke • Repaired swords for the Board of Ordnance, 1639. |  | 1630–1640 |
| Sword cutler | Glen (Henry) & Powell (Arthur John) • Successors to Arthur John Powell (see Arthur John Powell). | 18a London St. | 1909–1926 |
|  | Glen, Errington & Harrison • Successors to Errington & Harrison (see Errington & Harrison). | 18a London St. | 1927–1932 |
|  | David Glenny (see Richard Thresher) |  |  |
| Goldsmith Hilt maker Sword cutler | P. (Pierre) Glisier (Glasier) |  | 1748–1754 |
| Silversmith Hilt maker Sword cutler | B. & R. Godden | Hemmings Row | 1840–1853 |
| Cutler | John Godman • Master of the Cutlers Company, 1693. |  | 1683–1698 |
| Furniture dealer | S. (Samuel) Goff & Co. | 17–18 & 22 King St. Covent Garden | 1920–1922 |
| Portmanteau (leather luggage) maker | • Became a military outfitter in 1917. S. (Samuel) Goff & Co. | 28 Bedford St. & 17–18 & 22 King St. Covent Garden | 1923–1930 |
| Iron monger (dealer) |  |  |  |
| Iron worker |  |  |  |
| Equipment maker |  |  |  |
| Military outfitter |  |  |  |
| Sword cutler |  |  |  |
| Sword cutler | Peter Gogney |  | 1594–1590 |
| Sword cutler | George Goldmen |  | 1900–1902 |
| Goldsmith | Thomas Goldney | Bath | Bc1769–1791 |

| | | | |
|---|---|---|---|
| Jeweler<br>Hilt maker<br>Sword cutler | **Samuel Goldney** | Bath | Bc1760–1791 |
| | • Apprenticed to his brother-in-law Philip Rundell, 1784–1791.<br>• Married Eleanore Rundell, Philip Rundell's sister. | | |
| | **Thomas & Samuel Goldney** | | 1791–1793 |
| | **Neild (James) & Goldney<br>(Thomas & Samuel)**<br>(see James Neild) | 4 St. James St.,<br>near the Royal Exchange | 1794–1795 |
| | **T. (Thomas) & S. (Samuel) Goldney** | 4 St. James St.<br>near the Royal Exchange | 1796–1809 |
| | • Appointed sword cutlers to His Royal Highness the Prince of Wales and the royal family. | | |
| | **Thomas Goldney** | 4 St. James St.<br>near the Royal Exchange | 1810–1829 |
| | • James Morisset and Ray & Montaque mounted some of the presentation swords sold by Goldney, 1810–1829.<br>  (see James Morisset, Ray & Montaque) | | |
| | **Thomas Goldney**<br>• Retired in 1830. | Clifton Hill<br>Gloucestershire | 1830–D1856 |
| Goldsmith<br>Silversmith<br>Hilt maker<br>Sword cutler | **The Goldsmiths &<br>Silversmiths Co. Ltd.** | 112 Regent St.<br>46–47–48–49 Warwick St.<br>50 & 52 Glasshouse St. | 1890–1950 |
| | • Made a presentation sword for the city of London.<br>• Several silversmiths designed and supplied presentation swords for them,<br>  including Leslie G. Durbin, Hickleson & Philips, and Albert George Brooker. | | |
| Sword cutler | **Goodall & Graham** | 7 Conduit St. | 1910 |
| Silversmith<br>Goldsmith<br>Hilt maker<br>Sword cutler | **Samuel Goodbehere & Co.**<br>**(Godbehere)** | 86 Cheapside | 1784–1785 |
| | • Partner: James Bult.<br>• Successors to Bult & Sutton (see James Bult). | | |
| | **Goodbehere (Samuel) & Wigan<br>(Edward) & Co.** | 86 Cheapside | 1786–1799 |
| | • Partner: James Bult. | | |
| | **Goodbehere (Samuel),<br>Wigan (Edward) & Bult (James)<br>(S. Goodbehere & Co.)** | 86 Cheapside | 1800–1818 |
| | • Ray and Montaque mounted some presentation swords sold by them.<br>• Samuel Goodbehere was master of the Goldsmiths Company, 1804. | | |
| | **Goodbehere (Samuel) & Bult (James)**<br>(see Edward Wigan, James Bult) | 86 Cheapside | 1819 |
| | **John Goodbody**<br>(see Peter Grosjean) | | |
| Gold and silver<br>lacemen<br>Sword cutler | **Robert Gordon** | 68 Princes St. | 1859–1872 |
| Sword cutler | **W. (William) H. Gore** | | 1895 |
| Knife and dirk<br>maker<br>Sword belt<br>maker<br>Sword cutler | **Charles Goss**<br>**John Gross**<br>• Succeeded by Charles Oakden (see Charles Oakden). | 35 Brownlow St., Drury Lane<br>35 Brownlow St., Drury Lane | 1833–1838<br>1839–1840 |

| | | | |
|---|---|---|---|
| Sword cutler | **Algernon Graves** | | 1902 |
| Cutler | **Henry Graves** <br> • Master of the Cutlers Company, 1868, 1874, and 1877. | | 1858–1882 |
| Goldsmith <br> Hilt maker <br> Sword cutler | **Robert Gray (Grey)** <br> **Gray (Robert) & Son** <br> • Made small swords with gold inlayed steel hilts. | Bond St. <br> Bond St. | 1770–1800 <br> 1801–1805 |
| Gun maker <br> Sword cutler | **Samuel Gray (Grey)** | 10 Marshall St., Golden Square | 1850–1856 |
| Goldsmith <br> Silversmith <br> Jeweler <br> Enameler <br> Hardwaremen <br> (dealer) <br> Hilt maker <br> Sword cutler <br> Cutler | **Thomas Gray (Grey)** <br><br><br> **Gray (Thomas) & Constable (William)** <br> • Successors to William Constable (see William Constable). <br> **Thomas Gray** <br> • Succeeded by Delafons & Sons (see Delafons & Sons). <br> • James Morisset mounted some presentation swords for Gray & Constable (see James Morisset). | 42 Sackville St. <br><br><br> 41 & 42 Sackville St. <br><br> 41 & 42 Sackville St., off Piccadilly | Bc1760 <br> 1780–1793 <br><br> 1794–1801 <br><br> 1802–1825 |
| Sword cutler | **Green & Son** | Savile Row | 1850 |
| Goldsmith <br> Silversmith <br> Hilt maker <br> Sword cutler | **Green & Ward** <br> **Green, Ward & Green** <br> • Sold presentation swords mounted by James Morisset and Ray & Montagne. <br> (see James Morisset, Ray & Montagne) | | 1790–1801 <br> 1802–1820 |
| | **Samuel Green** <br> (see Clarke & Green) | | |
| Sword cutler | **Thomas Green** | 16 Clement's Lane, Temple Bar | 1791–1800 |
| Silversmith <br> Hilt maker <br> Sword cutler | **Francis Greene** <br> • In 1675, Greene was fined 40 shillings for making silver-hilted swords <br>   with low-grade cast silver. | | 1670–1680 |
| Cutler | **Jeremiah Greene** <br> • Master of the Cutlers Company, 1666. | | 1656–1671 |
| Sword cutler | **John Greene** | | 1616–1636 |
| Cutler | **Laurence Greene** <br> • Master of the Cutlers Company, 1563 and 1570. | | 1553–1575 |
| Cutler | **Reynold Greene** <br> • Master of the Cutlers Company, 1611–1612. | | 1601–1617 |
| Cutler | **Thomas Greene** <br> • Master of the Cutlers Company, 1594–1595. | | 1584–1601 |
| Goldsmith <br> Silversmith <br> Hilt maker | **George Greensill** <br> • Had apprentice Richard Kersill, 1736–1743 (see Richard Kersill). <br> **Joseph Greensill** | 36 Strand <br><br> 36 Strand | 1720–1759 <br><br> 1760–1778 |

| Profession | Name | Address | Dates |
|---|---|---|---|
| Sword cutler<br>Gun mounter | • Son of George Greensill.<br>**Joseph Greensill**<br>• Silver-mounted some pistols.<br>**Joseph & Edward Greensill**<br>• Employee: John Salter.<br>• Succeeded by John Salter (see John Salter). | 35 Strand<br><br>34 & 35 Strand | 1779<br><br>1780–1800 |
| Cutler | **Charles Greer**<br>• Master of the Cutlers Company, 1852. | | 1842–1857 |
| Cutler | **James Greer**<br>• Master of the Cutlers Company, 1888. | | 1878–1893 |
| Cutler | **John Greer**<br>• Master of the Cutlers Company, 1812. | | 1802–1817 |
| Cutler | **Robert Greer**<br>• Master of the Cutlers Company 1887. | | 1877–1892 |
| Gun maker<br>Sword cutler | **Griffen (Joseph) & Tow (John)**<br>• Succeeded by John Tow (see John Tow). | 10 New Bond St. | 1772–1782 |
| Sword cutler | **Hugh Griffen** | | 1578–1793 |
| Goldsmith<br>Jeweler<br>Hilt maker<br>Sword cutler | **John Griffen**<br>**Griffen (John) & Adams (John M.)**<br>**Griffen (John) & Adams (John M.)**<br>• Appointed sword cutler to His Royal Highness the Duke of Clarence, 1802.<br>**John M. Adams** | 17 Ludgate St.<br>17 Ludgate St.<br>76 Strand<br><br>76 Strand | 1793–1794<br>1795–1799<br>1800–1828<br><br>1829–1839 |
| Sword cutler | **William Griffen** | Parish of St. Botolph, Aldergate | 1580–D1603 |
| Sword cutler | **Griffiths, McAlister Ltd.** | 10 Warwick St., Regent St. | 1930–1940 |
| Military agent<br>Sword cutler | **Capt. Robert Melville Grindlay**<br>• Ship owner.<br>**Capt. Robert Melville Grindlay**<br>**Capt. Robert Melville Grindlay**<br><br>• Purchased arms and swords for the East India Co.<br>**Christian Grindlay & John Matthews**<br><br>**Grindlay & Co.**<br><br>**Grindlay & Co.** | 16 Cornhill<br><br>16 Cornhill, Trafalgar Place<br>16 Cornhill<br>St. Martins Place, Trafalgar Square<br><br>16 Cornhill,<br>St. Martins Place, Trafalgar Square<br>16 Cornhill,<br>St. Martins Place, Trafalgar Square<br>63 Cornhill,<br>St. Martins Place, Trafalgar Square<br>& 124 Bishopsgate within | 1831–1834<br><br>1835–1836<br>1837–1838<br><br><br>1839–1843<br><br>1844–1848<br><br>1849–1860 |
| Military tailor<br>Hatter<br>Sword cutler | **Peter Grosjean**<br>**Frederick Grosjean (Sr.)**<br>**Ann Grosjean**<br>• Widow of Frederick Sr.<br>**Grosjean (Ann) & Goodbody (John)**<br>**Frederick Grosjean (Jr.)**<br>**Frederick Grosjean (Jr.)** | 208 Piccadilly<br>208 Piccadilly<br>99 Quadrant, Regent St.<br><br>99 Quadrant, Regent St.<br>99 Quadrant, Regent St.<br>109 Regent St. | 1800–1819<br>1820–D1832<br>1832–1833<br><br>1834<br>1835–1847<br>1848–1856 |

| | | | |
|---|---|---|---|
| | Grosjean (Frederick Pierre Genaret) & Petherick (Henry) | 109 Regent St. | 1857–1870 |
| | Frederick Pierre Genaret Grosjean | 109 Regent St. | 1871–1873 |
| | Grosjean (Frederick Pierre Genaret) & Nelson (Samuel) | 109 Regent St. | 1874–1883 |

• Succeeded by Samuel Nelson (see Samuel Nelson).

| | | | |
|---|---|---|---|
| Sword cutler | Thomas Gruson | | 1760 |
| Cutler | Robert Grymes | | 1670–1685 |

• Master of the Cutlers Company, 1680.

| | | | |
|---|---|---|---|
| Silversmith Goldsmith | Richard Gurney | | Bc1703–1724 |

• Apprenticed to Richard Bayley, 1717–1724 (see Richard Bayley).

| | | | |
|---|---|---|---|
| Hilt maker | Richard Gurney | Foster Lane | 1725–1726 |
| Sword cutler | Richard Gurney & Co. | Foster Lane at the sign of the Golden Cup | 1727–1773 |

• Partner Thomas Cooke.
• Had apprentice William Preist Sr., 1740–1747 (see William Preist Sr.).

| | | | |
|---|---|---|---|
| Army tailor Sword cutler | Thomas Ansley Guthrie | 12 Cort St. | 1863–1889 |

• Successor to Hunter & Guthrie (see Hunter & Guthrie).

| | | | |
|---|---|---|---|
| | Guthrie (Thomas Ansley) & Valentine (Robert) | 12 Cort St. | 1890–1924 |
| | Guthrie (Thomas Ansley) & Valentine (Robert) | 46 Albemarle St. | 1925–1929 |
| | Guthrie (Thomas Ansley) & Valentine (Robert) | 17 Sackville St. | 1930–1943 |
| | Guthrie (Thomas Ansley) & Valentine (Robert) | 16 Sackville St. | 1944–1950 |

• Succeeded by Robert Valentine (see Robert Valentine).

| | | | |
|---|---|---|---|
| Sword cutler | John Haddam | Charing Cross | 1810–1826 |
| Hafter (sword hilt maker) | Richard Le Haftere | | 1300–1312 |
| | Soloman Le Haftere | | 1312–1320 |

• Son of Richard Le Haftere.

| | | | |
|---|---|---|---|
| Hafter (sword hilt maker) | William Le Haftere | | 1310 |
| Sword cutler | Henry Haggen | | 1620–D1649 |
| | Katherine Haggen | | 1649–1655 |

• Widow of Henry Haggen.
• Supplied swords to the Board of Ordnance, 1650.

| | | | |
|---|---|---|---|
| Sword cutler | Richard Hales | | 1606 |
| Sword cutler | James Hall | 17 Whithouse Yard, Drury Lane | 1827 |
| Sword cutler | Sir John Hall | | 1845 |
| Silversmith Hilt maker Sword cutler | Ralph Hall | | 1625–1635 |

• In 1632, the Cutlers Company confiscated some of Hall's swords with counterfeit cast silver hilts.

| | Samuel Hall
(See Richards & Hall) | | |
|---|---|---|---|
| Damascener
(Damascus
blade maker)
Sword cutler | **William Hall**
• Made sword with gilt blades. | | 1615 |
| Gold and silver
lacemen
Army and navy
outfitters
Army and navy
contractors
Sword cutler
Gun maker | **Hamburger & Co.**
**Hamburger, Harwood & Co.**
**Hamburger, Rogers & Co.**
• Sold M1821 light cavalry sabers.
• Partners: William and John Rogers.
• William and John Rogers purchased Hamburger, Rogers & Co. in 1870.
• Succeeded by Rogers (William and John) & Co. Ltd. in 1917 (see Rogers & Co. Ltd.). | 30 King St., Covent Garden
30 King St. | 1812–1826
1827–1839
1840–1917 |
| Goldsmith
Jeweler
Hilt maker
Sword cutler | **John Hamlet**
• Mounted Patriotic Fund swords. | Princes St.,
Leicester Square | 1810–1820 |
| Cutler | **Thomas Hammant**
• Master of the Cutlers Company, 1858. | | 1848–1863 |
| Military
outfitter | **Robert Hamond**
• In 1645, during the English Civil War, Hammond contracted for 550 Spanish pikes (15 feet long) at 4 shillings each for the "New Model" army (parliamentary forces) under Gen. Sir Thomas Fairfax. On July 26, 1645, delivered 150 Spanish pikes; on September 18, 1645, delivered 400 Spanish pikes.
• Also provided 220 shaphaune pistols on December 22, 1645.
• On December 22, 1645, contracted for 250 matchlock muskets. | Broad St. | 1642–1649 |
| Sword cutler | **Charles F. Hancock** | Bruton St. | 1850–1860 |
| Sword cutler | **Henry Hant** | | 1934–1941 |
| Cutler | **George Harberd**
• Master of the Cutlers Company, 1622. | | 1612–1627 |
| Cutler | **William Hardy**
• Master of the Cutlers Company, 1752. | | 1742–1757 |
| | **John Harker**
(see Robert Makepeace) | | |
| Armourer
Sword maker | **John Harmer**
• Supplied swords to the Board of Ordnance, 1621–1627. | | 1620–1630 |
| Sword cutler | **Victor Harold & Co.** | | 1850–1854 |
| Hardwareman
(dealer) | **Thomas Harper**
• Immigrated to Charleston, SC, USA, c. 1760. | Bristol | B1736–1760 |
| Jeweler
Goldsmith | **Thomas Harper**
• In 1773, advertised: *Working jeweler and goldsmith has opened a shop in Broad St.* | Broad St., Charleston, SC, USA | 1760–1778 |

| | | | |
|---|---|---|---|
| Silversmith<br>Hilt maker<br>Sword cutler | *where all kinds of work in the above branches will be completely executed upon the cheapest terms and with all possible dispatch.*<br>• In 1778, refused to take the oath of allegiance to the United States and immigrated to Eustatious, Dutch West Indies. | | |
| | Thomas Harper | Eustatious, Dutch West Indies | 1778–1781 |
| | • In 1781, after British captured Charleston, SC, Harper returned. | | |
| | Thomas Harper | Church St., Charleston, SC, USA | 1781–1782 |
| | • Immigrated back to London, England, 1782. | | |
| | Thomas Harper | 207 Fleet St., near Temple Bar | 1783–1809 |
| | • Made a presentation sword given to Vice Admiral Lord Collingwood, 1806.<br>• Ray & Montuque mounted some presentation swords for Harper.<br>• Also made masonic jewels. | | |
| | Thomas Harper | 29 Arundel St., Fleet St. | 1810–1831 |
| | Thomas Harper | 1 Featherstone Building, Holborn | 1831–D1832 |
| | • Thomas Harper died in 1832 at 96 years old. | | |
| | Richard Harris<br>(see Maynard & Co.) | | |
| Silversmith<br>Hilt maker<br>Sword cutler | William Harris | | Bc1681–1702 |
| | William Harris | White Friars, Temple Gate | 1702–1730 |
| | • Apprenticed to cutler Richard Lowe, c. 1695–1702. | | |
| Silversmith<br>Hilt maker<br>Sword cutler | George Harrison | | Bc1736–1759 |
| | • Apprenticed to John Watkins, 1751–1759 (see John Watkins). | | |
| | George Harrison | Fenchurch St., Goulston Square, Whitechapel | 1760–1775 |
| | • Also made silver gun mountings and watch cases. | | |
| Sword blade maker | J. (John) Harrison | | 1840–1870 |
| | • Sold sword blades to George W. Simons & Co., Philadelphia, PA, USA. | | |
| Bladesmith | Thomas Harrison | | 1480–1500 |
| | • Master of the Bladesmiths Company, 1490. | | |
| Sword cutler | Harrods Ltd. | Brompton St | 1950 |
| | • Sold swords made by the Wilkinson Sword Company. | | |
| Army and navy tailor<br>Sword cutler | Benjamin Hart | 26 Warwick St., Golden Square | 1860–1881 |
| | • Succeeded by Joseph & Co. (see Joseph & Co.). | | |
| Army tailor<br>Clothier<br>Army accoutrement maker<br>Sword cutler | Brien Hart | 34 Pall Mall | 1834 |
| | Henry Hart | 26 Pall Mall | 1835–1886 |
| | Henry Hart & Co. | 26 Pall Mall | 1887–1890 |
| | • Sold M1821 light cavalry sabers. | | |
| Cutler | Thomas Hart | | 1631–1646 |
| | • Master of the Cutlers Company, 1641. | | |
| Sales agent for a sword maker | H. Hartjen & Co. | 4 Falcon St., Aldersgate | 1883–1897 |
| | H. Hartjen & Co. | 1 & 2 Falcon St., Aldersgate | 1898–1903 |
| | H. Hartjen & Co. | 35 & 37 Noble St. | 1904–1913 |
| | H. Hartjen & Co. | 67 New Compton St. | 1914 |

| | | | |
|---|---|---|---|
| | • Sales agent for sword maker P.D. Luneschloss, Solingen, Prussia, 1883–1914. | | |
| Cutler | **William Hartwell** | | 1478–1500 |
| | • Master of the Cutlers Company, 1488, 1489, 1494, and 1495. | | |
| Sword and sword blade maker | **John Harvey** <br> • In 1638, Benjamin Stone of Hounslow obtained an order from the Office of Ordnance for 5,000 sword and rapier blades. All were subcontracted. John Harvey was one of three subcontractors (the other two were John Hayes and the Nicholas brothers). <br> • All blades delivered to Capt. William Legge, Master of Armoury (see Chapter 3). | | 1630–1645 |
| Pike maker | **Henry Haselforte** <br> • Supplied Morris pikes to the Drapers Company for the defense of their settlement in Calais, France, during the French War, 1557–1558. | | 1555–1560 |
| | **John Hassell** <br> (see Henry Gillott) | | |
| Gun maker <br> Sales agent for a sword maker | **Friedrich E.D. Hast** <br> **Friedrich E.D. Hast** <br> **Friedrich E.D. Hast** <br> • Sales agent for sword makers Samuel Hoppe & Co., Solingen, Prussia, 1861–1862. | 34 Broad St. Building <br> 18 Aldermansbury <br> 34 Jewin St., Cripplegate | 1847–1848 <br> 1849–1860 <br> 1861–1865 |
| Sword cutler | **Simon Hattfielde** <br> • Apprenticed to sword cutler Richard Hawes. | Fleet St., <br> Parish of St. Sepulchre without <br> Newgate, at the sign of the Kather & Wheel | 1560–D1576 |
| Sword cutler | **Lawrence Hattfielde** <br> • Brother of Simon Hatfielde. | Fleet St., <br> Parish of St. Sepulchre without Newgate | 1560–D1576 |
| Sword cutler | **Richard Hattfielde** | Fleet St., <br> Parish of St. Sepulchre without Newgate | 1585–D1592 |
| Sword cutler | **Christopher Hattfielde** <br> • Son of Lawrence Hattfielde. <br> • Master of the Cutlers Company, 1609–1610. | Fleet St., <br> Parish of St. Sepulchre without Newgate | 1595–D1619 |
| Cutler | **Joseph Hatton** <br> • Master of the Cutlers Company, 1769–1770. | | 1759–1775 |
| Sword cutler | **James Hawes** | | 1844 |
| Sword cutler | **Richard Hawes** <br> • A tenant of Thomas Buck (see Thomas Buck). <br> • Master of the Cutlers Company, 1590–1591. | Fleet St., <br> Parish of St. Sepulchar without Newgate | 1560–1600 |
| Silversmith <br> Hilt maker <br> Sword cutler | **John Hawgood** <br> • Brother of Thomas Hawgood. <br> • Made silver-hilted rapiers. <br> • Signed an advertisement warning sword purchasers about counterfeit cast silver hilts, 1661. | against Bedford House, <br> near the Savoy, Strand <br> at the sign of the Reindeer | 1645–1682 |
| | **John Hawgood** <br> • Master of the Cutlers Company, 1687. | Charing Cross <br> near Angel Court, at the top of Whitehall <br> near the statue of King Charles II on horseback <br> at the sign of King Charles II head | 1682–D1707 |

|  |  |  |  |
|---|---|---|---|
| | • Appointed sword cutler to the Board of Ordnance, November 6, 1693. | | |
| | • Supplied and repaired swords for the Board of Ordnance, 1700–1707. | | |
| | • Sold horseman's swords. | | |
| Silversmith<br>Hilt maker | **Robert Hawgood** | East end of Charing Cross<br>near Angel Court, near the river | 1650–1673 |
| Sword blade<br>importer | **Thomas Hawgood** | East end of Charing Cross<br>near Angel Court, near the river | 1673–1720 |
| Sword and<br>bayonet<br>cutler | • His brother, John Hawgood, moved to a location near him in 1682.<br>• Appointed sword cutler to His Majesty the King.<br>• Supplied plug bayonets to the Board of Ordnance, 1685–1711.<br>• Supplied swords to the Board of Ordnance, 1685–1711.<br>• On August 21, 1685, Hawgood petitioned the Board of Ordnance for a special liscense to import a large number of sword blades from Solingen, Germany, to fulfil his sword contract (import of munitions of war had been made illegal).<br>• The Board of Ordnance allowed him to import six chests of sword blades.<br>• Hawgood and partner John Prigg imported sword blades from Solingen, Germany, and sold them to the Board of Ordnance, 1685 and 1686.<br>• Also supplied plug bayonets to the Board of Ordnance. Appointed sword cutler to the Board of Ordnance on August 17, 1685, January 11, 1689, November 6, 1693, and March 15, 1714. | | |
| Military<br>accoutrement<br>maker | **Thomas Hawkes** | 17 Piccadilly | 1788–1796 |
| | **Thomas Hawkes** | 24 Piccadilly | 1797–1809 |
| | • Sword cutler to the Ordnance Department. | | |
| Hatter | • Appointed cap maker to the royal family, His Royal Highness the Prince of Wales, and several dukes. | | |
| Army cap,<br>hat and helmet<br>maker | **T. (Thomas) Hawkes,<br>Moseley (John) & Co.** | 22 Piccadilly | 1810–1820 |
| Military tailor | **T. (Thomas) Hawkes,<br>Moseley (John) & Co.** | 14 Piccadilly | 1821–1852 |
| Gun dealer | **Hawkes & Co.** | 14 Piccadilly | 1853–1890 |
| Sword cutler | • Partner: Henry Thomas White, 1859.<br>• In 1879, Henry Thomas White purchased Hawkes & Co. but kept the name.<br>• Sold M1821 light cavalry sabers. | | |
| | **Hawkes & Co. Ltd.** | 1 Savile Row | 1891–1980 |
| | • Hatters by appointment to the royal family and His Royal Highness the Prince of Wales.<br>• Sold swords made by the Wilkinson Sword Company. | | |
| | **Gieves & Hawkes Ltd.** | 1 Savile Row | 1981–2003 |
| | • Sold swords made by the Wilkinson Sword Company. | | |
| Sword cutler | **Benjamin Hawkins** | | 1642–1649 |
| | • On April 3, 1645, during the English Civil War, Hawkins contracted for 2,500 swords (infantry) with belts at 4 shillings, 8d each for the "New Model" army (parliamentary forces) under Gen. Sir Thomas Fairfax. | | |
| Sword cutler | **John Haydon** | Parish of St. Bride | 1610–1625 |
| Sword cutler | **George Haye (Hayes)** | | 1635–1645 |
| | • Supplied swords to the Board of Ordnance, 1639. | | |
| Sword blade<br>maker | **John Hayes** | | 1630–1645 |
| | • In 1638, Benjamin Stone of Hounslow obtained an order for 500 sword and rapier blades from the Office of Ordnance. All were subcontracted. John Hayes, John Harvey, and the Nicholas brothers were the subcontractors.<br>• All blades were delivered to Capt. William Legge, Master of Armoury.<br>(see Chapter 3) | | |

| | | | |
|---|---|---|---|
| Goldsmith<br>Jeweler<br>Toy (hardware) maker<br>Hilt maker<br>Sword cutler | **Haynes (Henry) & Kentish (John)**<br>• Successors to Kentish & Haynes.<br>(see John Kentish) | 18 Cornhill St. | 1790–1820 |
| | **Edward Hayson**<br>(see James Hole) | | |
| Sword maker | **Robert Heath**<br>• Supplied swords to the Board of Ordnance, 1639. | | 1635–1640 |
| Military ornament maker<br>Military helmet maker<br>Military outfitter<br>Sword cutler | **I. (Isaac) Hebbert**<br>**I. (Isaac) Hebbert & Co.**<br>**I. (Isaac) Hebbert & Co. Ltd.**<br>**Isaac Hebbert & Co. Ltd.**<br>• Succeeded by Charles Reeves (see Charles Reeves). | 8 Air St., Piccadilly<br>8 Air St., Piccadilly<br>8 Air St., Piccadilly<br>8 Air St., Piccadilly | 1824–1826<br>1827–1828<br>1829–1832<br>1833–1852 |
| Army helmet cap and accoutrement maker<br>Army uniform maker<br>Sword cutler | **Charles Hebbert**<br>**Hebbert (Charles) & Hume (William)**<br>• From 1824 on, cap and helmet maker to the Dukes of Cambridge and Gloucester.<br>**Hebbert (Charles) & Hume (William)**<br>**Hebbert (Charles) & Hume (William)**<br>**Charles Hebbert**<br>**Hebbert & Co.**<br>**Hebbert & Co.**<br>**Hebbert & Co.**<br>**Hebbert & Co.**<br>**Hebbert & Co.**<br>**Hebbert & Co.**<br>• Had branches at, Bradford, Chathan, Colchester, Leeds, Manchester, Portsmouth, and Glasgow. | <br>30 & 33 Princes St., Leicester Square<br><br>30 & 33 Princes St., Leicester Square<br>8 Pall Mall East<br>8 Pall Mall East<br>8 Pall Mall East<br>8 New Coventry St.<br>16 James St.<br>32 Bethnal Green Rd. & 16 James St.<br>32, 37 & 40 Bethnal Green Rd. &<br>24 Cecil Ct., Charing Cross<br>35 & 37 Bethnal Green Rd. | 1780–1813<br>1814–1827<br><br>1828<br>1829<br>1830–1849<br>1850–1864<br>1865–1867<br>1868–1886<br>1887–1899<br>1900–1911<br><br>1912 |
| Sales agent for a sword maker | **S. Hecht & Co.**<br>**S. Hecht & Co.**<br>**S. Hecht & Co.**<br>• Sales agent for Alexander Coppel & Co., Solingen sword makers, 1868–1878. | 19 Carter Lane<br>20 Noble St.<br>14 Hamsell St. | 1868–1870<br>1870–1877<br>1877–1878 |
| | **Heintzman**<br>(see Rochussen & Co.) | | |
| Goldsmith<br>Hilt maker<br>Sword maker | **Thomas Heming**<br>• Apprenticed to Peter Butcher, 1738–1745.<br>**Thomas Heming**<br>• Appointed goldsmith to His Majesty the King, 1760–1782.<br>**Thomas Heming** | <br><br>Piccadilly<br><br>New Bond St. | B1723–1745<br><br>1746–1762<br><br>1763–D1795 |
| Sword cutler | **Thomas Hendy** | Lady Place, North Underhill | 1710–D1721 |
| Silversmith<br>Goldsmith<br>Hilt maker<br>Sword cutler | **David Hennell (Sr.)**<br>**David Hennell (Sr.)**<br>**David (Sr.) & Robert (Sr.) Hennell** | <br>Kings Head Court, Gutter Lane<br>11 Foster Lane | B1717–1735<br>1736–1762<br>1763–1785 |

| | | | |
|---|---|---|---|
| | • Partner: Robert Hennell Sr. (son of David Hennell Sr.), born 1741. | | |
| | • David Hennell Sr. died in 1785. | | |
| | David (Jr.) & Robert (Sr.) Hennell | 11 Foster Lane | 1786–1801 |
| | • David Hennell Jr., son of Robert Sr., born 1767. | | |
| | David (Jr.) & Robert (Sr.) & Samuel Hennell | 11 Foster Lane | 1802 |
| | • Partner: Samuel Hannell, son of David Hennell Jr. | | |
| | • David Hennell Jr. died, 1802. | | |
| | Robert (Sr.) & Samuel Hennell | 11 Foster Lane | 1803–1811 |
| | • Robert Hennell Sr. died, 1811. | | |
| | Samuel Hennell | 11 Foster Lane | 1812–1813 |
| | Hennell (Samuel) & Terry (John) | 11 Foster Lane | 1814–1815 |
| | Samuel Hennell | 11 Foster Lane & 8 Aldernanbury St. | 1816 |
| | Samuel Hennell | 11 Foster Lane & 8 Charles St., Gowell St. | 1816–1817 |
| | Samuel Hennell | 5 Snowhill | 1818 |
| Silversmith Goldsmith Hilt maker Sword cutler | Richard Hennell | | 1877–1886 |
| Silversmith Goldsmith Hilt maker Sword cutler | Robert Hennell Jr. | Windmill Court, Smithfield | B1769–1784 |
| | • Son of Robert Hennell Sr. | | |
| | Robert Hennell Jr. | Windmill Court, Smithfield | 1785–1807 |
| | Hennell (Robert Jr.) & Nutting (Henry) | 38 Noble St., Foster Lane | 1808 |
| | Robert Hennell (Jr.) | 35 Noble St., Foster Lane | 1809–1816 |
| | Robert Hennell (Jr.) | 3 Lancaster Ct. | 1817–1827 |
| | Robert Hennell (Jr.) | 14 Northumberland St., Strand | 1828–1834 |
| Gold and silver lacemen Button maker Sword cutler | Herbert (George) & Boys (John) & Co. | 26 Bedford St., Covent Garden | 1855–1875 |
| | • Successors to E. Tyler & Co. (see E. Tyler & Co.). | | |
| | George Herbert & Co. | 26 Bedford St., Covent Garden | 1876–1921 |
| | Herbert & Co. | 31 Carnaby St. | 1922–1925 |
| | Herbert & Co. | 31 Fauberts Place, Regent St. | 1926–1948 |
| | • Herbert & Co. bought by Charles Pitt & Co. in 1948 (see Charles Pitt & Co.). | | |
| Cutler | William Alfred Herbert | | 1900–1962 |
| | • Master of the Cutlers Company, 1912. | | |
| Sword maker | John Herder (Herd) | St. Andrew, Holborn | 1590–D1605 |
| Goldsmith Silversmith Sword cutler | James Hervot | Rouen, France | 1670–1698 |
| | • Immigrated to London, 1698. | | |
| | • Apprenticed to Nicholas Faulcon, 1699–1706. | | |
| | James Hervot | Glasshouse Yard, | 1707–1721 |
| | | Parish of St. Anne, Blackfriars over Jackson court by Breakneck stairs | 1722–1740 |
| | Hervot (James) & Co. | Temple Bar | 1741–1750 |
| Blacksmith Sword maker | William Hethe (Hythe) | | 1479–D1535 |
| | • Registered his "W" mark (two Vs overlapping) in 1519. | | |
| Hatter Sword cutler | Hewson & Williams | | 1899–1910 |
| | • Successors to Busain & Smith (see Busain & Smith). | | |

| | | | |
|---|---|---|---|
| Silversmith<br>Hilt maker<br>Sword cutler | **Hickleton & Phillips**<br>• Decorated some presentation swords for the Goldsmiths & Silversmiths Co. Ltd.<br>(see Goldsmiths & Silversmiths Co. Ltd.) | | 1940–1950 |
| Sword cutler | **Hicks (Hugh) & Varnan (John)**<br>• On April 3, 1645, during the English Civil War, contracted for 1,000 swords<br>(infantry) with belts at 4 shillings, 8 d each for the "New Model" army<br>(parliamentary forces) under Gen. Sir Thomas Fairfax. | | 1642–1649 |
| Silversmith<br>Hilt maker<br>Sword cutler | **Francis Hide** | against the Royal Exchange | 1722–1730 |
| Cutler | **Thomas Higgs**<br>• Master of the Cutlers Company, 1840. | | 1830–1845 |
| | Henry Hilburn<br>(see Welch & Stalker) | | |
| Goldsmith<br>Silversmith<br>Hilt maker<br>Sword cutler | **Charles Hill**<br>**Charles Hill & Samuel Swain**<br>**Charles Hill**<br>**Charles Hill**<br>**Hill (Charles) & Yardley (Henry)**<br>**J. (John) Hill & Henry Yardley** | <br>New Gravel Lane, Southwark<br>New Gravel Lane, Southwark<br>3 Charing Cross<br>3 Charing Cross<br>3 Charing Cross | Bc1739–1759<br>1760–1773<br>1774–1789<br>1790–1799<br>1800–1809<br>1810–1813 |
| Military tailor<br>Sword cutler | **Henry Hill**<br>**Henry Hill**<br>**Henry Hill**<br>**Hill Bros.**<br>• Appointed military tailors to Her Majesty the Queen.<br>• Sold M1821 light cavalry sabers.<br>**Hill Bros.**<br>• Appointed military tailors to His Majesty the King.<br>**Hill Bros. Ltd.**<br>**Hill Bros. Ltd.**<br>• In 1939, merged with Henry Poole & Co. Ltd. (see Henry Poole & Co. Ltd.). | 14 Little Portland St.<br>53 York St., Westminster<br>3 Old Bond St.<br>3 Old Bond St.<br><br><br>3 & 4 Old Bond St.<br><br>3 & 4 Old Bond St.<br>25 Bruton St. | 1842–1844<br>1845–1849<br>1850–1855<br>1856–1889<br><br><br>1890–1922<br><br>1923–1930<br>1931–1939 |
| Sword and<br>bayonet cutler | **John Hill (Sr.)**<br>• Supplied plug bayonets to the Board of Ordnance, 1688–1691.<br>• John Hill Sr. became ill in 1691 and died in 1702.<br>• After his death, his goods were auctioned off at Mitre Coffee House, Mitre Court,<br>near St. Dunstan's Church.<br>• The catalog lists *"rich swords of all sorts, simiters, bayonets and belts, all of the newest fashion, some fit*<br>*for merchants exportation."*<br>• Hill's wife Mary ran the business, 1691–1702.<br>**Mary Hill**<br>• Widow of John Hill.<br>• Supplied plug bayonets to the Board of Ordnance, 1691–1702.<br>• Probably took over John Hill's business when he became ill in 1691 and continued until<br>her son, John Hill Jr., took over the business in 1702.<br>**John Hill (Jr.)**<br>• Son of John Hill Sr.<br>**John Hill (Jr.)** | behind the Royal Exchange<br><br><br><br><br><br><br>behind the Royal Exchange<br><br><br><br><br>behind the Royal Exchange<br><br>Princes St.<br>at the sign of the Cutlers Arms & Sword | 1680–D1702<br><br><br><br><br><br><br>1691–1702<br><br><br><br><br>1702<br><br>1702–1770 |

| | | | |
|---|---|---|---|
| | John Hill (Jr.) | 40 St. Katherines St. | 1771–1725 |
| Gun and pistol maker | John Hill | 168 Tooley St. | 1817–1832 |
| Sword cutler | John Hill | 76 Tooley St. | 1833–1851 |
| Gold and silver lacemen | Richard Hill | Strand | 1735–1742 |
| Sword cutler | Walter Turner, Richard Hill & Robert Pitter | 3 Strand at the sign of the White Heart | 1743–1761 |
| | • Successor to Walter Turner (see Walter Turner). | | |
| | Hill (Richard) & Pitter (Thomas) (Sr.) | 3 Strand | 1762–1779 |
| | Hill (Richard), Pitter (Thomas) & Son (Thomas Pitter Jr.) | 3 Strand | 1780–1794 |
| | • Succeeded by T. (Thomas) Pitter & Son (see T. Pitter & Son). | | |
| Accoutrement maker | Henry Hillman | Bond St. | 1690–1759 |
| Sword cutler | John Hillman | New Bond St., New Hanover Square at the sign of the Flaming Sword | 1760–1763 |
| | Edward Hillman | New Bond St., near Hanover Square | 1764–1775 |
| Silversmith Hilt maker Sword cutler | John Hilman | Russell St., Covent Garden | 1731–1733 |
| Sword cutler | Benjamin Hilton | 12 Whitcomb St., Charing Cross | 1804–1807 |
| Bladesmith | Robert Hinkley (Hynkeley) | | 1420–D1452 |
| | • In 1450, being very ill, Hinckley sold his "two crescent moons" mark to bladesmith John Morthe (see John Morthe). | | |
| Silversmith Hilt maker Sword cutler | John Hinton | | 1661 |
| | • Signed an advertisement warning sword purchasers against counterfeit cast silver hilts | | |
| | Thomas Hobbs (see James Taylor) | | |
| Military tailor Embroiderer | Hobson & Sons Ltd. | 94 Great Windmill St. Artillery Place, Woolwich | 1873–1877 |
| Accoutrement maker | Hobson & Sons Ltd. | 37–38 Great Windmill St. Artillery Place, Woolwich | 1878–1883 |
| Sword and dirk cutler | Hobson & Sons Ltd. | Little Windmill St. | 1884–1886 |
| | Hobson & Sons Ltd. | 1–3–5 Lexington St., Golden Square & 1–3 Longston St. | 1887–1901 |
| | Hobson & Sons Ltd. | 1–3–5 Lexington St., Golden Square & 156 to 164 Tooley St. & 1 to 7 Shand St. | 1902–1907 |
| | Hobson & Sons Ltd. | 1–3–5 Lexington St., Golden Square & 156 to 164 Tooley St. & 1 Brewer St. | 1908–1956 |
| | Hobson & Sons Ltd. | 1–3–5 Lexington St., Golden Square & 156 to 164 Tooley St. & 91 Lewisham Rd. | 1957–1970 |

|  |  |  |  |
|---|---|---|---|
|  | Hobson & Sons Ltd. | 1–3–5 Lexington St., Golden Square & 156 to 164 Tooley St. & 126 to 134 Creek Road | 1971–1972 |
|  | Hobson & Sons Ltd. | Greek Street west end, 156 to 164 Tooley St. 126 to 134 Creek Rd. | 1973–1986 |
|  | Hobson & Sons Ltd. | 156 to 164 Tooley St. 126 to 134 Creek Rd. | 1987–1992 |
|  | Hobson & Sons Ltd. | 126 to 134 Creek Rd. | 1993–1996 |
|  | Hobson & Sons Ltd. | Kenneth Road, Thundersley Benfleet, Essex | 1997–2003 |

- Subsidiary, Bernard Uniforms Ltd.
- The Hobson Company manufactures uniforms and military clothing, belts, slings, badges, holsters, buttons, webbing, patches, anklets, buckles, boots, sashes, epaulettes, caps, lanyards, plumes, leather goods, spurs, and ceremonial accoutrements.
- They also sell swords and sword accoutrements.

**Henry Holbeck**
(see John Lambert)

| | | | |
|---|---|---|---|
| Hatter Sword cutler | Thomas Holden | Ludgate St. at the sign of Two Eagles | 1745–1759 |

- Sold combination sword and pistol, many decorated with eagles.
- Succeeded by Daniel Fry (see Daniel Fry).

| | | | |
|---|---|---|---|
| Military tailor Sword cutler | James Hole | 43 Freeschool St., Horsley Down | 1860–1883 |
|  | Hole (James) & Hayson (Edward) | 16 London St. | 1884–1886 |
|  | Hole (James) & Hayson (Edward) | 109 Fenchurch St. | 1887–1889 |
|  | Francis Hole | 57 Fenchurch St. | 1890 |
|  | Francis Hole | 21 London St. | 1891–1899 |
|  | Francis Hole & Co. | 18a London St. | 1900–1903 |
|  | Hole (Francis) & Powell (Arthur John) | 18a London St. | 1904–1907 |

- Succeeded by Arthur John Powell (see Arthur John Powell).

| | | | |
|---|---|---|---|
| Sword maker | William Holland |  | 1693–1713 |

- Used the "star and crown" mark.

| | | | |
|---|---|---|---|
| Gun maker Gun part maker Sword and sword blade maker | Thomas Hollier | Whitechapel | 1700–1729 |

- Appointed sword cutler to the Board of Ordnance on June 10, 1716, September 24, 1722, and December 24, 1722.
- Sold the Board of Ordnance 1,000 naval cutlasses, 1726.
- Also rebladed naval cutlasses for the Board of Ordnance.

| | | | |
|---|---|---|---|
| | Thomas Hollier | Armoury Mills, Lewisham, Kent | 1730–D1754 |

- A Hollier employee (Thomas Nelson) supplied naval swords and bayonets to the Board of Ordnance, 1743.
- Made swords, sword blades, bayonets, guns, ramrods, iron work, and brass gun furniture.
- Succeeded by Richard Hornbuckle (see Richard Hornbuckle).

| | | | |
|---|---|---|---|
| Cutler | Henry Holmes |  | 1530–1545 |

- Master of the Cutlers Company, 1540.

| | | | |
|---|---|---|---|
| Military tailor Habit (riding clothes) maker Sword cutler | Holt (Thomas) & Plush (John) | 20 Sackville St. | 1822–1833 |
|  | Thomas Holt | 20 Sackville St. | 1834–1844 |
|  | Thomas Holt & Son | 20 Sackville St. | 1845–1871 |
|  | Holt (Thomas) & Son | 9 Old Burlington St. | 1872–1933 |

| | | | |
|---|---|---|---|
| Steel ornament maker | Thomas Hook | 157 Wardour St. | 1794–1806 |
| Steel button maker | Thomas Hook | 43 Brewer St., Golden Square | 1807–1815 |
| Steel buckle maker | | | |
| Steel sword hilt maker | | | |
| Silversmith | John Hooke | | 1429 |
| Hilt maker | | | |
| Sword cutler | | | |
| Cutler | Thomas Hooker<br>• Master of the Cutlers Company, 1677. | | 1667–1882 |
| Sword maker | John Hopkins<br>• Apprenticed to Nicholas Croucher, 1692–1699 (see Nicholas Croucher).<br>John Hopkins | | Bc1678–1699<br><br>1700–1720 |
| Sword and sword blade maker | Richard (Recardes) Hopkins<br>• Had a sword blade mill in Hounslow, c. 1655–1659, during the Commonwealth under Oliver and Richard Cromwell.<br>Richard (Recardes) Hopkins<br>• Master of the Cutlers Company, 1671.<br>• Rentor Warden of the Cutlers Company, 1667.<br>• Warden of the Cutlers Company, 1668–1669. | Hounslow<br><br><br>London | 1655–1659<br><br><br>1659–1675 |
| Cutler | Sir Richard Hopkins<br>• Master of the Cutlers Company, 1701 and 1723. | | 1691–1728 |
| Cutler | Thomas Hopkins<br>• Master of the Cutlers Company, 1655. | | 1645–1660 |

## THE HOPPE (HOPPIE, HOPPER, HOPPY) FAMILY
(see Chapters 3 and 4)

| | | | |
|---|---|---|---|
| Bladesmith | Walter Hoppe | Parish of St. Anne, Aldergate | 1392–D1413 |
| Sword blade maker | Johannes Hoppe (Hoppie) the Younger<br>• Immigrated to Greenwich, 1629.<br>Johannes Hoppe (Hoppie)<br>• Worked at the Greenwich Armouries.<br>Johannes Hoppe (Hoppie)<br>Johannes Hoppe (Hoppie)<br>Johannes Hoppe (Hoppie)<br>• Listed as John Hoppy) | Solingen, Prussia<br><br>Greenwich, England<br><br>Hounslow Heath<br>Oxford<br>London | Bc1600–1629<br><br>1629–1633<br><br>1633–1642<br>1642–c1649<br>c1649–1675 |
| Sword blade maker | Heinrich Hoppe (Hoppie) the Elder<br>• Immigrated to Hounslow in 1629.<br>Heinrich Hoppe (Hoppie)<br>Heinrich Hoppe (Hoppie)<br>Heinrich Hoppe (Hoppie)<br>• Listed as Henry Hoppe. | Solingen, Prussia<br><br>Hounslow Heath<br>Oxford<br>London | Bc1609–1629<br><br>1629–1642<br>1642–c1649<br>c1649–1675 |

|  |  |  |  |
|---|---|---|---|
|  | Henry Hoppe the Younger | Oxford | Bc1642–c1649 |
|  | • Son of Henry Hoppe the Elder. |  |  |
|  | Henry Hoppe the Younger | London | c1649–1685 |
|  | Henry Hoppe the Younger | Shotley Bridge | 1685–1702 |
|  | • Listed as Henry Hopper. |  |  |
|  | Joseph Hoppe | Hounslow Heath | Bc1629–1642 |
|  | • Son of Henry Hoppe the Elder. |  |  |
|  | Joseph Hoppe | Oxford | 1642–c1649 |
|  | Joseph Hoppe | London | c1649–1685 |
|  | • Listed as Joseph Hopper. |  |  |
|  | • Master of the Cutler Company. |  |  |
| Gun part maker | Richard Hornbuckle | Armoury Mills | 1754–D1779 |
| Sword and sword |  | Lewisham, Kent |  |
| blade maker | • Successor to Thomas Hollier (see Thomas Hollier). |  |  |
| Bayonet maker | • Made ramrods, swords, sword blades, and bayonets. |  |  |
| Cutler | Thomas Horne |  | 1722–1737 |
|  | • Master of the Cutlers Company, 1732. |  |  |
| Hilt maker | Thomas Horner |  | 1613 |
|  | • On March 15, 1613, fined by the Cutlers Company for making counterfeit sword hilts of pewter. |  |  |
| Silversmith | Elias Hosier |  | Bc1680–1701 |
| Hilt maker | • Apprenticed to Thomas Vicaridge, 1694–1701 (see Thomas Vicaridge). |  |  |
| Sword cutler | Elias Hosier | New St., near Shoe Lane | 1702–1710 |
| Cutler | Ephraim How |  | 1530–1545 |
|  | • Master of the Cutlers Company, 1540. |  |  |
| Goldsmith | James Howell & Co. | 5, 6 & 9 Regent St, Pall Mall | 1860–1875 |
| Silversmith | • Thomas Buxton Moorish made hilts and decorated some of their presentation swords. |  |  |
| Hilt maker | (see Thomas Buxton Moorish) |  |  |
| Sword cutler |  |  |  |
| Cutler | John Howes |  | 1671–1686 |
|  | • Master of the Cutlers Company, 1681. |  |  |
| Sword cutler | Thomas Howes | Parish of St. Dunston in the East | 1612 |
| Bladesmith | John Howesman |  | 1462–1470 |
|  | • Became freeman of the Bladesmiths Company, 1462. |  |  |
| Sword cutler | John Hubbert |  | 1622–1625 |
| Bladesmith | John Hubberd | Fleet St., Parish of St. Martin, Ludgate | 1448–1460 |
| Cutler | Robert Huddy |  | 1789–1804 |
|  | • Master of the Cutlers Company, 1799. |  |  |
| Sword cutler | Michael Hudson |  | 1550–1570 |

• In January 1558, Hudson supplied swords to the Drapers (clothing makers) Company for the defense of their settlement in Calais, France, during the French War.

| | | | |
|---|---|---|---|
| Sword cutler | W. (William) J. Huett | | 1840–1860 |
| Cutler | William Hullett Hughes<br>• Master of the Cutlers Company, 1849. | | 1839–1854 |
| Sword maker | Hulbeck & Sons | 4 New Bond St. | 1790–1800 |
| Hatter<br>Sword cutler | John Hull | 48 Fleet St. | 1772–1774 |
| Sword cutler | Joseph Hume | 13 Perey St. | 1829–1845 |
| | William Hume<br>(see Charles Hebbert) | | |
| Sword cutler | Richard Humphreys | Haymarket | 1875 |
| Sword cutler | Hunt & Collins | 30 Haymarket | 1822 |
| Goldsmith<br>Jeweler<br>Silversmith<br>Sword cutler<br>Hilt maker | Hunt (John) & Potter (Edward)<br>• Succeeded by Edward Potter (see Edward Potter). | 15 Vere St., Clare Market | 1805–1807 |
| Goldsmith<br>Jeweler<br>Silversmith<br>Hilt maker<br>Sword cutler | Hunt (John Samuel) & Roskell (Robert)<br>• Successors to Mortimer & Hunt (see Mortimer & Hunt).<br>Hunt (John Samuel) & Roskell (Robert)<br>Hunt (John Samuel) & Roskell (Robert)<br>• Appointed goldsmiths and silversmiths to Her Majesty the Queen.<br>• Taken over by J.W. Benson in 1880, but kept the company name (see J.W. Benson). | 156 New Bond St.<br><br>23 Old Bond St.<br>25 Old Bond St. | 1845–1910<br><br>1911–1922<br>1923–1968 |
| | Thomas Hunt<br>(see Henry Potts) | | |
| Goldsmith<br>Hilt maker<br>Sword cutler | William Hunt<br>• Apprenticed to John Hunt 1709–c. 1716.<br>• Journeyman to John Hunt, c. 1716–1720.<br>William Hunt<br>• Had apprentice James Hunt (son), 1742–1749.<br>William Hunt & Son (James)<br>• William Hunt died in 1760.<br>• Their trade card read:<br>*Makes & sells all sorts of rich work in gold, watch chains, Etwees,*<br>*Snuff boxes, sword hilts, cane heads, and buckles*<br>James Hunt<br><br>James Hunt<br>James Hunt | <br><br><br>Old Change<br><br>King St., Cheapside<br>at the sign of the Golden Lion<br><br><br><br>King St., Cheapside<br>at the sign of the Golden Lion<br>9 King St.<br>Iron Monger Lane | Bc1695–1720<br><br><br>1721–1749<br><br>1750–D1760<br><br><br><br><br>1760–1772<br><br>1773–1780<br>1781–D1782 |
| Army tailor<br>Sword cutler | George Hunter<br>• Successor to John Le Gassick (see John Le Gassick). | 12 Cork St., Bond St. | 1850–1852 |

| | | | |
|---|---|---|---|
| | George Hunter & Co. | 12 Cork St., Bond St. | 1853–1854 |
| | Hunter (George) & Guthrie (Thomas Ansley) | 12 Cork St., Bond St. | 1855–1862 |
| | • Succeeded by Thomas Ansley Guthrie (see Thomas Ansley Guthrie). | | |
| Sword cutler | **Huntley Boyd & Co.** | | 1840–1860 |
| Army tailor<br>Sword cutler | **Henry Huntsman**<br>• Successor to John Walker Taylor (see John Walker Taylor). | 126 New Bond St. | 1864–1897 |
| Goldsmith<br>Toy (hardware)<br>maker<br>Hilt maker<br>Sword cutler | **Henry Hurt**<br>• Successor to Thomas Chesson at 32 Ludgate Hill.<br>**Henry Hurt**<br>• Succeeded by Thead & Picket.<br>(see Thomas Chesson, Thead & Picket) | St. Pauls, Churchyard<br><br>32 Ludgate Hill<br>at the sign of the Golden Salmon | 1745–1757<br><br>1758–1760 |
| Cutler | **Edmund Hutchinson**<br>• Master of the Cutler Company, 1630. | | 1620–1635 |
| Cutler | **Francis Hyde**<br>• Master of the Cutler Company, 1734. | | 1724–1739 |
| Cutler | **Edward Hynson**<br>• Master of the Cutler Company, 1638. | | 1628–1643 |
| Cutler | **John Iland**<br>• Master of the Cutler Company, 1572. | | 1562–1577 |
| Silversmith<br>Hilt maker<br>Sword cutler | **Robert Innes**<br>**Robert Innes**<br>**Robert Innes**<br>• Made silver-hilted small swords. | Mays Building, St. Martins Lane<br>Drury Lane<br>Harrie Court, Strand | 1743–1753<br>1754<br>1755–1758 |
| Silversmith<br>Hilt maker<br>Sword cutler | **Edward Ironside (Sr.)**<br>• Apprenticed to Edward Beesley, 1688–1698.<br>**Edward Ironside (Sr.)**<br>**Ironside (Edward Sr.) & Belcher (James)**<br><br>**Ironside (Edward (Jr.),<br>Belcher (James) & How (Peter)**<br>• Edward Ironside (Jr.) died in 1754. | <br><br>Lombard St.<br>Lombard St.<br>at the sign of the Black Lion<br>Lombard St. | Bc1674–1698<br><br>1699–1728<br>1729–D1737<br><br>1737–1754 |
| Silversmith<br>Hilt maker<br>Sword cutler | **William Ironside**<br>• Mounted small swords. | | 1708–1709 |
| Sword cutler | **John Irwin** | | 1740–1750 |
| Goldsmith<br>Hilt maker<br>Sword cutler | **Charles Jackson (Jakson)**<br>• Apprenticed to John Ladyman, 1700–1707.<br>• Journeyman to John Ladyman, 1707–1712.<br>**Charles Jackson**<br>• Had apprentice William Toone, 1723–1725.<br>• Had apprentice Benjamin Corbett, 1718.<br>**Charles Jackson** | <br><br><br>Cannon St.<br><br><br>Tower St. | Bc1686–1712<br><br><br>1713–1727<br><br><br>1728–1738 |

| | Charles Jackson | St. Sweetings Lane at the sign of the Golden Cap | 1739–1740 |
|---|---|---|---|
| Fine steel worker<br>Military goods dealer<br>Silversmith<br>Sword cutler | John Jackson (Jakson)<br>• A trade card reads:<br>*Sells, knives, spurs, buckles, powder flasks, shot bags, fine steel work (sword hilts).* | Wood St., Cheapside at the sign of the Unicorn | 1679–1710 |
| Bladesmith | Thomas Jackson (Jakson)<br>• A bladesmith member of the Armourers Company, 1515.<br>• A bladesmith member of the Cutlers Company, 1517.<br>• Registered his "two crescent moons" mark with the Cutler Company, 1517. | | 1515–1525 |
| Goldsmith<br>Hilt maker<br>Buckle maker<br>Sword cutler | Joel Jacobson<br>Joel Jacobson<br>Joel Jacobson<br>Joel Jacobson & John Yardley<br>(see John Yardley) | 10 St. James Walk, Clerkenwell<br>88 High Holborn<br>37 Charles St., Hatton Garden<br>37 Charles St., Hatton Garden | 1776–1778<br>1779<br>1780<br>1780–1786 |
| Cutler | Samuel James<br>• Master of the Cutlers Company, 1691. | | 1681–1696 |
| Linen draper (dealer)<br>Hatter<br>Sword cutler | James Janaway<br>• Linen draper only.<br>James Janaway & Co.<br>• Became a sword cutler. | 30 Parliament St., Westminster<br>30 Parliament St., Westminster | 1805–1810<br>1811–1815 |
| | Japan Warehouses<br>• Obtained a royal patent for lacquering (jappaning) upon iron, guns, pistols, armour and sword blades (to preserve them from rusting). | West end of the Royal Exchange and upper end of St. James St. | 1692 |
| Goldsmith<br>Silversmith<br>Jeweler<br>Hilt maker<br>Sword and knife cutler | Daniel Jeffreys (Jeffris)<br>Daniel Jeffreys<br>• Appointed cutler to His Majesty the King.<br>George Jeffreys | <br>76 Strand<br><br>76 Strand | Bc1743–1762<br>1763–1776<br><br>1777–1801 |
| Goldsmith<br>Silversmith<br>Jeweler<br>Hilt maker<br>Sword and knife cutler | Henry Jeffreys<br>Jeffreys (Henry) & Co. (Jeffris)<br><br>Henry Jeffreys & Co.<br>Henry Jeffreys & Co. | <br>91 Fleet St.<br>at the sign of the Great Knife Case<br>91 & 96 Fleet St.<br>91 Fleet St. | Bc1766–1785<br>1786–1789<br><br>1790–1793<br>1794–1800 |
| Goldsmith<br>Silversmith<br>Jeweler<br>Hilt maker<br>Sword cutler | Nathaniel Jeffreys (Sr.) (Jeffris)<br>Nathaniel Jeffreys (Sr.)<br>• Appointed cutler to His Majesty the King.<br>• Mounted and sold small swords.<br>• In October 1759, delivered 3,500 iron basket hilt (M1757) broadswords with paste broad scabbards (with single-edged blades) to the Board of Ordnance at 50 shilling each.<br>• Issued to the 87th, 88th, and 89th Highland Regiments. | <br>32 Strand,<br>corner of Villiers St., Westminster | Bc1734–1753<br>1754–1769 |

|  |  |  |  |
|---|---|---|---|
| | • Some served in the French and Indian War in America, 1756–1763. | | |
| | **Jeffreys (Nathaniel Sr.) & Drury (Dru Jr.)** | 32 Strand, corner of Villiers St., Westminster | 1770–1779 |
| | • Successor to Dru Drury Jr. (see Dru Drury Jr.). | | |
| | • Supplied basket hilt broadswords to the Gordon Fencibles, 1778. | | |
| | • Supplied basket hilt broadswords to the 2nd Highland Battalion, 1779. | | |
| | • Nathaniel Jeffreys died in 1779. | | |
| | • Succeeded by Dru Drury (Jr.) & Son (see Dru Drury & Son). | | |
| Goldsmith Silversmith | **Nathaniel Jeffreys (Jeffris) (Jr.)** | | Bc1763–1783 |
| | • Son of Nathaniel Jeffreys Sr. | | |
| Jeweler | **Nathaniel Jeffreys (Jr.)** | 22 Piccadilly | 1784–1789 |
| Hilt maker Sword cutler | **Nathaniel Jeffreys (Jr.)** | 70–71 Piccadilly corner of Dover St. | 1790–1798 |
| | • Royal jeweler to King George III. | | |
| | • Ray & Montaque mounted some Jeffreys swords (see Ray & Montaque). | | |
| | • A 1791 receipt called for a "*sword hilt, sword & tassells entirely of Brilliants* [diamonds], *(cost 500 pounds).*" | | |
| | • Another receipt called for a diamond-mounted sword at a cost 6,300 pounds, a huge amount for that period. | | |
| | **Nathaniel Jeffreys (Jr.)** | 34 Pall Mall | 1799–1802 |
| | • Bought blades from Birmingham sword and blade makers Samuel Harvey I, II, and III. | | |
| Sword cutler | **Robert Jeffreys** | | 1642–1649 |
| | • On April 3, 1645, during the English Civil War, Jeffreys contracted for 2,000 swords (infantry) with belts at 4 shillings 8 denier each for the "New Model" army (parliamentary forces) under Gen. Sir Thomas Fairfax. | | |
| Goldsmith | **Thomas Jeffreys (Jeffris)** | | Bc1744–1764 |
| Silversmith | **Thomas Jeffreys** | Charing Cross | 1765–1768 |
| Jeweler | **Thomas Jeffreys** | 20 Cockspur St., near Charing Cross | 1769–1778 |
| Toy (hardware) maker | **Jeffreys (Thomas) & Jones (Samuel) & Co.** | 20 Cockspur St., near Charing Cross | 1779–1794 |
| | • Appointed goldsmiths and sword cutlers to His Majesty the King. | | |
| Sword hilt maker Sword cutler | **Jeffreys (Thomas), Jones (Samuel) & Gilbert (Philip)** | 20 Cockspur St., near Charing Cross | 1795–1797 |
| | **Jeffreys (Thomas) & Gilbert (Philip)** | 20 Cockspur St., near Charing Cross | 1798–1801 |
| | • Sold swords mounted by Ray & Montaque and James Morisset. | | |
| | • Succeeded by Gilbert & Jeffreys & Co. | | |
| | (see Samuel Jones; Ray & Montaque; James Morisset; Gilbert & Jeffreys & Co.) | | |
| Sword and knife maker | **John Jenks (Jenckes)** | Parish of St. Anne, Blackfriars | Bc1578–Dc1625 |
| | • A German immigrant. | | |
| | • Apprenticed to John Wyat, 1592–1599. | | |
| | • On January 8, 1595, married Sarah Fulwater, daughter of sword and knife maker Henry Fulwater (also a German), who immigrated to England from Solingen, Prussia, in 1566 and changed his name from Heinrich Vollenweidner (see Henry Fulwater). | | |
| | • On August 26, 1599, son Joseph Jenks was born. | | |
| | • In 1607, Jenks obtained the "thistle" mark from knife maker Thomas Wills. | | |
| Sword and knife maker | **Joseph Jenks (Sr.) (Jenckes)** | Parish of St. Anne, Blackfriar | B1599–1625 |
| | • Son of John Jenks. | | |
| | • Moved to Hammersmith in 1625. | | |
| | **Joseph Jenks (Sr.)** | Hammersmith, County of Middlesex | 1625–1629 |
| | • Hammersmith is 3 miles from Blackfriars, London. | | |
| | • On November 5, 1627, married Joan Hearne of Horton (a county of Buckinghamshire on the Colne River 14 miles from Hammersmith). | | |

- On October 12, 1628, son Joseph Jenks Jr. was born at Horton (hometown of his mother, Joan).
- In 1629, Jenks and his family moved to Isleworth at Hounslow Heath, County of Middlesex, on the Isleworth River (see Chapter 3).

| | | |
|---|---|---|
| **Joseph Jenks (Sr.)** | **Isleworth (Hounslow Heath) County of Middlesex** | **1629–1641** |

- On February 29, 1635, his wife Joan died.
- In 1639, Jenks petitioned Algernon, the Earl of Northumberland, for the lease of some waste ground near Worton Bridge (over the Isleworth River) at Isleworth for the construction of a sword blade mill. The petition was granted and he set up his mill.
- Jenks signed his blades: "JOSEPH JENCKES ME FECIT HOUNSLOE"
- In 1641, Jenks and his family immigrated to New Hampshire, USA, eventually settling in Lynn, Massachusetts, in 1645.
- Jenks brought some German bladesmiths with him.

| | | |
|---|---|---|
| **Joseph Jenks (Sr.)** | **New Hampshire, USA** | **1641–1645** |

- Settled on the York River temporarily but moved to Lynn in 1645.

| | | |
|---|---|---|
| **Joseph Jenks (Sr.)** | **Lynn, MA, USA** | **1645–D1683** |

- Jenks became partner with a John Winthrop Jr. (major stockholder) and built a furnace and edged tool mill called the Hammersmith Iron Works (named after Hammersmith, England).
- The works were located at Saugus, Massachusetts, near Lynn, Massachusetts.
- Also had a saw mill.
- Jenks became an edged tool maker.
- In 1645, Jenks was granted a patent by the English parliament (the Commonwealth, under Cromwell) for 14 years for the building of new type of mill machinery. It was the first "engine" (mill machinery) patent granted to English immigrants in America.
- His petition to Parliament read:

    *Whereas the Lord hath been pleased to give me knowledge in the making and erecting of engines of mills* [mill machinery] *to go* [run by] *by water* [water wheel] *for the speedy dispatch of much work, with few mens labor, in little time. My desire is to improve this talent for the public good and benefit and service of this country, to which end my intention and purpose (If God Permits) is to build a mill for the making of sythes and also a new invented saw mill and diverse other engines for making of diverse sorts of edged tools, whereby the country may have such necessities in short time at far cheaper rates than now they can.*

- Jenks made scythes and other edged tools such as knives, long axes, and belt axes (hatchets).
- Also cut dies for the pine tree shilling (English coin).

| | | | |
|---|---|---|---|
| | **Joseph Jenks (Jenckes) (Jr.)** | Horton near Hammersmith | B1628–1629 |

- His family moved in 1629.

| | | | |
|---|---|---|---|
| Knife and edged tool maker | **Joseph Jenks (Jr.)** | Isleworth (Hounslow Heath) County of Middlesex | 1629–1641 |
| | **Joseph Jenks (Jr.)** | New Hampshire, USA | 1641–1645 |

- He would have apprenticed with his father, Joseph Jenks Sr., at 14 years of age for seven years (1642–1649).

| | | | |
|---|---|---|---|
| | **Joseph Jenks (Jr.)** | Lynn, MA, USA | 1645–1649 |

- Joseph Jenks Jr. moved to Pawtucket, Rhode Island, in 1649.

| | | | |
|---|---|---|---|
| | **Joseph Jenks (Jr.)** | Pawtucket, RI, USA | 1649–Dc1710 |

- Made edged tools such as scythes, knives, long axes, and belt axes (hatchets) and regular tools such as hammers.
- Also made ship anchors.

| | | | |
|---|---|---|---|
| Sword cutler | **Jennens & Co. Ltd.** | 56 Conduit St. | 1900–1920 |

- Sold Wilkinson Sword Co. swords.

| | | | |
|---|---|---|---|
| Silversmith Hilt maker Sword cutler | **Henry Jennings** | | 1670–1695 |

- Sold small swords with silver hilts.
- Master of the Cutlers Company, 1686.

| | | | |
|---|---|---|---|
| Hardwaremen (dealer)<br>Cutlery dealer<br>Sword cutler<br>Military goods dealer | Benjamin Jewson | Crooked Lane,<br>near the Monument<br>at the sign of the Green Man | 1764–1779 |
| Hardwaremen (dealer)<br>Cutlery dealer<br>Sword cutler<br>Gun and flint dealer | Joseph Jewson<br>Jewson (Joseph) & Burgon (John)<br>• Succeeded by Burgon & Jewson (see Burgon & Jewson). | 16 Fish Street<br>16 Fish Street | 1765–1785<br>1786–1787 |
| Sword cutler | Johnson & Sadler<br>Johnson & Folcard<br>Johnson & Pegg | 2 Clifford St.<br>2 Clifford St.<br>2 Clifford St. | 1850–1870<br>1871–1909<br>1910–1920 |
| Ornament maker<br>Military feather and hair plume maker<br>Sword cutler | Henry Johnson<br>Henry Johnson<br>• Succeeded by John Lilly (see John Lilly). | 4 Warwick St., Golden Square<br>13 Archer St., Great Windmill St. | 1820–1835<br>1836–1844 |
| Ornament maker<br>Military feather and hair plume maker<br>Sword cutler | John Johnson (Sr.)<br>John Johnson (Sr.)<br>J. (John) Johnson (Jr.) & Son<br>John Johnson (Jr.) & Sons<br>John Johnson (Jr.) & Sons | 25 Pulteney St.<br>4 Great Vine St., Regent St.<br>4 Great Vine St., Regent St.<br>4 Great Vine St., Regent St.<br>42 Gerrard St. | 1809–1821<br>1822–D1832<br>1832–1834<br>1835–1858<br>1859–1868 |
| Gold and silver laceman<br>Sword cutler | Johnson & Simpson<br>Johnson, Simpson & Simons<br>Johnson, Simpson & Simons<br>• Succeeded by Simpson & Rook (see Simpson & Rook). | 10 Little Britain<br>10 Little Britain<br>9 & 10 Little Britain | 1840–1848<br>1849–1855<br>1856–1871 |
| Gold wire maker | Joseph Johnson<br>• Wire used for sword hilts. | 12 Princes St., Barbican | 1799–1810 |
| Cutler | James Johnston<br>• Master of the Cutlers Company, 1831. | | 1821–1836 |
| Hardwareman (dealer)<br>Sword cutler<br>Scabbard maker | Joshua Johnston (Sr.) (Johnstone)<br>Joshua Johnston (Sr.)<br>Joshua Johnston (Sr.)<br>• Retired in 1793, died in 1801.<br>Richard & Joshua Johnston (Jr.)<br>• Sons of Joshua Johnston (Sr.), Joshua Jr. born c. 1764, Richard born c. 1774.<br>• In 1798, Richard Johnston joined Robert Foster (see Robert Foster).<br>Joshua Johnston (Jr.)<br>• Made bayonets for Board of Ordnance.<br>J. (Joshua Jr.) W. (William) &<br>T. (Thomas) Johnston<br>• William and Thomas were sons of Joshua Johnston Jr.<br>J. (Joshua Jr.), W. (William) &<br>T. (Thomas) Johnston | <br>Blackmore Lane<br>8 Newcastle, Strand<br><br>8 Newcastle, Strand<br><br><br>8 Newcastle, Strand<br><br>8 Newcastle, Strand<br><br><br>12 Newcastle, Strand | Bc1764–1783<br>1784–1790<br>1791–1793<br><br>1794–1797<br><br><br>1798–1800<br><br>1801–1805<br><br><br>1806–1825 |

|  |  |  |  |
|---|---|---|---|
|  | • William Johnston died in 1825. |  |  |
|  | • Thomas Johnston opened his own shop in 1825 (see Thomas Johnston). |  |  |
|  | **Joshua Johnston (Jr.)** | 12 Newcastle, Strand | 1826–1839 |
| Army and navy accoutrement maker | **Richard Johnston (Johnstone)** |  | Bc1774–1793 |
|  | **Richard & Joshua (Jr.) Johnston** | 8 Newcastle, Strand | 1794–1797 |
|  | • Sons of Joshua Johnston (Sr.). |  |  |
| Gun maker | **Robert Foster & Richard Johnston** | 68 St. James St., Pall Mall | 1798 |
| Sword belt maker | • Robert Foster died in 1798. |  |  |
|  | **Richard Johnston** | 38 & 68 St. James St., Pall Mall | 1798–1806 |
| Sword cutler | • Appointed sword cutler and belt maker to His Majesty the King. |  |  |
|  | **Woolley (James), Deakin (Thomas), Dutton (John & Joseph) & Johnston (Richard)** | 38 & 68 St. James St., Pall Mall & 10 Bush Lane Cannon St. | 1806–1811 |
|  | (see James Woolley's London and Birmingham listings; John Dutton's London listing; Thomas Deakin's Birmingham listing) |  |  |
|  | **Richard Johnston** | 38 & 68 St. James St., Pall Mall | 1811–D1838 |
|  | • Master of the Cutlers Company, 1819 and 1820. |  |  |
|  | • Richard Johnston died in 1838. |  |  |
|  | **James Johnston** | 68 St. James St., Pall Mall | 1838–D1861 |
|  | • Son of Richard Johnston. |  |  |
|  | **J. (John) B. (Bryan) Johnston** | Sackville St. | 1861–1897 |
|  | • Son of James Johnston. |  |  |
| Hardwareman (dealer) | **Thomas Johnston** | 44 Stanhope, St. Clare Market | 1825–1826 |
|  | • Son of Joshua Johnston Jr. |  |  |
| Sword cutler | • Left J. W. & T. Johnston in 1825. |  |  |
|  | (see Joshua Johnston) |  |  |
|  | **Thomas Johnston** | 41 Stanhope, St. Clare Market | 1827–1828 |
|  | **Thomas Johnston** | 39 Stanhope, St. Clare Market | 1829–1848 |
|  | • Master of the Cutlers Company, 1832. |  |  |
| Cutler | **William Grenville Johnston** |  | 1861–1876 |
|  | • Master of the Cutlers Company, 1871. |  |  |
| Sword cutler | **Jones, Chalk, & Co.** |  | 1910–1912 |
| Sword cutler | **David Jones** | Parish of St. Martin in the Fields | 1617–1620 |
| Gun maker | **Edward Jones** | near the Minories | 1605–D1644 |
| Sword cutler | • Had apprentice Richard Birch, 1605–1614 (see Richard Birch). |  |  |
|  | **Ellis Jones** (see Thomas & Jones) |  |  |
| Sword cutler | **John Jones** | Parish of St. Dunstan in the West | 1593–1608 |
| Military tailor | **John Jones** | 8 Regent St. | 1828–1831 |
| Army uniform maker | **John Jones** | 6 Regent St. | 1832–1868 |
|  | **Jones (John) & Co.** | 6 Regent St. | 1869–1893 |
| Army clothier | • Sold swords made by the Wilkinson Sword Company. |  |  |
| Sword cutler | **John Jones & Co.** | 6 Regent St. | 1894–1923 |
|  | **J. (John) Jones & Co. Ltd.** | 33 St. James Square | 1924–1960 |
|  | • Succeeded by Rogers & Co. Ltd. (see Rogers & Co. Ltd.). |  |  |

| Trade | Name | Address | Dates |
|---|---|---|---|
| Goldsmith<br>Hilt maker<br>Buckle maker<br>Sword cutler | Samuel Jones<br>• Succeeded by Jeffreys, Jones & Co.<br>(see Thomas Jeffreys) | 2 Halsey Court,<br>Blackman St., Borough | 1770–1778 |
| Silversmith<br>Hilt maker<br>Buckle maker<br>Sword cutler | Thomas Jones<br>Thomas Jones | 10 Crown Court, Fleet St.<br>Buckley Court, Clerkenwell | 1766–1791<br>1792–1796 |
| Goldsmith<br>Hilt maker<br>Buckle maker<br>Sword cutler | Thomas Jones<br>• Apprenticed to Goldsmith John Threadway, 1748.<br>• Apprenticed to tin plate worker John Wray, 1748–1755.<br>Thomas Jones<br>Thomas Jones<br>Thomas Jones<br>Thomas Jones | <br><br><br>Rupert St.<br>Bells Building, Salisbury Court<br>1 Great Rider St., Parish of St. James<br>36 Duke St., Parish of St. James | Bc1734–1755<br><br><br>1756–1769<br>1770–1781<br>1782–1805<br>1806–1811 |
| Gold and silver<br>lacemen<br>Embroiderer<br>Sword cutler | William Jones & Co.<br>• Successors to Lambert & Brown (see Lambert & Brown).<br>William Jones & Co. | 236 Regent St.<br><br>7 Golden Square | 1859–1888<br><br>1889–1896 |
| Army tailor<br>Sword cutler | Joseph (Barnet) & Co.<br>• Successors to Lambert & Brown (see Lambert & Brown).<br>Joseph (Barnet) & Co.<br>Alfred Barnet Joseph & Co.<br>(see Portsea listing) | 26 Warwick St., Golden Square<br><br>150 Regent St.<br>150 Regent St. | 1862<br><br>1862–1863<br>1864–1865 |
| Gun maker<br>Sword cutler | William I. Jover (Sr.)<br>Nock (Henry), Jover<br>(William I. Sr.) & Green (John)<br>Jover (William I. Sr.) & Son (William I. Jr.)<br>William I. Jover (Jr.)<br>William I. Jover (Jr.) & Son<br>William I. Jover (Jr.) & Son | <br>10 Ludgate St.<br><br>337 Oxford St.<br>67 New Bond St.<br>65 New Bond St.<br>26 Mortimer St. | 1750–1774<br>1775–1784<br><br>1785–1792<br>1793–1794<br>1795–1801<br>1802–1810 |
| Military<br>outfitter<br>Sword cutler | Daniel Judd<br>• In 1645, during the English Civil War, provided 33 tons of musket shot, 1 ton of pistol shot, 32 tons of match, and 200 cassocks and breeches for the "New Model" army (parliamentary forces) under Gen. Sir Thomas Fairfax.<br>• On July 5, 1645, provided 400 swords (infantry) with belts with "Dutch" (Solingen, Prussia) blades at 4 shillings, 6 denier each, to parliamentary forces. | East side of Pudding Lane<br>New Fish St.<br>Parish of St. Margaret | 1638–1650 |
| Sword cutler | John Jupp | Strand | 1768 |
| Sword cutler | Thomas Justice | | 1650–1660 |
| Knife, dagger<br>and sword cutler | John Jyker<br>• On January 22, 1558 supplied swords and daggers to the Drapers Company (cloth makers) for their settlement at Calais, France, during the French War in defense of Calais. | | 1550–1560 |
| Silversmith | Charles Kandler (Sr.) & James Murray | St. Martins Lane | 1727 |

| | | | |
|---|---|---|---|
| Goldsmith | **Charles Kandler** | St. Martins Lane | 1728–1734 |
| Hilt maker | **Charles Frederick Kandler** | Jermyn St. | 1735–1766 |
| Sword cutler | • Son of Charles Kandler. | Parish of St. James, Westminister | |
| | • Made a silver-mounted basket hilted broadsword for Prince Charles Edward Stuart in 1741. | | |
| | **Charles Frederick Kandler** | corner of Jermyn St. & Harman St. | 1767–1777 |
| | **Charles Kandler (Jr.)** | 1000 Jermyn St. | 1778–1793 |
| | • Son of Charles Frederick Kandler. | | |
| Cutler | **Richard Kay** | | 1704–1719 |
| | • Master of the Cutlers Company, 1714. | | |
| Merchant | **Malcolm Kearton (Kirton) & Co.** | 28 Fenchurch St. | 1880–c1955 |
| Colonial agent | • Founder George James Malcolm Kearton (B1859–D1902) started as general merchant and east and west India colonial agent. | | |
| Military | • Around 1897, became military outfitter. | | |
| Outfitter | • Sons and partners: James Linton Graham Kearton (B1889–D1917), Arthur Noel Malcolm Kearton (B1895–Dc1955). | | |
| | • James took over the company in 1902. | | |
| | • Arthur took over the company in 1917. | | |
| | • Bought swords from George Pillin (see George Pillin). | | |
| | **John Kelke** (see John Kirke) | | |
| Saddle and | **J. & S. Kelly** | 139 Strand | 1786–1815 |
| harness maker | **T. Kelly** | 139 Strand | 1816 |
| Whip maker | | | |
| Leather goods maker | | | |
| Sword cutler | | | |
| Sword maker | **John Kempe** | | 1670–D1690 |
| | • Successor to Jean de Latre as furbisher of arms (including swords) at the Palace of Whitehall, 1689–1690. | | |
| | • Repaired swords for the Board of Ordnance, 1690. | | |
| | **William Kempe** | | 1690–1710 |
| | • Son of John Kempe. | | |
| | • Succeeded his father as furbisher of arms at the palace of Whitehall, 1690. | | |
| Sword cutler | **James Kennedy** | Covent Garden | 1870 |
| Gold and silver | **Kenning (George) & Co.** | 18 Little Britain | 1845–1860 |
| lacemen | **George Kenning & Co.** | 18 Little Britain | 1861–1862 |
| Embroiderer | **Kenning (George) &** | 4 & 18 Little Britain | 1863 |
| Sword cutler | **McKierman (James)** (see James McKierman) | | |
| | **Kenning (George) & McKierman (James) & Co.** | 4 Little Britain | 1864–1865 |
| | **George Kenning** | 4 Little Britain | 1866–1870 |
| | **George Kenning** | 2 & 4 Little Britain | 1871 |
| | **George Kenning** | 4 Little Britain & 198 Fleet St. | 1872–1874 |
| | **George Kenning** | 1 to 3 Little Britain 175 Aldersgate St. & 198 Fleet St. | 1875–1880 |
| | **George Kenning** | 1 to 4 Little Britain 197 Aldersgate St. & 198 Fleet St. | 1881–1882 |

| | | | |
|---|---|---|---|
| | George Kenning | 1 to 4 Little Britain<br>197 Aldersgate St. & 16 Great Queen St. | 1883–1884 |
| | George Kenning & Son | 1 to 4 Little Britain<br>197 Aldersgate St. & 16 Great Queen St. | 1885–1908 |
| | George Kenning & Son | 1 to 4 Little Britain<br>16 Great Queen St. | 1909–1930 |
| | George Kenning & Son | Eagle Wharf, Great Queen St.,<br>Little Britain St. | 1931–1955 |
| | Kenning & Spencer | 1 to 4 Eagle Wharf<br>19 to 21 Great Queen St. | 1956 |
| | Kenning & Spencer | 19 to 21 Great Queen St. | 1957 to c1990 |
| | Toye, Kenning & Spencer Ltd. | New Town Rd.<br>Bedworth, Warwickshire | c1990–2003 |

• Ceremonial and service uniforms and accoutrements, accessories and equiptment.
(see Glasgow, Scotland, listing in Chapter 15)

| | | | |
|---|---|---|---|
| Bladesmith | John Kent | | 1376 |
| Sword cutler | John Kent | Parish of St. Margaret, Westminster | 1606–1621 |
| Silversmith<br>Hilt maker<br>Sword cutler | William Kentember (Kentesber) | | 1780–1790 |

• Made silver-hilted cavalry sabers with two branches.

| | | | |
|---|---|---|---|
| Goldsmith<br>Jeweler | John Kentish | Johnson Ct., Fleet St.<br>opposite Royal Exchange | 1748–1757 |
| Toymen<br>(hardware | John Kentish | Popes Head Alley, Cornhill<br>at the sign of the Star | 1758–1775 |
| dealer)<br>Hilt maker | Kentish (John) & Turner (Samuel) | 18 Cornhill St. | 1776–1779 |
| Sword cutler | Kentish (John) & Haynes (Henry) | 18 Cornhill St. | 1780–1789 |

• Sold naval officers hangers and small swords.
• Succeeded by Haynes & Kentish (see Haynes & Kentish).

| | | | |
|---|---|---|---|
| Silversmith<br>Hilt maker | John Kenton | | Bc1669–1700 |

• Apprenticed to cutler Henry Panton, 1683–1700 (see Henry Panton).

| | | | |
|---|---|---|---|
| Sword cutler | John Kenton | Lombard St. | 1701–1711 |
| Gold encruster<br>Sword inlayer | Johann Andreas Kern | | 1760–1775 |

• Did work for Matthew Boulton, Birmingham (see Matthew Boulton).
• Did gold inlay work on swords.

Richard Kerry
(see James Forrest)

| | | | |
|---|---|---|---|
| Silversmith<br>Silver plater | Richard Kersill | | Bc1722–1743 |

• Apprenticed to George Greenhill, 1736–1743 (see George Greenhill).

| | | | |
|---|---|---|---|
| Hilt maker<br>Sword cutler | Richard Kersill | Foster Lane | 1744–D1747 |
| | Ann Kersill | Foster Lane | 1747–1748 |

• Widow of Richard Kersill.

| | | | |
|---|---|---|---|
| | William Kersill | Gutter Lane, Newport St. | 1749–1771 |

• Son of Richard.

| | | | |
|---|---|---|---|
| | William Kersill | 21 Aldergate St. | 1772–1777 |

• Made silver-hilted small swords.
• Made silver hilts for Thomas Bibb (see Thomas Bibb).

| | | | |
|---|---|---|---|
| Sword cutler | **H. Kettle** | 41 Conduit St., Hanover Square | 1840–1850 |
| | • Sold children's swords with horsehead pommels. | | |
| Silversmith | **William Kidney** | | Bc1716–1732 |
| Hilt maker | • Apprenticed to David Willaume, 1723–1732. | | |
| Assayer | **William Kidney** | 6 Bell Court, Fosterlane | 1733–1750 |
| Plate worker | | | |
| Sword cutler | | | |
| Sword maker | **Johann Kindt** | | |
| | (see Chapter 3) | | |
| Button and | **King & Preston** | 33 George St., Hanover Square | 1840–1842 |
| accoutrement | • Succeeded by William Smith & Co. (see William Smith & Co.). | | |
| maker | | | |
| Sword cutler | | | |
| Cutler | **Edward King** | | 1639–1654 |
| | • Master of the Cutlers Company, 1649. | | |
| | **Henry King** | | |
| | (see Samuel Firmin) | | |
| Silversmith | **James King** | | Bc1748–1767 |
| Hilt maker | **William Portal & James King** | Orange St., near Leicester Square | 1768–1771 |
| Sword cutler | | | |
| Bladesmith | **John King (Kynge)** | | 1452 |
| Silversmith | **John King** | | Bc1736–1756 |
| Goldsmith | • Apprenticed to Henry Brind, 1750–1751. | | |
| Hilt maker | • Apprenticed to William Shaw, 1751–1756. | | |
| Sword cutler | **John King** | 73 Little Britian | 1757–1784 |
| | • Son Joseph King apprenticed and worked for him, 1780–1800. | | |
| | **John King** | 1 Moor Lane, Fore St. | 1785–1800 |
| | • Also made silver gun furniture. | | |
| Army and navy | **Thomas King** | 53 Parliament St., Westminster | 1807–1808 |
| tailor | **Thomas King** | 22 Charing Cross | 1809–1828 |
| Military | | | |
| warehouse | | | |
| Sword cutler | | | |
| Silversmith | **William King** | Charing Cross | 1777–1778 |
| Hilt maker | | | |
| Sword cutler | | | |
| Silversmith | **William Kinman** | East Harding St. | 1759–1775 |
| Brass founder | **William Kinman** | 9 New Street Square | 1776–1797 |
| Hilt maker | • Made silver and silver gilt hilts for small swords, military swords, hangers, | | |
| Sword cutler | and hunting swords for many London cutlers. | | |
| Armourer | **John Kirke (Kelke)** | | 1552–1635 |
| Sword and | • Armourer at the Greenwich Royal Armouries. | | |
| sword blade | • Master of the Greenwich Royal Armouries, 1567–1575. | | |
| maker | • Obtained a royal patent for making sword and rapier blades, c. 1630. | | |

| | | | |
|---|---|---|---|
| Cutler | **John Kittebuter** | | 1682–1697 |
| | • Master of the Cutlers Company, 1692. | | |
| | • Used "moon and P" mark. | | |
| Armourer | **Richard Klinge** | | 1660 |
| Sword maker | • A German. | | |
| | • Member of the Armourers Company. | | |
| Silversmith | **William Knight** | New St., Shoe Lane | 1680–1700 |
| Hilt maker | • Made the the "Lichfield" Bearing sword (ceremonial town sword used | | |
| Sword cutler | in parades, etc.) in 1685. | | |
| | • Made plug bayonets. | | |
| Cutler | **Roger Knowlls** | | 1574–1589 |
| | • Master of the Cutlers Company, 1584. | | |
| Hatter | **Richard Knowlton** | 2 Fleet St. | 1784–1795 |
| Hosier | **Knowlton (Richard) & Deans (James)** | 2 Fleet St. | 1796–1810 |
| Haberdasher | **Charles William Knowlton** | 2 Fleet St. | 1811–1815 |
| of hats | | | |
| Sword cutler | | | |
| Gun maker | **John Knubley** | Otley, Yorkshire | 1766–1771 |
| Pistol | **John Knubley** | 11 Charing Cross | 1772–1793 |
| warehouse | **John Knubley** | 7 Charing Cross | 1794–1797 |
| Sword cutler | • John Knubley died in 1795. | | |
| | • Samuel Brunn bought the controlling interest in Knubley business in 1795 | | |
| | but kept the Knubley name (see Samuel Brunn). | | |
| | • John Mallet bought out Samuel Brunns interest in Knubley in 1797 (see John Mallet). | | |
| Sword and sword | **Johann Konigs (Connyne, Conine)** | Berg & Solingen, Prussia | Bc1587–1607 |
| blade maker | • Immigrated to London, 1607. | | |
| | • Changed his name to John Connyne. | | |
| | • Brought his German sword maker's mark (crossed arrows) with him, which | | |
| | was approved by the Cutlers Company. | | |
| | **John Connyne** | London | 1607–1629 |
| | • Became a naturalized citizen. | | |
| | • Moved to Hounslow Heath sword blade center near London, 1629 (see Chapter 3). | | |
| | **John Connyne** | Hounslow Heath | 1629–1642 |
| | • Operated a sword blade mill. | | |
| | • In 1642, when parliamentary forces took over Hounslow and converted many | | |
| | of the sword and blade mills to powder mills, Connyne moved back to London. | | |
| | **John Connyne** | London | 1642–1659 |
| Bladesmith | **Thomas Kyngeston (Kingston)** | | 1500–1510 |
| | • Moved from the Armourers Company to the Bladesmiths Company, 1506. | | |
| Gun maker | **J. (John) D. (David) Lacy** | 67 Threadneedle St., Royal Exchange | 1811–1812 |
| Sword cutler | • Successor to Bennett & Co. (see John Bennett). | | |
| | **John George Lacy** | 67 Threadneedle St., Royal Exchange | 1813–1814 |
| | **Lacy (John George) & Witton** | 67 Threadneedle St., | 1815–1826 |
| | **(David William)** | Royal Exchange & 63 Fenchurch St. | |
| | • Succeeded by Gameson & Williams at 67 Threadneedle St. (see Gameson & Williams). | | |
| | **Lacy (John George) & Co.** | 63 Fenchurch St. | 1827–1828 |
| | • Partner: David William Witton. | | |

|  |  |  |  |
|---|---|---|---|
|  | Lacy (John George) & Witton (David William) | 63 Fenchurch St. | 1829 |
|  | Lacy (John George) & Witton (David William) | 13 Camomile St. | 1830–1835 |
|  | Lacy (John George) & Reynolds (Henry) | 16 Camomile St. | 1836–1837 |
|  | • Partner: David William Witton. |  |  |
|  | Lacy (John George) & Reynolds (Henry) | 21 Great St., Bishopsgate Parish of St. Helens | 1838–1852 |
|  | • Partner, David William Witton. |  |  |
|  | Lacy & Co. | 21 Great St., Bishopsgate Parish of St. Helens | 1853 |
|  | • Partner, David William Witton. |  |  |
|  | • Succeeded by David William Witton (see David William Witton). |  |  |

Edward Lake
(see Linney & Ulton)

| | | | |
|---|---|---|---|
| Bladesmith | Robert Lambe | | 1400 |
| Gold and silver lacemen Embroiderer Sword cutler | John Lambert | 436 Strand | 1802–1803 |
| | John Lambert | 14 Coventry St. | 1804 |
| | Lambert (John) & Burden (William) | 14 Coventry St. & 2 Whitcomb St. | 1805–1811 |
| | John Lambert & Co. | 2 Whitcomb St. | 1812 |
| | Lambert (John) & Maclauren (George) | 2 Whitcomb St. | 1813–1818 |
| | Lambert (John) & Maclauren (George) | 26 Princes St. | 1819–1823 |
| | Lambert (John) & Maclauren (George) | 236 Regent St. | 1824–1828 |
| | Lambert (John) & Holbeck (Henry) | 236 Regent St. | 1829–1832 |
| | Samuel Lambert | 236 Regent St. | 1833–1834 |
| | Lambert (Samuel) & Brown (James) | 236 Regent St. | 1835–1842 |
| | Lambert (Samuel) & Brown (James) & Co. | 236 Regent St. | 1843–1850 |
| | Lambert (Samuel) & Brown (James) | 236 Regent St. | 1851–1852 |
| | Lambert (Samuel) & Brown (James) & Co. | 236 Regent St. | 1853–1858 |
|  | • Succeeded by William Jones (see William Jones). |  |  |
| Sword cutler | William Lambert | | 1610–1623 |
| Bladesmith | Thomas Lamyan | | 1517–1527 |
|  | • Registered his "three crescent moons" mark with the Cutlers Company, 1517. |  |  |
| Military tailor Sword cutler | Landon & Co. | 17 Jermyn St., Parish of St. James | 1835–1847 |
| | Landon & Morland | 17 Jermyn St., Parish of St. James | 1848–1856 |
| | Landon, Morland & Landon | 7 New Burlington St. | 1857–1863 |
| | Landon & Morland | 7 New Burlington St. | 1864–1866 |
| | Landon & Co. | 7 New Burlington St. | 1867–1903 |
| | Carr, Son & Landon | 7 New Burlington St. | 1904–1939 |

John Langdale
(see Samuel Firming)

| | | | |
|---|---|---|---|
| Haberdasher of hats Sword cutler | Nicholas Langford | Fleet St., near Salisbury Court | 1730–1740 |
| | Nicholas Langford | Fleet St., corner of the entrance to Serjent's Inn, at the sign of the Angel & Oxford Arms | 1740–D1767 |
| | Thomas Langford | 50 Fleet St. | 1767–1790 |
|  | • Son and successor to Nicholas Langford. |  |  |
|  | • Master of the Cutlers Company, 1776. |  |  |
| | Langford (Thomas) & Son | 50 Fleet St. | 1791–1793 |
| | Thomas Langford & Son | 50 Fleet St. | 1794–D1797 |
|  | • The Langfords sold small swords, many with steel hilts made by Matthew Boulton of Birmingham, and hollow ground blades made in Solingen, Prussia. |  |  |

| Sword cutler | **William Langford** | Parish of St. James Clerkenwell, Middlesex | 1604–1626 |
|---|---|---|---|
| Sword cutler | **Thomas Langton** | Parish of St. Bride, Fleet St. | 1604–1619 |
| Sword cutler | **William Langton** | Westminster | 1617–1632 |
| | **J.F. Latham** (see Henry Wilkinson) | | |
| Sword maker | **Jean de Latre**<br>• Furbisher of arms (swords) at the Palace of Whitehall.<br>• Succeeded by John Kempe. | | 1680–1689 |
| Goldsmith<br>Hilt maker | **Benjamin Laver**<br>• Apprenticed to Thomas Heming, 1751–1763 (see Thomas Heming). | | Bc1737–1763 |
| Sword cutler | **Benjamin Laver** | New Bond St.<br>Parish of St. George, Hanover Square | 1764–1780 |
| | **Benjamin Laver & Co.** | Barlows Mews<br>Bruton St., near Bond St. | 1781–1782 |
| | • Partner and son: Thomas Laver.<br>**Benjamin Laver** | 4 Bruton St., Berkeley Square | 1783–1800 |
| | • Appointed goldsmith to His Royal Highness the Duke of Clarence.<br>• Sold presentation swords mounted by James Morisset (see James Morisset). | | |
| Sword maker | **Samuel Law**<br>• Furbisher at the Tower of London, proving swords at the salary of 30 pounds a year (1649–1652).<br>• Supplied plug bayonets and swords to the Board of Ordnance, 1649–1666.<br>• Also repaired swords for the Board of Ordnance. | | 1630–D1668 |
| | **Edith Law**<br>• Widow of Samuel Law.<br>• In 1663, with Joseph Audley and Robert Steadman, repaired and cleaned 500 short swords/bayonets for the Board of Ordnance.<br>• Repaired swords for the Board of Ordnance, 1668–1669. | | 1668–1675 |
| Hatter | **James Lawson** | 234 Strand | 1750–1773 |
| Hosier<br>Sword cutler | **James Lawson** | 181 Strand | 1774–1784 |
| Sword hilter | **Lawson (James) & Ward (William)** | 5 Hatton Garden | 1851–1894 |
| | **John Leaton** (see Chapter 4) | | |
| Tailors<br>Sword cutler | **John Leckie & Co.** | 117 Leadenhall St. | 1875–1883 |
| | **John Leckie & Co.** | 14 St. Mary's Way | 1884–1897 |
| | **John Leckie & Co.** | 84 Fore St. | 1898–1931 |
| | **John Leckie & Co.** | 56 Kingsway | 1932–1935 |
| | **John Leckie & Co.** | 6 Love Lane, Wood St. | 1936–1941 |
| | **John Leckie & Co.** | 15–16 Aldermanbury | 1942–1944 |
| | **John Leckie & Co.** | 1-2-3 St. Pauls Churchyard | 1945–1947 |
| | (see Glasgow, Scotland, listing in Chapter 15) | | |

| | | | |
|---|---|---|---|
| Sword cutler | James Lee | | 1850 |
| Sword cutler | John A. Lee | | 1611 |
| Military tailor<br>Sword cutler | John LeGassick<br>• Successor to Davidson & Co. (see Davidson & Co.).<br>• Succeeded by George Hunter (see George Hunter). | 12 Cork St., Bond | 1845–1849 |
| Sword cutler | J. Legg | Brighton | 1840 |
| Sword cutler | W.E. Legge | 43 Station Rd. | 1914 |
| Blade maker<br>Armourer | David Le Hope<br>• In 1321, King Edward II sent LeHope to Paris, France, to study French methods of blade making. | | 1300–1330 |
| Goldsmith<br>Silversmith<br>Hilt maker<br>Sword cutler | John Hugh LeSage<br>• Apprenticed to Lewis Cuncy, 1708–1717.<br>John Hugh LeSage<br>John Hugh LeSage<br>• Had apprentice Edward Wakelin, 1730–1738 (see Edward Wakelin).<br>John Hugh LeSage<br>• Had a journeyman Edward Wakelin, 1738–1746. | <br><br>Little St. Martins Lane near Long Acre<br>corner of Suffolk St.<br><br>Great Suffolk St. near the Hay market | Bc1694–1717<br><br>1718–1721<br>1722–1738<br><br>1739–1746 |
| Jeweler<br>Sword cutler | John Lengran<br>• Royal jeweler to King Henry VIII. | | 1524–1530 |
| Sword cutler | W. Leonard & Co. | | 1826 |
| Cutler | Robert Lepper<br>• Master of the Cutlers Company, 1784. | | 1774–1789 |
| Sword cutler | Edmund Lever | Parish of St. Magnas | 1596 |
| Army and navy outfitter<br>Sword cutler | Samuel Levy<br>Samuel Levy | Hemmings Row, St. Martins Lane<br>28 King William St., Strand | 1827–1838<br>1839–1865 |
| Military tailor<br>Gold and silver lacemen<br>Embroiderer<br>Sword cutler | George Lewis<br>Nathan Lewis<br>N. (Nathan) Lewis | 10 St. James St.<br>37 St. James St.<br>33 St. James St. | 1817–1821<br>1822–1825<br>1826–1839 |
| Military tailor<br>Sword cutler | Lewis, Sons & Partners | Parish of St. James | 1900 |
| Cutler | John Leycester<br>• Master of the Cutlers Company, 1558. | | 1548–1563 |
| Bladesmith | John Leyner (Layner)<br>• Master of the Bladesmiths Company, 1424, 1428, and 1441. | | 1420–1445 |
| Sword cutler | R. Lill & Co.<br>• Sold Wilkinson Sword Company swords. | 1 Maddox St. | 1950 |

| | | | |
|---|---|---|---|
| Military ornament maker<br>Sword cutler | John Lilly<br>• Successor to Henry Johnson (see Henry Johnson).<br>• Succeeded by Charles Rowley (see Charles Rowley). | 13 Archer St., Haymarket | 1845–1863 |
| | David Lincohn<br>(see James Lock) | | |
| Sword cutler | Richard Lincohn | | 1694–1696 |
| Sword cutler | George Lindly | | 1887 |
| Sword cutler | Samuel Lingard | | 1714 |
| Military tailor<br>Sword cutler | Linney (Amos) & Ulton (Henry) | 112 Jermyn St., Parish of St. James | 1828–1829 |
| | Amos Linney | 112 Jermyn St. | 1830–1834 |
| | Amos Linney | 23 Regent St. | 1835–1843 |
| | Amos Linney & Son | 23 Regent St. | 1844–1849 |
| | James & George Linney | 21 Regent St. | 1850–1857 |
| | James Linney | 21 Regent St. | 1858–1861 |
| | James Linney & Co. | 21 Regent St. | 1862–1882 |
| | Linney (James) & Lake (Edward) | 26 Regent St. | 1883 |
| | Linney (James) & Co. | 26 Regent St. | 1884–1885 |
| | • Succeeded by Sibley (William) & Linney (James) (see Sibley & Linney). | | |
| Silversmith<br>Hilt maker<br>Sword cutler | George Littleboy<br>• Apprenticed to Thomas Cooke, 1712–1719.<br>• Journeyman to Thomas Cooke, 1719–1730. | | Bc1998–1730 |
| | George Littleboy | Noble St.,<br>near the coach makers | 1731–1740 |
| Cutler | Horace Charles Lloyd<br>• Master of the Cutlers Company, 1889. | | 1879–1894 |
| Cutler | Richard Lloyd<br>• Master of the Cutlers Company, 1856. | | 1846–1861 |
| Cutler | Richard Gore Lloyd<br>• Master of the Cutlers Company, 1883. | | 1873–1888 |
| Silversmith<br>Hilt maker<br>Dagger maker<br>Sword cutler | William Lobe | Fleet Street,<br>next to Hercules Pillars Alley<br>at the sign of Three Daggers | 1675–1685 |
| | • Sold silver-hilted rapiers with hollow ground blades,<br>some with silver and gold wire on the handles. | | |
| Hatter<br>Sword cutler | James Lock | 6 St. James St. | 1759–1812 |
| | Lock (James) & Lincoln (David) | 6 St. James St. | 1813–1820 |
| | G. (George) & J. (James) Lock | 6 St. James St. | 1821–1829 |
| | J. (James) & G. (George) Lock | 6 St. James St. | 1830–1850 |
| | Lock (James) & Co. | 6 St. James St. | 1851–1927 |
| | Lock (James) & Co. Ltd. | 6 St. James St. | 1928–1954 |
| Gun maker<br>Sword cutler | W. (Walter) Locke,<br>Watts (William) & Co.<br>(Adams Patent Small Arms Mfg. Co.) | 391 Strand | 1866–1893 |

| | | | |
|---|---|---|---|
| Sword maker | **London & Morland** | Jermyn St., Parish of St. James | 1835–1845 |
| Gold and silver laceman<br>Sword cutler | **Edward Longdon**<br>• Succeeded by Bodley & Elley (see Bodley & Elley). | 31 Lombard St. | 1736–1783 |
| Jeweler<br>Silversmith<br>Engraver<br>Hilt maker<br>Sword cutler | **John & Richard Longman**<br>**John & Richard Longman &**<br>**John Strongilharm**<br>• Successors to John Strongitharm (see John Strongitharm).<br>**Longman (John & Richard) &**<br>**Strongitharm (John)**<br>• Merged with Widdowson & Veale in 1877 (see Widdowson & Veale).<br>**Longman & Strongitharm Ltd.**<br>**Longman & Strongitharm Ltd.**<br>**Longman & Strongitharm Ltd.**<br>• Strongitharm was an engraver. | 1 Waterloo Place, Pall Mall<br>1 Waterloo Place, Pall Mall<br><br>1 Waterloo Place, Pall Mall<br><br>1 Waterloo Place, Pall Mall<br>1 Albemarle St.<br>13 Dover St. | 1852–1855<br>1856–1876<br><br>1877–1909<br><br>1910–1921<br>1922–1935<br>1936–1968 |
| Gold and silver laceman<br>Sword cutler | **Thomas Lonsdale**<br>**Thomas Lonsdale**<br>**Lonsdale (Thomas) & Son**<br>**Thomas Lonsdale & Son**<br>• Appointed laceman to His Majesty the King, 1811–1831.<br>**Charles & Thomas Lonsdale**<br>**C. (Charles) & T. (Thomas) Lonsdale**<br>• Succeeded Benjamin Fox (see Benjamin Fox).<br>**Lonsdale (Charles & Thomas) & Tyler (E.)**<br>• Succeeded by Tyler & Rogers (see Tyler & Rogers). | 35 Bedford St., Covent Garden<br>32 King St.<br>32 King St.<br>32 King St.<br><br>32 King St.<br>26 Bedford St, Covent Garden<br><br>26 Bedford St., Covent Garden | 1784–1787<br>1788–1799<br>1800–1805<br>1806–1831<br><br>1832<br>1833–1834<br><br>1835–1845 |
| Cutler | **James Looker**<br>• Master of the Cutlers Company, 1783 and 1798. | | 1773–1803 |
| Sword cutler | **Henry Looker** | Parish of St. Andrew, Holborn | 1595–1615 |
| Cutler | **Henry William Looker**<br>• Master of the Cutlers Company, 1833. | | 1823–1838 |
| Silversmith<br>Sword hilter<br>Sword cutler | **Richard Loveit (Lovett, Lovell)** | Red Lion St. | 1701–1720 |
| Silversmith<br>Sword hilter<br>Sword cutler | **Moses Loveit (Lovett, Lovell)**<br><br>**Moses Loveit**<br>• Sold silver-hilted small swords<br>• His customers could buy the silver hilt only and trade in old silver hilts.<br>• Supplied swords to the Board of Ordnance, 1715. | St. Martins Lane,<br>next to Mores Yard, Strand<br>near Dukes Court | 1696–1709<br><br>1710–1720 |
| Silversmith<br>Sword hilter<br>Sword cutler | **Nicholas Loveit (Lovett, Lovell)**<br>• Probably a realative of Moses Loveit.<br>• Master of the Cutlers Company, 1705. | Strand | 1696–1709 |
| Silversmith<br>Hilt maker<br>Sword cutler | **George Loveit (Lovett, Lovell)**<br>• Signed an advertisement warning sword purchasers about counterfeit cast silver hilts. | | 1661 |

| Trade | Name | Address | Dates |
|---|---|---|---|
| Sword cutler | **Richard Lowe** | | 1670–1700 |
| | • Had apprentice William Harris, c. 1695–1702. | | |
| Steel hilt maker<br>Steel wire drawer<br>Steel button maker | **L. Lowther** | Great Earl St.<br>at the sign of the Seven Dials | 1784–1800 |
| | • Made steel sword hilts and steel sword hilt wire wrappings. | | |
| Gun and accoutrement maker<br>Sword cutler | **J.S. Lowther** | | 1811–1815 |
| Hatter<br>Sword cutler | **William Loxham** | Corner of St. Sweeting's Alley<br>and the Royal Exchange, Cornhill | 1740–1754 |
| | • Master of the Cutlers Company, 1742. | | |
| | **William & Edward Loxham** | Corner of St. Sweeting's Alley<br>and the Royal Exchange, Cornhill | 1755–1776 |
| | • Edward Loxham, Master of the Cutlers Company, 1758. | | |
| | **William & Edward Loxham** | 88 Cornhill | 1777–1780 |
| | • Made bayonets for the Board of Ordnance.<br>• William Loxham died in 1780. | | |
| | **Edward Loxham** | 88 Cornhill | 1781–D1787 |
| | **Robert Loxham** | 88 Cornhill | 1787–1791 |
| | • Son of Edward. | | |
| | **Robert Loxham** | 88 Cornhill & 61 Cannon St. | 1792–1816 |
| | • Master of the Cutlers Company, 1797 and 1811. | | |
| | **R.C. Lukin**<br>(see James Morisset) | | |
| Military tailor<br>Sword cutler | **Macken (James) & Co.** | 42 Conduit St. | 1843 |
| | **James Macken & Co.** | 42 Conduit St. | 1844–1849 |
| | **James Macken & Co.** | 10 Little Ryder St. | 1850–1864 |
| | **James Macken & Co.** | 25 Ryder St. | 1865–1883 |
| | **Macken (James) & Co.** | 25 Ryder St. | 1884–1900 |
| Silversmith<br>Goldsmith<br>Hilt maker<br>Sword cutler | **Robert Makepeace (Sr.)** | Maiden Lane, Wood St. | 1755–1766 |
| | **Robert Makepeace (Sr.)** | Serle St., Lincoln's Inn | 1767–1776 |
| | **Robert Makepeace (Sr.) & Richard Carter** | Bartholomew Close | 1777–1778 |
| | • Sons: Robert Makepeace (Jr.) and Thomas Makepeace<br>  apprenticed in 1777–1778 (see Richard Carter). | | |
| | **Robert Makepeace (Sr.)** | 6 Searle St., Lincoln Inn Fields | 1779–1783 |
| | **Robert Makepeace (Sr.) & Sons**<br>**(Thomas & Robert)** | 6 Searle St., Lincoln Inn Fields | 1784–1794 |
| | • Robert Makepeace (Sr.) died in 1794. | | |
| | **Robert (Jr.) & Thomas Makepeace** | 6 Searle St., Lincoln Inn Fields | 1794–1795 |
| | • Sons of Robert Makepeace Sr.<br>• Thomas Makepeace died in 1795. | | |
| | **Robert Makepeace (Jr.)** | 6 Searle St., Lincoln Inn Fields | 1795–1808 |
| | • Sold presentation swords mounted by Ray & Montaque and James Morisset.<br>• Made presentation sword that the city of London presented to Admiral Lord Viscount<br>  Horatio Nelson after the Nile campaign.<br>• In 1799, made the presentation sword for Viscount Duncan after Camperdown.<br>(see Ray & Montaque, James Morisset) | | |

| | | | |
|---|---|---|---|
| | Makepeace (Robert Jr.) & Harker (John) | 6 Searle St. | 1809–1818 |
| | Makepeace (Robert Jr.) & Harker (John) | 5 Searle St. | 1819–1821 |
| | Makepeace (Robert Jr.) & Harker (John) | 3-4-5 Searle St. | 1822–1824 |
| | Makepeace (Robert Jr.) & Harker (John) | 4-5 Searle St. | 1825–1827 |
| | • Robert Makepeace died in 1827. | | |
| | William & Robert Henry Makepeace | 5 Searle St. | 1827–1840 |
| | • Sons of Robert Makepeace. | | |
| | Makepeace (William & Robert Henry) & Walford (Henry) | 5 Searle St. | 1841–1863 |
| Gun maker Sword cutler | John Mallet (Malet) | 7 Charing Cross | 1798–1803 |
| | • Successor to John Knubley (see John Knubley) | | |
| Bladesmiths | Robert Malteby | | 1360–1370 |
| | Robert Malteby & Roger Mark | corner of Seacoal Lane & Cock Lane, Parish of St. Sepulchre without Newgate | 1371–1381 |
| | Robert Malteby (see Roger Mark) | corner of Seacoal Lane & Cock Lane, Parish of St. Sepulchre without Newgate | 1382–1406 |
| Cutler | John Manifold | | 1648–1663 |
| | • Master of the Cutlers Company, 1658. | | |
| Sword cutler | Henry Manning & Co. | 124 New Bond St. | 1840–1900 |
| Silversmith Sword and knife hilt maker Sword and knife cutler | Samuel Mansell | | Bc1730–1761 |
| | • Apprenticed to Dru Drury Sr., 1754–1761 (see Dru Drury Sr.). | | |
| | Samuel Mansell | 48 Strand | 1762–1773 |
| | Samuel Mansell | Orange Court, Leicesterfields | 1774–1780 |
| Gun maker Sword cutler | John Manton & Co. | 6 Dover St. | B1752, 1802–1832 |
| | • Foreman for gun maker Twigg until 1801. | | |
| | J. (John) Manton & Son | 6 Dover St. | 1833–1839 |
| | • John Manton died in 1834. | | |
| | J. (John) Manton & Son | 6 Jermyn St. | 1840–1868 |
| | J. (John) Manton & Co. | 6 Jermyn St. | 1869–1874 |
| | J. (John) Manton & Son | 6 Jermyn St. | 1874–1877 |
| | • In Calcutta, India, also. | | |
| Silversmith Sword and knife hilt maker | Mappin (John Newton) & Co. | 77–78 Oxford St. | 1859–1862 |
| | Mappin (John Newton) & Webb (Henry) | 71 & 72 Cornhill & 77 & 78 Oxford St. | 1863–1867 |
| Sword and knife cutler | Mappin (John Newton) & Webb (Henry) | 1 Winsley St., Oxford St., 71–72 Cornhill & 76 to 78 Oxford St. | 1868–1871 |
| Jeweler Goldsmith | Mappin (John Newton) & Webb (Henry) | 1 Winsley St., Oxford St., 76–78 Oxford St. | 1872–1881 |
| Religious silverware maker | Mappin (John Newton) & Webb (Henry) | 1 Winsley St., Oxford St., 158–162 Oxford St. | 1882–1909 |
| | • In 1902, bought out Mappin Bros. | | |
| | Mappin (John Newton) & Webb (Henry) | 1 Winsley St., Oxford St., 156 to 162 Oxford St., 220 Regent St. | 1910–1915 |
| | Mappin (John Newton) & Webb (Henry) | 1 Winsley St., Oxford St., 156 to 162 Oxford St., 172 Regent St. | 1916–1956 |
| | Mappin (John Newton) & Webb (Henry) | 1 Winsley St., Oxford St., 172 Regent St. | 1957–1961 |

| | | | |
|---|---|---|---|
| | Mappin (John Newton) & Webb (Henry) | 1 Winsley St., 106 & 170 Regent St. | 1962–c1978 |
| | Mappin (John Newton) & Webb (Henry) | 23 Albermarle St. | c1978–2003 |
| Silversmith | Joseph Mappin | 15 Fore St. | 1845–1850 |
| Goldsmith | Mappin Bros. | 37 Moorgate St. | 1851–1855 |
| Jeweler | Mappin Bros. | 67 & 68 King William St. | 1856–1861 |
| Sword and knife hilt maker | Mappin Bros. | 222 Regent St. | 1862–1872 |
| | Mappin Bros. | 200 Regent St. | 1873–1888 |
| Sword and knife cutler | Mappin Bros. | 220 Regent St. & 35 St. Pauls Churchyard | 1889–1890 |
| | Mappin Bros. | 220 Regent St. & 66 Cheapside | 1891–1910 |

- Bought out by Mappin & Webb, 1903 (see Mappin & Webb).
- Name of Mappin Bros. kept until 1910.
(see Sheffield listing)

| | | | |
|---|---|---|---|
| Bladesmith | John Marhan | Fleet St., Parish of St. Bride | 1376–D1407 |
| Bladesmith | Roger Mark | | 1360–1370 |
| | Robert Malteby & Roger Mark | corner of Seacoal Lane & Cock Lane, Parish of St. Sepulchre without Newgate | 1371–1381 |
| | Roger Mark | corner of Cock Lane & Hosier Lane, Parish of St. Sepulchre without Newgate | 1382–D1407 |
| Bladesmith | William Marlar | | 1515–1525 |

- Bladesmith member of the Armourers Company, 1515.
- Bladesmith member of the Cutlers Company, 1517.
- Registered his "cross and crescent" mark with the Cutlers Company, 1517.

| | | | |
|---|---|---|---|
| Silversmith Goldsmith Hilt maker Sword cutler | Jeremiah Marlow Jr. | | 1722–1740 |

- Had apprentice Joseph Clare Jr., 1732–1739 (see Joseph Clare Jr.).

| | | | |
|---|---|---|---|
| Hatter | Marr (Charles) & Third (Henry) | Arundel St., Strand | 1765–1769 |
| Sword cutler | Charles Marr | Arundel St., Strand | 1770–1772 |
| Army accoutrement maker Sword cutler | J. (John) Marsh | 18 Brewer St., Golden Square | 1800–1820 |
| Sword cutler | Marshall & Co. | St. James St. | 1840 |
| Silversmith Hilt maker Sword cutler | Henry Marshall | Fleet Street, near Inner Temple Gate | 1695–1710 |

- Made silver-hilted small swords and hunting sabers some with gold inlayed blades.

| | | | |
|---|---|---|---|
| Silversmith Hilt maker Sword cutler | Robert Martin | | 1661 |

- Signed an advertisement warning sword purchasers about counterfeit cast silver hilts.

| | | | |
|---|---|---|---|
| Gun maker Sword maker | William Martindale | Chiswell St., near the artillery ground, at the sign of the Stirrup | 1654–D1666 |

- Gunsmith to King Charles II.
- In 1661, advertised pocket pistols, cannons, all sorts of guns, and swords with pistols affixed in the *Mercurius Publicus*.

| | | | |
|---|---|---|---|
| Sword cutler | James Master | | 1640–1676 |
| Sword cutler | Carey Matcham | | 1586 |
| | John Matthew<br>(see Capt. Robert Melville Gridlay) | | |
| Sword and knife maker | **Richard Matthew (Sr.)** | Fleet Bridge, Parish of St. Bride | 1557–D1591 |

- Cutler to Queen Elizabeth I.
- In 1562, made the sword of state for the city of London.
- In 1563, Queen Elizabeth granted Matthew exclusive patent to make knives and knife hilts in England (Matthew had petitioned for the patent in 1558).
- In 1569–1570, supplied swords to the Drapers Company (cloth makers) for their Calais, France, settlement.
- In 1571, Queen Elizabeth granted Matthew exclusive patent to make turky hafts (Turkish-style hilts made of pieces of horn separated by yellow or white plates), 1571–1577.
- In 1571, granted the "half moon with face" mark.
- Master of the Cutlers Company, 1585, 1588, and 1589.
- Had apprentice Thomas Beckwith, 1584–1589 (see Thomas Beckwith).
- Richard Matthew Sr. died in 1591 and his sons Nathaniel, Richard Jr., and Paul took over the business.

| | | | |
|---|---|---|---|
| | **Nathaniel, Richard (Jr.) & Paul Matthew** | Fleet Bridge, Parish of St. Bride | 1591–1599 |

- Used their father's "half moon with face" mark.

| | | | |
|---|---|---|---|
| | **Nathaniel Matthew** | Fleet Bridge, Parish of St. Bride | 1600–1610 |

- Appointed cutler to His Majesty the King and His Royal Highness the Prince of Wales.
- In 1610, Henry Moorehead (employee, 1591–1620) obtained their "half moon with face" mark.

| | | | |
|---|---|---|---|
| | Henry G. Matthews<br>(see J. Gieve & Sons Ltd.)<br>(see Portsmouth listing) | | |
| Sword cutler | William Matthews | Parish of St. Sepulchre without Newgate | 1570–1580 |
| Sword cutler | John Maxsted | 4 Morris St., Parish of St. James | 1822–1825 |
| Army and navy Outfitter and agent | Maynard (Robert) & Co. | 27 Poultry St. | 1832–1837 |
| | Robert Maynard & Co. | 27 Poultry St. | 1838–1844 |
| | Maynard (Robert) & Harris (Richard) | 27 Poultry St. | 1845–1847 |
| Sword cutler | Maynard (Robert) & Harris (Richard) | 126 Leadenhall St. | 1848–1868 |
| | Maynard (Robert), Harris (Richard) & Grice (Henry) | 126 Leadenhall St. | 1869–1878 |
| | Henry Maynard & Co. | 126 Leadenhall St. | 1879–1890 |

- Son of Robert.

| | | | |
|---|---|---|---|
| Sword cutler | T. (Thomas) McBride | St. James St. | 1857–1890 |
| | T. (Thomas) McBride | Haymarket | 1890–1900 |
| Sword cutler | McCallan & Co. | St. James St. | 1850 |
| Gold and silver laceman | James McKiernan | 12 & 15 St. Johns Lane, Clerkenwell | 1866 |
| Sword cutler | • Successor to Kenning & McKiernan (see Kenning & McKiernan).<br>(see George Kenning) | | |
| | James McKiernan | 7 Albion Place & 15 St. Johns Lane, Clerkenwell | 1867–1869 |
| | James McKiernan | 62 St. Johns Square | 1870–1880 |

| | | | |
|---|---|---|---|
| Cutler | Thomas Henry McLean | | 1914–1929 |
| | • Master of the Cutlers Company, 1924. | | |
| Sword cutler | R. (Robert) McQueen & Son | New Castle | 1910 |
| Gold and silver laceman | John Medlicott | 17 Conduit St., Hanover Square | 1806–1811 |
| | John Medlicott | 154 Piccadilly | 1812 |
| Sword cutler | J. (John) Medlicott | 164 Piccadilly | 1813–1816 |
| Lace, jewel, and bead warehouse | F.W. Medlicott | 164 Piccadilly | 1817–1820 |
| | J.M. Medlicott | 163 Piccadilly | 1821–1826 |
| Bladesmith | John Meire (Meyer) | | 1376 |
| Sword maker | Thomas Melcher (Melchior, Melser) | Solingen, Prussia | Bc1570–1608 |
| | • Immigrated to London, 1608. | | |
| | Thomas Melcher | London | 1608–D1622 |
| | • Brought the "pincers" mark with him from Solingen, Germany. | | |
| Sword maker | Jonas Melcher | Fleet St., Parish of St. Bride | 1622–1640 |
| | • Son of Thomas Melcher. | | |
| | • Obtained "dolphin" mark, 1622. | | |
| | • Gave up his father's "pincers" mark to his brother John Melcher, 1622. | | |
| Sword maker | (Johann) John Melcher | Fleet St., Parish of St. Bride | 1622–1640 |
| | • Son of Thomas Melcher. | | |
| | • Obtained "pincer" mark from his brother Jonas Melcher, 1622. | | |
| Bladesmith | John Mercer | | 1515–1525 |
| | • Bladesmith member of the Armourers Company, 1515. | | |
| | • Bladesmith member of the Cutlers Company, 1517. | | |
| | • Registered his "orchid" mark with the Cutlers Company, 1517. | | |
| Sales agent for a sword maker | Joseph Louis Merfield | 17 Savage Garden | 1869–1891 |
| | • Sales agent for sword maker Herman Boker & Co., of Solingen, Prussia, 1869–1891. | | |
| Silversmith Hilt maker | John Merridon (Sr.) | Parish of St. Mary Abchurch, Candlewick ward | 1691–1695 |
| Sword cutler | William Merridon | Parish of St. Mary Abchurch, Candlewick ward | 1696–1700 |
| | • Master of the Cutlers Company, 1696. | | |
| | J. (John) Merridon (Jr.) | under the Royal Exchange, Cornhill, at the sign of the Black Moors Head | 1701–1704 |
| | • Master of the Cutlers Company, 1708. | | |
| | • Made silver-hilted small swords with hollow ground blades. | | |
| | John Merridon (Jr.) | under the Royal Exchange, Cornhill | 1705–1716 |
| Cutler | Walter Allen Merridon | | 1817–1848 |
| | • Master of the Cutlers Company, 1827 and 1843. | | |
| Gold and silver lacemen Fringe and trim maker | Meyer (Edward Simeon) & Mortimer (John) | 13 New Bond St. | 1812–1821 |
| | • Succeeded by Storr & Mortimer (see Paul Storr). | | |
| Embroiderer | Edward Simeon Meyer | 22 Bedford St. | 1822–1831 |
| Sword cutler | Meyer (Edward Simeon) & Clark (John) | 22 Bedford St. | 1832–1840 |
| | • Became sword cutlers. | | |

|  |  |  |  |
|---|---|---|---|
|  | Edward Simeon Meyer | 22 Bedford St. | 1841–1847 |
|  | Meyer (Edward Simeon) & Brownsmith (Henry) | 22 Bedford St. | 1848–1851 |
| Sales agent for a sword maker | William Meyerstein & Co. | Glove Lane, Blood St. | 1870–1888 |
|  | William Meyerstein & Co. | 5 London Wall Ave. | 1888–1895 |

• Sales agent for sword maker F.W. Holler of Soligen, Prussia, 1870–1895.

| | | | |
|---|---|---|---|
| Sword cutler | John Milborne | | 1644–1660 |

• Immigrated from Germany, 1644.
• Granted "flaming sword with three half moons" mark in 1644.

| | | | |
|---|---|---|---|
| Military outfitter Sword cutler | Miller & Sons | 9 London St., Fenchurch St. | 1904–1905 |
|  | Miller & Sons | 17 London St., Fenchurch St. | 1906–1916 |
| Cutler | Joseph Miller | | 1752–1781 |

• Master of the Cutlers Company, 1762–1776.

| | | | |
|---|---|---|---|
| Sales agent for a sword maker | Nathaniel Mills & Co. | 25 Mary Ann St. | 1865–1867 |

• Sales agent for sword maker Clemen & Jung of Solingen, Prussia, 1865–1867.

| | | | |
|---|---|---|---|
| Sword cutler | Philip Mist | | 1695–1720 |

• Supplied swords and plug bayonets to the Board of Ordnance, 1703–1711.
• Also repaired swords for the Board of Ordnance.
• Master of the Cutlers Company, 1710.

| | | | |
|---|---|---|---|
| Goldsmith Silversmith Hilt maker Sword cutler | Colin Mitchell | Canongate | 1720–1750 |

• Mounted small swords.
• Made silver-plated basket hilts for Walter Allen of Stirling, Scotland (1732–1760).

| | | | |
|---|---|---|---|
| Goldsmith Jeweler Hilt maker Sword cutler | V. (Viet) Mitchell | 6 Cornhill | 1768–1771 |
|  | V. (Viet) Mitchell | 9 Cornhill | 1772–1777 |

• Sold swords with chiseled steel hilts.

| | | | |
|---|---|---|---|
| Sword maker | Robert Mole & Son (see Birmingham listing) | 11 Great Chapel St. | 1895–1906 |
| Jeweler Gold, silver, pearl, diamond and bullion dealer | Montaque (James) & Co. | 22 Denmark St. | 1821–1828 |
|  | Montaque (James) & Co. | 13 Charlotte St., Bedford Square, Bloomsbury | 1829–1832 |

• Successor to Ray & Montaque at this address (see Ray & Montaque).

| | | | |
|---|---|---|---|
| Sword cutler Armourer | George Moody | | 1720–1750 |

• Armourer to King George I and II.

| | | | |
|---|---|---|---|
| Army cap, hat and accoutrement maker Sword cutler | Charles Moore | 39 St. James St. | 1830–1831 |
|  | Moore (Charles) & Co. | 39 St. James St. | 1832–1838 |
|  | Charles Moore | 39 St. James St. | 1839–1845 |
|  | Charles Moore | 120 Pall Mall | 1846–1847 |

| | | | |
|---|---|---|---|
| Gun maker<br>Sword cutler | **Charles Moore** | 77 St. James St. | 1818–1851 |
| Sword cutler | **George Moore**<br>• Master of the Cutlers Company, 1637.<br>• Supplied swords to the Board of Ordnance, 1626–1639. | | 1610–1645 |
| Army hat,<br>glove, hose,<br>accoutrement<br>maker<br>Sword cutler | **J. (John) Moore**<br>• Successor to Bicknell & Moore (see Bicknell & Moore).<br>**William Moore**<br>• John Moore's son.<br>• Succeeded at this address by Christy & Co. (see Christy & Co.).<br>**William Moore** | 1 Old Bond St.<br><br>1 Old Bond St.<br><br><br>40 Rupert St., Haymarket | 1838–1847<br><br>1848–1851<br><br><br>1852–1855 |
| Cutler | **Percival Moore**<br>• Master of the Cutlers Company, 1648. | | 1638–1653 |
| Silversmith<br>Hilt maker<br>Sword cutler | **William Moore**<br>**William Moore**<br>• Sold presentation swords mounted by Ray & Montaque (see Ray & Montaque). | 5 Ludgate St.<br>33 Crispin St., Spital Fields | 1810–1826<br>1827–1830 |
| Sword and knife<br>maker | **Henry Moorehead**<br>• Worked for sword and knife makers Richard Nathaniel (Jr.) and Paul Matthew,<br>  1591–1616 (see Paul Matthew).<br>• In 1610, obtained the Matthew family "half moon with face" mark.<br>**Henry Moorehead**<br>• Master of the Cutlers Company, 1610. | <br><br><br><br>Fleet Bridge, Parish of St. Bride | Bc1571–1610<br><br><br><br>1610–1623 |
| Silversmith<br>Goldsmith<br><br>Jeweler<br>Enameler<br>Hilt maker<br>Sword cutler | **James Morisset**<br>• Morisset probably apprenticed and worked for his brother-in-law, Louis Toussaint<br>  (see Louis Toussaint), 1752–1762.<br>• Morisset worked for Edward Wakelin and John Parker (see Wakelin & Parker), 1762–1766.<br>**James Morisset & Louis Toussaint**<br><br>• Did enamel work for Wakelin & Parker, 1767–1772.<br>• Morisset registered his mark in 1770.<br>• Toussaint retired in 1775.<br>**Morisset (James) & Wingman (Gabriel)**<br><br>• Succeeded by Gabriel Wingman (see Gabriel Wingman) at this address in 1779.<br>**Morisset (James) & Lukin (Robert & Charles)**<br>• Located across the street from 11 Denmark Street.<br>• Morisset retired in 1800 and died in 1815.<br>• Morisset mounted presentation swords for many other goldsmiths and<br>  retailers, including Robert Makepeace, Jeffries & Gilbert, T. & S. Goldney,<br>  Rundell & Bridge, Laver & Co., Gray & Constable, James Shrapnell, Thomas<br>  Ayres, Henry Osborne, Green & Ward, and Jeffries & Jones.<br>• Morisset made many high-quality presentation small swords (some presented by the<br>  Honorable East India Company).<br>• He also made snuff boxes. His "freedom boxes" (presentation or award snuff boxes) were of the highest quality.<br>  Several were presented to heroes of the war against France in the 1790s.<br>• Made the first four swords presented by the corporation of London to naval heroes<br>  of the war against France (1790s).<br>• Succeeded by Ray & Montaque at 22 Denmark Street (see Ray & Montaque). | <br><br><br><br>11 Denmark St., Parish of St. Giles<br>in the Fields, Holborn<br><br><br><br>11 Denmark St., Parish of St. Giles<br>in the Fields, Holborn<br><br>22 Denmark St. | B1738–1766<br><br><br><br>1767–1775<br><br><br><br><br>1776–1778<br><br><br>1779–1800 |

| | | | |
|---|---|---|---|
| Sword cutler | **Morris & Co.** | Pall Mall | 1845–1855 |
| Goldsmith<br>Silversmith<br>Hilt maker | **Thomas Buxton Morrish**<br>• Made hilts and decorated some presentation swords for James Howell & Co.<br>  (see James Howell & Co.). | | 1860–1876 |
| Bladesmith | **John Morthe (Morth)**<br>• Bought the "two crescent moons" mark from bladesmith Robert<br>  Hinkley (Hynkeley), 1450. | | 1450–1475 |
| Goldsmith<br>Silversmith<br>Hilt maker<br>Sword cutler | **Mortimer (John) & Hunt (John Samuel)**<br>• Successor to Storr & Mortimer (see Storr & Mortimer).<br>• Succeeded by Hunt & Roskell (see Hunt & Roskell).<br>(see Edinburgh, Scotland, listing in Chapter 15) | 156 New Bond St.<br>17 & 18 Harrison St., Grays Inn Road | 1839–1844 |
| Cutler | **Joseph Morton**<br>• Master of the Cutlers Company, 1892. | | 1882–1897 |
| Cutler | **Joseph Underwood Morton**<br>• Master of the Cutlers Company, 1896. | | 1886–1901 |
| Government<br>contractor<br>Sword cutler | Samuel Moses | 17 Great Alie St. & 65 Mansell St. | 1880–1885 |
| | Samuel Moses | 17 & 18 Great Alie St. | 1886–1890 |
| | Samuel Moses & Sons | 65 Mansell St. & 17 Great Alie St.<br>& Tenter St. (North) & Tenter St. (West) | 1891–1907 |
| | Samuel Moses & Sons Ltd. | 65 Mansell St. & 17 Great Alie St.<br>& Tenter St. (North) & Tenter St. (West) | 1908–1917 |
| | Samuel Moses & Sons Ltd. | 65 to 67 Mansell St. &<br>17 Great Alie St. & Tenter St.<br>(North) & Tenter St. (West) | 1918–1924 |
| | Samuel Moses & Sons Ltd. | 65 to 67 Mansell St. &<br>15 to 21 Tenter St. (West)<br>& 18 Tenter St. (North) | 1925–1930 |
| | Samuel Moses & Sons Ltd. | 18 North Tenter St. | 1931–1968 |
| Sword cutler | **Moss Brothers & Co.** | 21 King St., Covent Garden | 1850 |
| Silversmith<br>Goldsmith<br>Hilt maker<br>Sword cutler | **Thomas Moulden**<br>• Apprenticed to Dru Drury Sr., 1713–1720 (see Dru Drury Sr.). | | Bc1699–1721 |
| | Thomas Moulden<br>• Had apprentice Benjamin Corbett, 1722–1725 (see Benjamin Corbett). | Dolphin Ct., Ludgate Hill | 1722–1732 |
| | Thomas Moulden | Lower end of Cheapside<br>at the sign of the Three Crowns | 1733–1739 |
| Cutler | **William Multone**<br>• Master of the Cutlers Company, 1422. | | 1412–1427 |
| Cutler | **John Munt**<br>• Master of the Cutlers Company, 1420. | | 1410–1425 |
| Sword maker | **Peter Munsten**<br>(see Chapter 3)<br>(see Peter English) | | |

| | | | |
|---|---|---|---|
| Sword cutler | **William Murden** | Cripplegate | 1625 |
| Sword blade maker | **Thomas Murrey** | | 1610–1625 |

- In early 1621, King James I granted Thomas Murrey (cutler and secretary to the "Prince's Highness," probably the Prince of Wales' wardrobe supplier) a patent (monopoly) for the sole manufacture of sword and rapier blades.
- On June 13, 1621, by order of the Cutlers Company of London, master John Porter, past master Thomas Chesshire, and royal cutler Robert South inspected a sword blade mill (presumably Thomas Murrey's mill) and were to report their findings to the court at the next meeting.
- In July 1621, Thomas Murrey presented his first group of sword blades to the Cutlers Company for inspection.
- The company rejected them, saying they needed much more work and they were too expensive.

| | | | |
|---|---|---|---|
| Silversmith<br>Goldsmith<br>Jeweler | **George Sigismund Natter**<br>• Apprenticed to Charles Wright, 1765–1772. | | Bc1751–1772 |
| Hilt maker<br>Sword cutler | **George Sigismund Natter** | 185 Fleet St. | 1772 |
| | **Bland (Cornelius) & Natter (George Sigismund)**<br>(see Cornelius Bland) | 185 Fleet St. | 1773–1781 |
| Goldsmith<br>Jeweler | **William Neild** | 4 St. James St.,<br>near the Royal Exchange | 1770–1775 |
| Hilt maker<br>Sword cutler | **James Neild** | 4 St. James St.,<br>near the Royal Exchange | 1776–1793 |
| | • Used "bright cut" engraving on some of his hilts. | | |
| | **Neild (James) & Goldney (Thomas & Samuel)**<br>• Succeeded by T. & S. Goldney.<br>(see Thomas & Samuel Goldney) | 4 St. James St.,<br>near the Royal Exchange | 1794–1795 |
| Military tailor | **Samuel Nelson** | 109 Regent St. | 1884–1886 |
| Sword cutler | **Samuel Nelson** | 13 Hanover St. | 1887–1909 |
| | • Successor to Frederick Grosjean (see Frederick Grosjean). | | |
| Bladesmith | **John Neuby** | | 1376 |
| Silversmith | **John Neville** | | Bc1716–1737 |
| Hilt maker<br>Sword cutler | **Ann Craig & John Neville**<br>• Ann Craig died in 1745. | corner of Norris St. &<br>the Haymarket, Parish of St. James | 1738–1745 |
| | **John Neville**<br>(see John Craig) | corner of Norris St. &<br>the Haymarket, Parish of St. James<br>at the sign of the Hand and Ring | 1745–1752 |
| Cutler | **Symon Newenton** | | 1482–1504 |
| | • Master of the Cutlers Company, 1492, 1493, 1498, and 1499. | | |
| Gold and silver lacemen<br>Hatter<br>Hosier<br>Sword cutler | **Newham (James) & Bingham (Charles)**<br>• Successor to Thomas Street (see Thomas Street).<br>**Newham (James) & Thresher (Richard)**<br>• Succeeded by Richard Thresher (see Richard Thresher). | 152 Strand, next to Somerset house<br><br>152 Strand, next to Somerset house | 1767–1776<br><br>1777–1779 |

| | | | |
|---|---|---|---|
| Tailor | William Newton | 242 Piccadilly | 1817–1820 |
| Gaiter (spats) maker | William Newton | 228 Piccadilly | 1821–1822 |
| | William Newton & Son | 228 Piccadilly | 1823–1829 |
| Habit (riding clothes) maker | William Newton & Son | 7 Maddox St., Bond St. | 1830–1837 |
| | William Newton & Son Ltd. | 7 Maddox St., Bond St. | 1838–1848 |
| | William Mears Newton | 7 Maddox St., Bond St. | 1849–1852 |
| Sword cutler | Newton (William Mears) & Chambers (William) | 8 Maddox St., Bond St. | 1853–1857 |
| | Newton (William Mears) & Chambers (William) | 43 Maddox St., Bond St. | 1858–1859 |
| | Newton (William Mears) & Chambers (William) | 7 Hanover Square | 1860–1862 |
| | William Mears Newton & Co. | 7 Hanover Square | 1863–1869 |

- Appointed tailor to His Royal Highness the Prince of Wales.
- Appointed habit maker to Her Majesty the Queen.

| | | | |
|---|---|---|---|
| | Newton (William Mears) & Co. | 7 Hanover Square | 1870–1887 |
| | Newton (William Mears) & Bean (Henry) | 7 Hanover Square | 1888–1900 |
| Sword blade maker | Nicholas Brothers | | 1630–1648 |

- In 1638, Benjamin Stone of Hounslow obtained an order for 5,000 sword and rapier blades from the Office of Ordnance. All were subcontracted to the Nicholas Brothers, John Hayes, and John Harvey. All blades were delivered to Capt. William Legge (Master of Armoury) under Stone's name.

| | | | |
|---|---|---|---|
| Military tailor | Nicoll (Henry) & Franklin (Joseph) | 12 Conduit | 1824–1825 |

- Successors to Pulford & Nicoll (see Pulford & Nicoll).

| | | | |
|---|---|---|---|
| Sword cutler | Nicoll (Henry), Franklin (Joseph) & Rowed (Anthony) | 12 Conduit | 1826–1828 |
| | Nicoll (Henry), Rowed (Anthony) & Greenshields (James) | 12 Conduit | 1829 |
| | Nicoll (Henry), Rowed (Anthony) & Co. | 12 Conduit | 1830–1831 |
| | Nicoll (Henry) & Sandilands (John) | 12 Conduit | 1832–1839 |

- Succeeded by Sandilands & Nicoll (see Sandilands & Nicoll).

| | | | |
|---|---|---|---|
| Goldsmith | John Nichols (Nicholls) | 6 Dukes Ct., St. Martins Lane | 1825–1829 |
| Silversmith | John Nichols | 40 Castle St., Leicester Square | 1830–1831 |
| Hilt maker Sword cutler | John Nichols | 42 Jermyn St., Parish of St. James | 1832–1850 |
| Gold and silver laceman Sword cutler | Thomas Nixon | 60 Lombard St., Birchin St. at the sign of the Artichoke | 1763–1765 |
| Silversmith Hilt maker Sword cutler | William Nixon | Over the sewer, next to the church of St. Clement Danes, Strand | 1745–1810 |

- Made a presentation sword for Vincount Admiral Horatio Nelson.
- Several of Nixon's swords are in the Windsor Castle Royal Armoury.

| | | | |
|---|---|---|---|
| Gun maker | Henry Nock | | B1741–1771 |
| Sword cutler | Henry Nock | 10 Ludgate St. | 1772–1774 |
| | Nock (Henry), Jover (William I. Sr.) & Green (John) | 10 Ludgate St. | 1775–1784 |
| | Henry Nock & James Wilkinson | 10 Ludgate St. | 1785–1804 |

- Henry Nock died in 1804.
- Appointed gun maker to His Majesty the King.
- Succeeded by James Wilkinson (see James Wilkinson).
- Nock willed his company to James Wilkinson.

| | | | |
|---|---|---|---|
| | Samuel Nock | 180 Fleet St. | 1804–1823 |

| | | | |
|---|---|---|---|
| | Samuel Nock | 43 Regent St. | 1824–1852 |
| | Samuel Nock | 116 Jermyn St. | 1853–1858 |

• Appointed gun maker to Her Majesty the Queen.
• Succeeded by John Wallis (see John Wallis).

| | | | |
|---|---|---|---|
| Goldsmith | John Nodes | near Craven St., Strand | 1763–1776 |
| Jeweler | William Nodes | New Bond St., near Grosvenor Square | 1777–1785 |
| Toy (hardware) | William Nodes | 126 Bond St. | 1786–1794 |
| maker | Nodes (William) & Sydenham (J.H.) | 126 Bond St. | 1795–1796 |
| Hilt maker | • Succeeded by J.H. Sydenham (see J.H. Sydenham). | | |
| Sword cutler | | | |

| | | | |
|---|---|---|---|
| Sword cutler | Edward Norman | | 1679 |

| | | | |
|---|---|---|---|
| Sword cutler | Alexander Normington | | 1640–1660 |

• Supplied swords to the Board of Ordnance, 1649–1651.

| | | | |
|---|---|---|---|
| Silversmith | Benjamin North | New Whay St., Parish of | 1783–1800 |
| Hilt maker | | St. Thomas, Southwark | |
| Sword cutler | | | |

| | | | |
|---|---|---|---|
| Silversmith | Thomas North | | Bc1702–1723 |
| Hilt maker | • Apprenticed to George Wilcox, c. 1718–1723 | | |
| Sword cutler | (see George Wilcox). | | |
| | Thomas North | Dean St., Fetterlane | 1724–1740 |
| | • Had apprentice William Wilson, 1724–1725. | | |

| | | | |
|---|---|---|---|
| Sword cutler | Daniel Nowell | 25 Stangate St., Lambeth | 1811–1820 |

| | | | |
|---|---|---|---|
| Sword cutler | Charles Oakden | 35 Brownlow St., Drury Lane | 1841–1846 |
| Knife and dirk | • Successor to John Goss (see John Goss). | | |
| maker | | | |
| Knife and dirk | | | |
| belt maker | | | |

| | | | |
|---|---|---|---|
| Gold and silver | J. (John) & B. (Bennett) Odell | 114 New Bond St. | 1800–1802 |
| lacemen | • Bennett Odell moved in 1803. | | |
| Hatter | J. (John) Odell | 114 New Bond St. | 1803–1804 |
| Sword cutler | J. (John) Odell | 85 New Bond St. | 1804–1806 |
| | J. (John) Odell | 4 Pall Mall | 1807–1808 |
| | J. (John) Odell | 57 Piccadilly | 1809–1810 |

| | | | |
|---|---|---|---|
| Gold and silver | Bennett Odell | 6 Mill St., Hanover Square | 1803–1804 |
| laceman | Bennett Odell | 9 Lower Brook St. | 1805 |
| Embroiderer | Bennett Odell | 4 Pall Mall | 1806–1810 |
| Military hat | Bennett Odell | 17 Old Bond St. | 1811–1818 |
| and cap maker | • Gold and silver laceman to the Princess of Wales, 1814–1833. | | |
| Jeweler | Bennett Odell | 71–72 Burlington Arcade, Piccadilly | 1819–1824 |
| Sword cutler | • Sold M1796 light and heavy cavalry officers sabers. | | |
| | Bennett Odell | 72–73 Burlington Arcade, Piccadilly | 1825–1828 |
| | Bennett Odell | 71–72–73 Burlington Arcade, Piccadilly | 1829–1833 |
| | Odell (Bennett) & Atherly (John) | 71–72–73 Burlington Arcade, Piccadilly | 1834–1847 |
| | • Succeeded by John Athley (see John Athley). | Piccadilly | |
| | Adam Ohlig | | |

|  |  |  |  |
|---|---|---|---|
|  | Robert Ohlig | | |
|  | William Ohlig | | |
|  | Nicholas Ohlig | | |
|  | William Ohlig | | |
|  | Christopher Ohlig | | |
|  | (see Chapter 4) | | |
| Bladesmith | **Thomas Okys (Oks)** | | 1515–1520 |
|  | • Member of the Armourers Company, 1515. | | |
|  | • Member of the Cutlers Company, 1517. | | |
|  | • Registered his mark "S" as a bladesmith with the Cutlers Company, 1517. | | |
| Cutler | **William Adam Oldaker** | | 1862–1884 |
|  | • Master of the Cutlers Company, 1872 and 1879. | | |
| Silversmith<br>Hilt maker<br>Sword cutler | **William Oldrenshawe** | | 1600–1620 |
|  | • Member of the Cutlers Company. | | |
|  | • In 1607, the Cutlers Company fined him for selling a rapier and dagger at the Sturbridge Fair, telling his customer the hilt was hatched (silver applied to the hilt by the process of hatching) when it was actually plain silvered (a cheaper method of applying silver to the hilt using less silver). He sold the rapier and dagger for 26 shillings, 8 denier. | | |
| Hatter<br>Army cap, hat, accoutrement maker<br>Sword cutler | James (Sr.) Oliphant | St. James St., Haymarket | 1763–1788 |
|  | James (Sr.) Oliphant & Son | 14 Cockspur St. | 1789–1794 |
|  | James (Sr.) Oliphant & Son Ltd | 14 Cockspur St. | 1795–1798 |
|  | James (Sr.) Oliphant & Co. | 14 Cockspur St. | 1799–1807 |
|  | C. (Campbell) & R. (Richard) Oliphant (Sr.) & Co. | 14 Cockspur St. | 1808 |
|  | Campbell Oliphant | 14 Cockspur St. | 1809–1814 |
|  | Oliphant (Campbell) & Son (James Jr.) | 14 Cockspur St. | 1815 |
|  | James Oliphant (Jr.) | 14 Cockspur St. | 1816–1820 |
|  | James Oliphant (Jr.) & Co. | 14 Cockspur St. | 1821–1825 |
|  | J. (James Jr.) & R. (Richard) Oliphant (Jr.) | 14 Cockspur St. | 1826–1836 |
|  | James Oliphant III | 14 Cockspur St. | 1837–1856 |
|  | James Oliphant III | 5 Warwick St., Charing Cross | 1857–1862 |
| Tailor<br>Naval outfitter<br>Sword cutler | James Ollivier (Sr.) | 6 Bolton St. | 1835–1837 |
|  | Ollivier (James) & Brown (Robert Sr.) | 6 Bolton St. | 1838–1849 |
|  | James Ollivier (Jr.) | 6 Bolton St. | 1850 |
|  | Ollivier (James Jr.) & Brown (Robert Jr.) | 37 Sackville St. | 1851–1906 |
|  | Ollivier, Brown & Scholfield Ltd. | 37 Sackville St. | 1907–1912 |
| Sales agent for a sword maker | **S. Oppenheim & Son** | 4 Bread St., Cheapside | 1859–1863 |
|  | • Sales agent for sword maker Carle Klonne & Co. of Suhl, Germany. | | |
| Sword cutler | P. (Peter) Orr & Sons | | 1850 |
| Miniature sword maker | Ortner & Haule | 3 St. James St. | 1900–1915 |
| Sword cutler<br>Accoutrement maker<br>Sword and gun maker | **Henry Osborne** | 82 Pall Mall | 1802–1807 |
|  | • Appointed sword cutler and accoutrement maker to His Majesty the King. | | |
|  | **Osborne (Henry) & Gunby (John)** | 82 Pall Mall | 1808–1818 |
|  | • John Gunby was a metal worker and sword and gun maker. | | |
|  | • They began to make swords, sword blades, and guns in 1808. | | |

|  |  |  |  |
|---|---|---|---|
|  | • Made officers swords for the 10th Light Dragoons. | | |
|  | • James Morisset mounted some presentation swords sold by them. | | |
|  | • In July 1812, took an order from Henry Upson for 1,000 gilt and plated officers and commissioned officers swords. | | |
|  | • By August 1812, Osborn & Gunby had shipped 200 swords to Upson. | | |
|  | • They sold sword blades to John Salter. | | |
|  | **Osborne (Henry) & Gunby (John)** | 48 Thames St. | 1819–1820 |
|  | • Appointed sword cutler and accoutrement maker to His Majesty the King, His Royal Highness the Prince of Wales, and the Honorable East India Company. | | |
|  | **Henry Osborne** | 48 Thames St. | 1821–1838 |
|  | • Henry Osborne died in 1827. His son Thomas took over the company but kept the Henry Osborne name. | | |
|  | (see Birmingham listing) | | |
| Cutler | **Thomas Otehill** | | 1845–1861 |
|  | • Master of the Cutlers Company, 1855 and 1856. | | |
| Silversmith<br>Hilt maker<br>Sword cutler | **John Outlaw** | | 1661 |
|  | • Signed an advertisement warning sword purchasers about counterfeit cast silver hilts. | | |
| Hatter<br>Hosier<br>Sword cutler | **Thomas Ovey** | 17 Fleet St. | 1778 |
|  | **Thomas Ovey** | 41 Fleet St. | 1779–1802 |
| Sword cutler | **Daniel Owen** | | 1620–1635 |
| Silversmith<br>Hilt maker<br>Sword cutler | **Sarah Owen** | 12 Magpie Alley, Fetter Lane | 1779–1829 |
| Merchant<br>Sword and<br>blade<br>importer | **George Page** | | 1630–1640 |
|  | • On July 6, 1639, Benjamin Stone of Hounslow bought 200 swords at 7 shillings each from Page to resell to the Board of Ordnance. | | |
| Sword cutler | **Henry Palmer** | St. James St. | 1790–1800 |
|  | • Appointed cutler to His Majesty the King. | | |
|  | • Exclusive agent of Thomas Gill's (Birmingham) Swords in London. | | |
| Goldsmith<br>Silversmith<br>Hilt maker | **Esaie Pantin (Sr.) (Pontin, Pointon, Panton)** | Rouen, France | Bc1630–c1650 |
|  | • Immigrated to London, c. 1650. | | |
|  | **Esaie Pantin (Sr.)** | London | c1650–Dc1708 |
|  | • Sons: Esaie Pantin (Jr.) and Simon Pantin (Sr.). | | |
|  | • His son Esaie Pantin (Jr.) was born in 1660 and apprenticed and worked for his father, c. 1674–1708. | | |
|  | **Esaie Pantin (Jr.)** | | 1708–1710 |
| Silversmith<br>Hilt maker<br>Sword cutler | **Henry Pantin (Pontin, Pointon, Panton)** | Rouen, France | Bc1630–c1650 |
|  | • Immigrated to London, c. 1650. | | |
|  | **Henry Pantin** | London | c1650–Dc1710 |
|  | • In 1661, signed an advertisement warning sword purchasers about counterfeit cast silver hilts. | | |
|  | • Had apprentice John Kenton, 1683–1700 (see John Kenton). | | |
|  | • Had apprentice Hugh Brawne, c. 1703–1709 (see Hugh Brawne). | | |

| | | | |
|---|---|---|---|
| Toy (hardware) maker<br>Goldsmith<br>Jeweler<br>Silversmith<br>Hilt maker<br>Sword cutler | **Simon Pantin (Sr.) (Pontin, Pointon, Panton)**<br>• Family came from Pouen, France.<br>• Son of Esaie Pantin Sr.<br>• Apprenticed to Peter Harache, c. 1692–1698. | | Bc1678–1698 |
| | **Simon Pantin (Sr.)**<br>• Became a naturalized citizen in 1709. | Peacock St., St. Martins Lane<br>Parish of St. Martins in the Field<br>at the sign of the Peacock | 1699–1716 |
| | **Simon Pantin (Sr.)** | Peacock St., Castle St. near Leicester Field<br>Parish of St. Martins in the Field | 1717–D1728 |
| | **Simon Pantin (Jr.)** | Peacock St., Castle St. near Leicester Field<br>Parish of St. Martins in the Field | 1729–1730 |
| | **Simon Pantin (Jr.)** | Green St., Castle St., Leicester Field<br>Parish of St. Martins in the Field | 1731–D1733 |
| | **Mary Pantin** | Green St., Castle St., Leicester Field<br>Parish of St. Martins in the Field | 1733 |
| | • Widow of Simon Pantin Jr.<br>**Lewis Pantin (Sr.)**<br>• Son of Simon Pantin Jr. | Green St., Castle St., Leicester Field,<br>Parish of St. Martins in the Field | Bc1713,<br>1733–D1767 |
| | **Lewis Pantin (Jr.)**<br>• Son of Lewis Pantin Sr. | 45 Fleet St. | Bc1746,<br>1767–1781 |
| | **Lewis Pantin (Jr.)** | 36 South Hampton St., Strand | 1782–1791 |
| | **Lewis Pantin (Jr.)** | 8 Sloane Square, Chelsea | 1792–1794 |
| | **Lewis Pantin (Jr.)** | 6 Crown St., Westminster<br>at the sign of the Crown and Sceptre | 1795–1799 |
| | • Son Lewis III apprenticed and worked for him, c. 1781–1797. | | |
| | **Lewis Pantin Jr.** | 17 Alfred Place, Newington Causeway | 1800–1801 |
| | **Lewis Pantin Jr.** | 30 Marsham St. | 1802–1804 |
| | **Lewis Pantin Jr.** | 5 Canterbury Place, New Manor Place | 1805–1810 |
| Toy (hardware) maker<br>Silversmith<br>Hilt maker<br>Sword cutler | **Lewis Pantin III**<br>• Son of Lewis Pantin Jr. | 62 St. Martins, Le Grand | 1798–1830 |
| Bayonet maker | **Thomas Pantin (Pontin, Pointon, Panton)**<br>• Made plug bayonets.<br>• Used the "diamond and star" mark. | | 1664–1680 |
| Steel manufacturer<br>Sword maker | **James Parish** | 144 Strand | 1784–1790 |
| | **James Parish** | 114 Fleet St. | 1791–1794 |
| | **John Parker**<br>(see Edward Wakelin) | | |
| | **William Parker**<br>(see John Field) | | |
| Silversmith<br>Goldsmith<br>Jeweler<br>Hilt maker<br>Sword cutler | **John Parker**<br>• Apprentice and journeyman to George Wickes, 1751–1761 (see George Wickes). | | Bc1737–1761 |
| | **John Parker & Edward Wakelin** | Panton St., two doors<br>from the Haymarket, Parish of St. James<br>at the sign of the Kings Arm | 1762–1777 |

| | | | |
|---|---|---|---|
| | • Successors to Wickes & Wakelin and Wickes & Netheron (see George Wickes). | | |
| | **John Parker** | | 1777–D1792 |
| Hatter | **Charles Parr** | | 1738–1758 |
| Hosier | **Adams (John) & Parr (Charles)** | 48 Fleet St. near Serjeants Inn | 1759–1766 |
| Sword cutler | **Adams (John) & Parr (Charles)** | 146 Fleet St. | 1767 |
| | **Charles Parr** | 149 Fleet St. | 1768–1770 |
| Cutler | **Thomas Parsons** | | 1640–1655 |
| | • Master of the Cutlers Company, 1650. | | |
| Cutler | **Gabriel Partridge** | | 1632–1641 |
| | • Master of the Cutlers Company, 1642. | | |
| Sword cutler | **John Partridge** | Parish of St. Bride | 1659–1664 |
| Sword cutler | **Thomas Partridge** | Parish of St. Bride | 1612–1627 |
| Bladesmith | **John Pascall** | | 1515–1520 |
| | • A bladesmith member of the Armourers Company, 1515. | | |
| | • A bladesmith member of the Cutlers Company, 1517. | | |
| | • Registered "three leaf clover" mark with Cutlers Company, 1517. | | |
| | **John Passman** (see Evan Evans) | | |
| Cutler | **William Patrick** | | 1512–1527 |
| | • Master of the Cutlers Company, 1522. | | |
| Cutler | **Edward Patterson** | | 1702–1717 |
| | • Master of the Cutlers Company, 1712. | | |
| Cutler | **Peter Patterson** | | 1838–1853 |
| | • Master of the Cutlers Company, 1848. | | |
| Cutler | **William Paxton** | | 1665–1680 |
| | • Master of the Cutlers Company, 1675. | | |
| Sword cutler | **Peter Pearsall** | 202 Holborn | 1790 |
| | **Peter Pearsall** | 231 High Holborn | 1791–1795 |
| Sword cutler | **Pearse (J. & B.) & Co.** | Hart St., Covent Garden | 1840–1845 |
| | **J. & B. Pearse & Co.** | Floral St., Covent Garden | 1846–1889 |
| Sword cutler | **Jeremy Peasely** | | 1645 |
| | **John Peat** (see Gibson & Son) | | |
| Gun maker Bayonet maker | **James Peddell** | | 1685–1710 |
| | • Supplied plug bayonets to the Board of Ordnance, 1703–1704. | | |
| Sword knot maker | **Henry Penel** | 10 Englefield Rd. | 1672–1675 |

| | | | |
|---|---|---|---|
| Sword cutler | | | |
| Cutler | **Henry Penharger**<br>• Master of the Cutlers Company, 1472 and 1473. | | **1462–1478** |
| Sword cutler | **Thomas Pennington**<br>• Master of the Cutlers Company, 1684.<br>• Supplied swords to the Board of Ordnance, 1686–1692. | Langbourne Ward<br>Parish of St. Mary Woolnoth | **1680–1692** |
| Cutler | **William Pepys**<br>• Master of the Cutlers Company, 1792. | | **1782–1797** |
| Cutler | **William Hasledine Pepys**<br>• Master of the Cutlers Company, 1822 and 1828. | | **1812–1833** |
| Cutler | **William Price Pepys**<br>• Master of the Cutlers Company, 1914. | | **1904–1919** |
| Sword maker | **Peter Perkins**<br>• Used "bunch of grapes" mark. | | **1677–1703** |
| Cutler | **Robert Perkins**<br>• Master of the Cutlers Company, 1895. | | **1885–1900** |
| Sword cutler | **Edward Perry**<br>• Master of the Cutlers Company, 1802.<br>• His swords are marked "Perry's Warranted."<br>• Exported swords to the USA. | | **1781–1815** |
| Cutler | **Robert Perry**<br>• Master of the Cutlers Company, 1757. | | **1747–1762** |
| Silversmith<br>Hilt maker<br>Sword cutler | **Thomas Perry**<br>**James Perry**<br>**James Perry**<br>**James Perry**<br>**James Perry** | Shoe Lane<br>Shoe Lane<br>Holborn<br>131 Chancery Lane<br>10 Crown Court, Fleet Street | **1727–1762**<br>**1763–1764**<br>**1765–1772**<br>**1773–1776**<br>**1777–1780** |
| Sword and pistol maker | **Charles Peters** | | **1725–1735** |
| | **Henry Petherick**<br>(see Peter Grosjean) | | |
| Iron monger (dealer)<br>Bladesmith | **Richard Petique att Nax**<br>• Son of Simon Petique atte Nax. | Parish of St. Benedict Shevehog | **1356–1400** |
| Iron monger (dealer)<br>Bladesmith | **Simon Petique atte Nax** | Fleet St., Parish of St. Brides | **1344–D1390** |
| Iron monger (dealer)<br>Bladesmith | **William Petique atte Nax** | Cornhill St.,<br>Parish of St. Andrew Undershaft | **1366–1380** |

| | | | |
|---|---|---|---|
| Sword cutler | **Phelps & Co.**<br>• Sold Wilkinson Sword Company swords. | Walbrook | c1950 |
| Armourer<br>Sword maker | **John Phillipes**<br>• Made falchions.<br>• Member of the Armourers Company of London.<br>• Used the "bell" mark. | | 1570–1590 |
| Sword blade<br>maker | **Edmond Phillips**<br>• Accused of selling defective sword blades to the Board of Ordnance, 1632. | | 1630–1635 |
| Sword cutler | **Keenan Phillips & Co.** | Maddex St. | 1830–1835 |
| Tailor<br>Habit<br>(riding clothes)<br>maker<br>Military outfitter<br>Sword cutler | **Lawrence Phillips**<br>**Lawrence Phillips**<br>**Lawrence Phillips & Son**<br>**L. (Lawrence) D. (David) &**<br>**W. (William) Phillips**<br>• Succeeded by James William Dove (see James William Dove). | 140 St. Martins Lane<br>28 Strand<br>13 George St., Hanover Square<br>13 George St., Hanover Square | 1826–1829<br>1830–1863<br>1864–1875<br>1876–1882 |
| Cutler | **John Phipps**<br>• Master of the Cutlers Company, 1790. | | 1780–1795 |
| Button and<br>accoutrement<br>maker | **Phipson (Peter),**<br>**Doughty (Benjamin) & Co.**<br>• Successor to Walter Williams (see Walter Williams). | 103 St. Martins Lane | 1829–1830 |
| Sword cutler | **Benjamin Doughty**<br>**Benjamin Doughty & Co.**<br>**Benjamin Doughty & Co.** | 103 St. Martins Lane<br>103 St. Martins Lane<br>109 St. Martins Lane | 1831–1836<br>1837–1857<br>1858–1876 |
| Goldsmith<br>Jeweler<br>Sword and knife<br>hilt maker<br>Sword and knife<br>cutler | **Pickett (William) & Rundell (Philip)**<br>• Successors to Thead & Picket<br>(see Thead & Picket).<br>**William Pickett & Philip Rundell**<br>• Succeeded by Philip Rundell (see Philip Rundell).<br>• William Pickett retired in 1785 and died in 1796. | 32 Ludgate Hill<br>at the sign of the Golden Salmon<br>32 Ludgate Hill<br>at the sign of the Golden Salmon | 1773–1774<br>1775–1784 |
| Sword cutler | **William Pierson** | Parish of St. Clement Danes,<br>Middlesex | 1612–1628 |
| Cutler | **Richard Piggott**<br>• Master of the Cutlers Company, 1713.<br>• Used the "flaming sword and ball" mark in 1679 and "tobacco pipe" mark in 1682. | | 1679–1718 |
| Sword maker | **John Burqoyne Pillin**<br>**John Burqoyne Pillin** | 3 Featherstone Building<br>19 Frith St., Soho | 1850–1853<br>1854 |
| Sword maker | **G. (George) A. (Alfred) Pillin**<br>**G. (George) A. (Alfred) Pillin**<br>**John Septimus Pillin**<br>**George Pillin**<br>• Sold swords to London military outfitter Malcohm Kearton (see Malcohm Kearton).<br>• Wilkinson Sword Company bought out George Pillin in 1922. | 3 Poland St.<br>31 Gerrard St., Soho<br>31 Gerrard St., Soho<br>31 Gerrard St., Soho | 1840–1861<br>1862–1880<br>1881–1919<br>1920–1922 |

| | | | |
|---|---|---|---|
| Sword cutler | **Pipe & McGill** | 27 Maiden Lane, Bedford St., Strand | 1900–1916 |
| Gold and silver lacemen Button maker Sword cutler | **Charles Pitt & Co.** **Charles Pitt & Co.** • Bought out Herbert & Co. in 1948. | 50 St. Martins Lane 31 Maddox St. | 1875–1895 1896–1968 |
| | **Robert Pitter** (see Richard Hill) | | |
| Gold and silver lacemen | **T. (Thomas Sr.) Pitter & Son (Thomas Jr.)** • Successor to Hill, Pitter, & Son. (see Richard Hill) | 3 Strand | 1795–1799 |
| Sword cutler | **T. (Thomas Jr.) Pitter** | 48 Pall Mall | 1800–1810 |
| Gold and silver lacemen | **J. (John) Pitter** • Successor to Amery & Pitter (see Amery & Pitter). | 26 Bedford St., Covent Garden | 1805–1809 |
| Embroiderer Sword cutler | **Pitter (John) & Fox (Benjamin)** • John Pitter died 1829. • Succeeded by Benjamin Fox (see Benjamin Fox). | 26 Bedford St., Covent Garden | 1810–1829 |
| | **George Pizey** (see Samuel Whitford) | | |
| Sales agent for a sword maker | **Lucas Platzhoff & Co.** **Lucas Platzhoff & Co.** • Sales agent for sword maker A. & E. Holler of Solingen, Prussia, 1865–1869. | 3 Church Passage, Guild Hall 30 Monkwell St., Cripplegate | 1865–1868 1869 |
| Cutler | **Oliver Pluckett** • Master of the Cutlers Company, 1602 and 1603. | | 1592–1608 |
| | **John Plush** (see Holt & Plush) | | |
| Cutler | **Alfred Pocock** • Master of the Cutlers Company, 1880 and 1909. | | 1870–1914 |
| Cutler | **Ebenezer Pocock** • Master of the Cutlers Company, 1886. | | 1876–1891 |
| Cutler | **George Pocock** • Master of the Cutlers Company, 1913. | | 1903–1918 |
| Cutler | **Percy Rogers Pocock** • Master of the Cutlers Company, 1919. | | 1909–1924 |
| Cutler | **Thomas Pocock** • Master of the Cutlers Company, 1850 and 1851. | | 1840–1856 |
| Cutler | **Thomas Gotch Pocock** • Master of the Cutlers Company, 1870 and 1878. | | 1860–1883 |
| | **Henry Pointon** (see Henry Panton) | | |

| | | | |
|---|---|---|---|
| Sword cutler | John Ponder | Duke St., Parish of St. James | 1856–1860 |
| Silversmith | John Pont (Sr.) | | Bc1699–1731 |
| Sword and knife hilt maker | • Apprentice and journeyman to Dru Drury Sr., c. 1713–1731 (see Dru Drury Sr.). | | |
| | John Pont (Sr.) | Staining Lane | 1732–D1761 |
| Sword and knife cutler | John Pont (Jr.) | Maiden Lane | 1761–1775 |
| Army and navy outfitter | James Poole | Shropshire | Bc1785–1805 |
| | • Moved to London in 1806. | | |
| Draper (cloth maker) | James Poole | Everett St., Brunswick Square | 1806–1810 |
| | • Linen draper (maker). | | |
| Sword cutler | James Poole | Great Ormonde St. | 1811–1821 |
| | • Draper and military tailor. | | |
| | • Began tailoring when he made tunics for Napoleon's Volunteer Corps after Napoleon escaped from Elba. | | |
| | James Poole | 171 Regent St. | 1822 |
| | Poole (James) & Cooling (Richard) | 4 Old Burlington St., corner of Savile Row | 1823–1829 |
| | James Poole | 4 Old Burlington St., corner of Savile Row | 1830–1834 |
| | James Poole | 4 Old Burlington St., 32 Savile Row | 1835–1842 |
| | • Opened an entrance on Savile Row. | | |
| | James Poole & Son (Henry) | 4 Old Burlington St., & 32 Savile Row | 1843–D1846 |
| | • Civilian tailoring became a large part of their business. | | |
| | • James Poole died in 1846. | | |
| | Henry Poole & Co. | 4 Old Burlington St., & 32 Savile Row | 1846–1865 |
| | • Became tailor to the court of Emporer Louis Napoleon. | | |
| | • Over the years, Poole & Co. became tailors to many dignitaries and royal households all over Europe and the United States. | | |
| | Henry Poole & Co. | 4 Old Burlington St., & 37 to 39 Savile Row | 1866 |
| | Henry Poole & Sons | 36 to 39 Savile Row | 1867–D1876 |
| | • Henry Poole died in 1876. | | |
| | Henry Poole & Co. | 36 to 39 Savile Row | 1876–1891 |
| | • Company directors: Samuel Cundey (1877–1887), Howard Cundey (1887–1927). | | |
| | Henry Poole & Co. | 36 to 39 Savile Row & 21 Clifford St. | 1892–1904 |
| | • Tailor to King Edward VII (1903–1910). | | |
| | Henry Poole & Co. | 37 to 39 Savile Row & 21 Clifford St. | 1905–1910 |
| | • Became army, navy, and civil service outfitter. | | |
| | • Had a branch in Paris, France, at 10 Rue Tronchet, 1904–1940. | | |
| | • Tailor to King George V (1910–1936). | | |
| | Henry Poole & Co. | 37 to 39 Savile Row | 1911–1946 |
| | • Became the largest tailor in the world, employing more than 300 tailors and cutlers. | | |
| | • Company directors: S. & H. Lundey (1927–1954), A. Mead (1927–1954). | | |
| | • Tailor to the Prince of Wales, 1930. | | |
| | • Absorbed tailor Hill Brothers, 1939. | | |
| | • Tailor to King George VI (1936–1952). | | |
| | Henry Poole & Co. | 37 & 39 Savile Row | 1947–1960 |
| | • Tailor to Queen Elizabeth II (1952–1998). | | |
| | Henry Poole & Co. | 11 & 12 Cork St. | 1961 |
| | Henry Poole & Co. | 10 & 12 Cork St. | 1962–1964 |
| | Henry Poole & Co. | 10 Cork St. | 1965–1981 |

| | | | |
|---|---|---|---|
| | • Absorbed tailor E.C. Squires, 1976. | | |
| | • Absorbed tailor Sullivan Wooley & Co., 1980. | | |
| | Henry Poole & Co. Ltd. | 15 Savile Row | 1982–2003 |
| Goldsmith | Guillaume (William) Portal (Sr.) | Bagnolssar Ceze Lanquedos, Holland | Bc1665–c1685 |
| Hilt maker | • A Huguenot who fled to London, c. 1685. | | |
| Sword cutler | William Portal (Sr.) | Parish of St. Helen, Abington | c1685–c1720 |
| | • A member (goldsmith) of the wardrobe (keepers of the royal clothes, weapons, equipment, utensils, etc.) of King William III (1689–1702) and Queen Anne (1702–1714). | | |
| | William Portal (Sr.) | Prittlewell, County of Essex | c1720–c1730 |
| | William Portal (Sr.) | South Hambridge, County of Essex | c1730–c1750 |
| Goldsmith | Abraham Portal | Prittlewell, County of Essex | Bc1726–1748 |
| Jeweler | • Son of Guilleume (William) Portal. | | |
| Toy (hardware) | • Apprenticed to Paul De La Merie, goldsmith to the king, (1740–1748). | | |
| maker | Abraham Portal | Rose St., Soho, | 1749 |
| Hilt maker | | at the sign of the Salmon and Pearl | |
| Sword cutler | Abraham Portal | Precinct of the Savoy | 1750–1752 |
| | Abraham Portal | Strand | 1753–1762 |
| | • Had apprentice William Portal Jr. (brother), 1750–1760. | | |
| | Abraham Portal & Harry Gearing | 34 Ludgate Hill | 1763–1798 |
| | Abraham Portal | Castle St., Holborn | 1779–D1809 |
| Goldsmith | William Portal (Jr.) | South Hambridge, County of Essex | Bc1736–1760 |
| Jeweler | • Son of Guilleume (William) Portal. | | |
| Toy (hardware) | • Apprentice to brother Abraham Portal, 1750–1760. | | |
| maker | William Portal | Orange St., near Leicester Square | 1761–1767 |
| Sword cutler | William Portal & James King | Orange St., near Leicester Square | 1768–1771 |
| | William Portal | Orange St., near Leicester Square | 1772–D1815 |
| | (see James King) | | |
| Sword cutler | John Porter | | 1580–1640 |
| | • Became Freeman of the Cutlers Company, 1603. | | |
| | • Master of the Cutlers Company, 1619–1621. | | |
| | • Supplied swords to the Board of Ordnance, 1621–1625. | | |
| | • On June 13, 1621, by order of the court of the Cutlers Company of London, master John Porter, past master Thomas Chesshire, and royal cutler Robert South were to inspect a sword blade mill, believed to be the mill of Thomas Murrey (see Thomas Murrey), and report their findings to the court at the next meeting. | | |
| Cutler | Thomas Porter | | 1582–1612 |
| | • Master of the Cutlers Company, 1592, 1593, 1606, and 1607. | | |
| Silversmith | E. (Edward) Potter (Sr.) | 15 Vere St., Clare Market | 1808–1826 |
| Goldsmith | • Successor to Hunt & Potter (see Hunt & Potter). | | |
| Jeweler | Thomas Potter | 15 Vere St., Clare Market | 1827–1830 |
| Gilders | Thomas & Edward (Jr.) Potter | 15 Vere St., Clare Market | 1831–1833 |
| Hilt maker | Thomas Potter | 15 Vere St., Clare Market | 1834–1835 |
| Sword cutler | Thomas Potter | 26 New Castle St., Strand | 1836–1837 |
| Gun maker | Thomas Potts | 70 Minories | 1828–1838 |
| Goldsmith | • Successor to Brander & Potts (see Brander & Potts). | | |
| Sword cutler | Thomas Potts | Hayden Square, Minories | 1839–1849 |
| | • Shop manager: William Scott. | | |

| | Thomas Potts | Hayden Square, Minories | 1850–1852 |
|---|---|---|---|
| | • Sword cutler and gun maker to the Board of Ordnance and Honorable East India Company. | | |
| Gun maker<br>Goldsmith<br>Sword cutler | Henry Potts | 32 & 33 Leman St.<br>& 27 Tentor Ground,<br>Goodmans Fields | 1845–1853 |
| | Potts (Henry) & Hunt (Thomas) | 32 & 33 Leman St.<br>& 27 Tentor Ground,<br>Goodmans Fields | 1854–1875 |
| | Thomas Hunt | 27 Tenter St. East | 1876–1880 |
| Sword cutler | William Poulton | Parish of St. Dunston in the West | 1625–1635 |
| Tailor<br>Sword cutler | Arthur John Powell | 18a London St. | 1908 |
| | • Successor to Hole & Powell (see Hole & Powell). | | |
| | • Succeeded by Glen & Powell (see Glen & Powell). | | |
| Sword cutler | Capt. Thomas Powell | | 1690 |
| Jeweler<br>Goldsmith | Thomas Powell | Parish of St. Martin Le Grand<br>Mouldmaker Row | 1756 |
| Silversmith | Thomas Powell | New Court, New St., Fetter Lane | 1757 |
| Hilt maker | Thomas Powell | Bolt Court, Fleet St. | 1758–1759 |
| Sword cutler | Thomas Powell | Peacock St., Gutter Lane, Cheapside | 1760–1772 |
| | Thomas Powell | Craigs Court, Charing Cross | 1773–1800 |
| Cutler | William Powell | | 1707–1782 |
| | • Master of the Cutlers Company, 1717. | | |
| Army clothier | William Prater | 7 Charing Cross | 1781–1783 |
| Linen draper | William Prater | 6 Charing Cross | 1784–1799 |
| (maker) | William & Charles Prater | 6 Charing Cross | 1800–1827 |
| Accoutrement | Charles Prater | 2 Charing Cross | 1828–1835 |
| maker | Charles Prater & Son | 2 Charing Cross | 1836–1849 |
| Sword cutler | C. (Charles) Prater & Co. | 2 Charing Cross | 1850 |
| Cutler | Benjamin Pratt | | 1894–1909 |
| | • Master of the Cutlers Company, 1904. | | |
| Goldsmith<br>Hilt maker<br>Sword cutler | James Preist | Parish of St. Sepulchre<br>without Newgate | 1621–1636 |
| Armourer<br>Brasier | William Preist (Sr.) | | 1710–1740 |
| | • Father of William Preist (Jr.) and James Preist. | | |
| Goldsmith<br>Hilt maker<br>Sword cutler | William Preist (Jr.) | | Bc1726–1747 |
| | • Son of William Preist Sr. | | |
| | • Apprentice to Richard Gurney, 1740–1747 (see Richard Gurney). | | |
| | William Preist (Jr.) & William Shaw<br>(see William Shaw) | Maiden Lane | 1748–1750 |
| | William Preist (Jr.) & William Shaw | corner of Lad Lane & Wood St. | 1751–1762 |
| | William Preist (Jr.) | corner of Lad Lane & Wood St. | 1763 |
| | William (Jr.) & James Preist | 30 Whitecross St. | 1764–1781 |
| | • James Preist, son of William Preist Sr., born c. 1736. | | |

| | William Preist (Jr.) | Hackney St. | 1782–D1802 |
| --- | --- | --- | --- |
| | James Preist | Watling St. | 1782–1810 |
| | • Had apprentice James Hammond Preist (son), 1783–1790. | | |
| Sword cutler | William Preist | 93 Albany St., Regents Park | 1829–1830 |
| Military tailor Sword cutler | Preston (William) & Son (John) | 126 New Bond St. | 1842–1848 |
| | • Successors to King & Preston (see Thomas King). | | |
| | William & John Preston | 126 New Bond St. | 1849–1856 |
| | • Appointed tailor to His Majesty the King and His Royal Highness the Prince of Wales. | | |
| | Preston (William & John), Taylor (John Walker) & Forder (James) | 126 New Bond St. | 1857–1860 |
| | Preston (William & John), Taylor (John Walker) & Co. | 126 New Bond St. | 1861 |
| | • Succeeded by John Walker Taylor (see John Walker Taylor). | | |
| Sword cutler | Edward Price | | 1620 |
| Silversmith Hilt maker Sword cutler | Thomas Price | | 1661 |
| | • In 1661, signed an advertisement warning sword purchasers about counterfeit cast silver hilts. | | |
| | John Price (see Richard Bryan) | | |
| Silversmith Hilt maker Sword cutler | Thomas Price | 13 Weston St., Pentonville | 1802–1815 |
| | • Mounted a sword for Rundell, Bridge & Rundell, which was awarded to Lt. Gen. Vicount Wellington by the city of London for the Peninsula Campaign, 1811. (see Rundell, Bridge & Rundell) | | |
| Sword cutler | William Price | 8 New St., Shoe Lane | 1790–1792 |
| | William Price | 44 Fetter Lane, Fleet St. | 1793–1797 |
| | John Prigg (see Thomas Hawgood) | | |
| Silversmith Knife and] sword hilter Knife and sword cutler | Joseph Pritchard | 4 Swinton Place, Bagnage Road | 1825–1826 |
| | Joseph Pritchard | 28 Steward St., Goswell St. | 1827–D1831 |
| | Sarah Pritchard | 28 Steward St., Goswell St. | 1831–1835 |
| | • Widow of Joseph Pritchard. | | |
| Gun maker Sword cutler | R. (Richard) Ellis Pritchett | 37 Chamber St. | 1814–1831 |
| | R. (Richard) Ellis Pritchett | 59 Chamber St., Goodmans Fields | 1832–1851 |
| | • Retired in 1851 died in 1866. | | |
| | Robert Taylor Pritchett | 86 St. James St. & 59 Chamber St. & 24 Prescot St., Goodmans Fields | 1851–1860 |
| | Robert Taylor Pritchett | 4 St. James St. & 59 Chamber St. & 24 Prescot St., Goodmans Fields | 1863–1865 |
| | • Retired in 1865 and died in 1907. | | |
| Sword belt maker | John Prosser (Sr.) | | B1769–1790 |
| | • Probably the son of sword cutler Thomas Prosser. | | |
| Army cap, helmet and accoutrement | John Prosser (Sr.) | Parish of St. James, Westminster | 1790–April 1795 |
| | • In April 1795, Mary Cullum, widow of Thomas Cullum, in poor health, hired Prosser as manager of her shop at Charing Cross. | | |

| | | | |
|---|---|---|---|
| maker<br>Goldsmith<br>Laceman<br>Sword cutler<br>Gun dealer | **John Prosser (Sr.)**<br>• Manager of Mary Cullum's shop.<br>• Mary Cullum, before she died, sold her shop to John Prosser in June. | 9 Charing Cross,<br>Parish of St. Martins in the Field,<br>Westminster | April 1795–June 1795 |
| | **John Prosser (Sr.)** | 9 Charing Cross,<br>Parish of St. Martins in the Field,<br>Westminster | June 1795–D1837 |

- Shop located down the street from the Admiralty.
- On July 9, 1795, appointed sword cutler and belt maker to King George III.
- In April 1796, Prosser registered his mark "I.P." in the small workers book at Goldsmith Hall.
- Appointed sword cutler to the Board of Ordnance.
- Sold M1814 household cavalry swords.
- Sold M1821 light and heavy cavalry sabers to the Board of Ordnance with J.J. Runkel blades.
- Sold M1796 foot officers swords.
- Sold half-basket guard officers swords for Windsor Castle lifeguards (king's guards).
- Repaired and remounted swords.
- In November 1807, made 27 swords for the officers of the 10th Light Dragoons.
- Sold large quantities of army and navy officers swords.
- Sold presentation swords with silver, gold, and cast brass hilts.
- Sold flintlock dueling pistols.
- Goldsmith Richard Teed mounted some of Prosser's presentation swords.
- An early invoice (1790s) read:
    *John Prosser, sword cutler to the King and the Royal Family &*
    *His Royal Highness Lord High Admiral*
- An 1814 invoice read:
    *John Prosser, Manufacturer of Arms & accoutrements to the*
    *King and their Royal Highnesses the Prince of Wales*
    *The Dukes of York, Kent & Cumberland*
- The Prince of Wales (son of King George III) owned more than 60 swords made by Prosser, who was a colonel and supplier of swords, uniforms, etc. to his regiment, the 10th Light Dragoons (Hussars).
- An 1825 invoice read:
    *John Prosser, Army Clothier, cap, accoutrement and helmet maker,*
    *Sword manufacturer & laceman to the King, the Royal Family, officer's*
    *Regimentals and other equipment for the Cavalry, Artillery, Infantry and*
    *East India cadets*
    *Sells    Soldiers knapsacks*
    *Soldiers shoes*
    *Soldiers shirts*
    *Soldiers gaiters*
    *Soldiers trowsers*
    *Soldiers forage caps*
    *Soldiers black socks*
    *And every article of necessaries*
- Had a warehouse at No. 1 Trinity Place, Charing Cross, 1814–1838.
- Sword cutler and belt maker to King George III, King George IV, and Queen Victoria.
- John Prosser (Sr.) died in 1837.

| | | | |
|---|---|---|---|
| Sword belt maker<br>Army cap helmet and accoutrement | **John Prosser (Jr.)**<br>• Son of John Prosser Sr.<br>• Sword cutler to the royal family.<br>• Also made children's swords. | 9 Charing Cross,<br>Parish of St. Martin in the Field,<br>Westminster | B1806, 1837–1852 |
| | **John Prosser (Jr.)** | 37 Charing Cross | 1853–D1860 |

| | | | |
|---|---|---|---|
| maker<br>Goldsmith<br>Laceman<br>Sword cutler | • Tennant of Henry Tatham Jr. (see) at 37 Charing Cross, but operated his own business.<br>• John Prosser Jr. died in 1860. | | |
| Steel worker<br>Sword cutler | **Thomas Prosser**<br>• Probably father of John Prosser Sr. | 21 Hyde St., Holborn | 1705–1795 |
| Cutler | **John Pryor**<br>• Master of the Cutlers Company, 1859. | | 1849–1862 |
| Cutler | **William Pryor**<br>• Master of the Cutlers Company, 1816 and 1830. | | 1806–1835 |
| Cutler | **William Squire Pryor**<br>• Master of the Cutlers Company, 1857. | | 1847–1852 |
| Draper<br>(cloth maker)<br>Military<br>tailor<br>Army and navy<br>contractor<br>Sword cutler | **Pulford (Robert) & Nicoll (Henry)**<br>**Robert Pulford & Co.**<br>**Robert Pulford**<br>**Robert Pulford**<br>**Robert Pulford**<br>**Robert Pulford & Son (Frederick)**<br>**Robert Pulford & Sons (Frederick & Frank)**<br>**Robert Pulford & Sons (Frederick & Frank)**<br>**Robert Pulford & Sons (Frederick & Frank)**<br>• At 33 Bruton Street, Pulford was successor to Roger & Co. Ltd. and John Jones & Co. (see Roger & Co. Ltd.; John Jones & Co.). | 26 St. James St.<br>23 St. James St.<br>23 St. James St.<br>65 St. James St.<br>65 St. James St.<br>65 St. James St.<br>65 St. James St.<br>19 Albemarle St.<br>33 Bruton St. | 1822–1823<br>1824–1825<br>1826–1827<br>1828–1838<br>1839–1843<br>1844–1897<br>1898–1932<br>1933–1961<br>1962–1965 |
| Gun maker<br>Sword cutler | **James Purdey (Sr.)**<br>• Worked for Joseph Manton (Forsyth Patent Gun Co.), 1808–1812.<br>**James Purdey (Sr.)**<br>**James Purdey (Sr.)**<br>**James Purdey (Sr.)**<br>• James Purdey (Sr.) retired in 1857 and died in 1863.<br>**James Purdey (Jr.) & Sons**<br>**James Purdey (Jr.) & Sons**<br>**James Purdey (Jr.) & Sons**<br>• James Purdey Jr. died in 1909. | <br><br>15 New Grey Coat Place, Tothill Fields<br>4 Princes St., Leicester Fields<br>314 1/2 Oxford St.<br><br>314 1/2 Oxford St.<br>314 1/2 Oxford St.<br>Audley House, 58 South Audley St. | B1784–1812<br><br>1813–1814<br>1815–1827<br>1828–1857<br><br>1857–1876<br>1877–1881<br>1881–1909 |
| Cutler | **Robert Pykemere**<br>• Master of the Cutlers Company, 1468, 1469, 1476, 1477, 1482, and 1483. | | 1458–1488 |
| Sword cutler | **J.F. Raab** | | 1806 |
| Bladesmith | **Thomas Racton**<br>• Master of the Bladesmiths Company, 1425. | | 1420–1430 |
| Goldsmith<br>Hilt maker<br>Sword cutler | **John Radburn** | New St., Fetter Lane | 1762–1773 |
| Sword cutler | **Andrew Ragdale** | | 1698–1725 |
| Sword cutler | **John Raines**<br>• Supplied swords to the Board of Ordnance, 1656. | | 1650–1660 |

| | | | |
|---|---|---|---|
| Bladesmith | **Adam Ramme** | | 1411 |
| Cutler | **Robert Rastrick**<br>• Master of the Cutlers Company, 1670. | | 1660–1675 |
| Goldsmith<br>Hilt maker<br>Sword cutler | **Andrew Raven**<br>• Mounted small swords and hunting swords. | Parish of St. Martins Le Grand | 1697–1700 |
| Sword cutler | **Ravenscroft & Sons** | | 1827 |
| Accoutrement maker<br>Medalist<br>Gun dealer<br>Sword cutler | **William Rawle**<br>• Sold silver-mounted pistols aand powder flasks. | Corner of Castle Ct., Strand | 1769–D1789 |
| Silversmith<br>Hilt maker<br>Sword cutler | **Valentine Rawle** | 23 Great Suffolk St., Charing Cross | 1792–1806 |
| Silversmith<br>Hilt maker<br>Sword cutler | **Richard Rawlings** | 2 Sparrows Rents, Pearpoole Lane | 1777–1787 |
| Goldsmith<br>Jeweler<br>Diamond and pearl dealer<br>Hilt maker<br>Sword cutler | **Ray (John) & Montaque (James)**<br>**(John Ray & Co.)**<br>• Successors to Morisset & Lukin (see James Morisset).<br>• Large maker of presentation small swords and "freedom boxes" (award/presentation snuff boxes), especially for the war against France. Many were presented by the city of London.<br>• Mounted many presentation swords sold by other goldsmiths and retailers, including R. Makepeace; Thomas Ayres; Ayres & Bennett; John White & Co.; James Birt; Jeffreys & Gilbert; Edinburgh; Thomas Harper; Rundell & Bridge; T. & S. Goldney; Charles Aldridge; Rundell, Bridge & Rundell; W. Moore; Goodbehere, Wigan, Bult; Green & Ward; C. Aldridge; Green, Ward & Green; and R. Clarke.<br>• Succeeded by James Montaque at 22 Denmark Street (see James Montaque). | 22 Denmark St., Soho | 1801–1821 |
| Jeweler<br>Sword cutler | **John Ray** | 25 Hart St., Parish of St. George, Bloomsbury | 1824–1832 |
| Sword cutler | **Thomas Raymond** | | 1430–D1461 |
| Goldsmith<br>Hilt maker<br>Buckle maker<br>Sword cutler | **Thomas Read**<br>**Thomas Read & Thomas Smith**<br>**Thomas Read**<br>**Thomas Read**<br>**Thomas Read** | <br>17 Bartholomew Close<br>17 Bartholomew Close<br>Caley Court, Holborn<br>4 Parliament St. | Bc1750–1770<br>1771–1773<br>1774–1775<br>1776–1800<br>1801–1820 |
| Silversmith<br>Hilt maker<br>Sword cutler | **John Reason**<br>• Brother of Joseph Reason.<br>• Apprentice to cutler John Whitton, 1678–1685.<br>• Journeyman to cutler John Whitton, 1685–1697. | | Bc1664–1697 |

|  |  |  |  |
|---|---|---|---|
|  | John Reason | Heart St., Covent Garden | 1698–1710 |
| Silversmith<br>Hilt maker<br>Sword cutler | Joseph Reason<br>• Brother of John Reason.<br>• Journeyman to Daniel Wilson, c. 1690–1696 (see Daniel Wilson).<br>Joseph Reason<br>• Had apprentice Thomas Bass, c. 1700–1707 (see Thomas Bass). | <br><br><br>Burleigh St., Strand | Bc1676–1696<br><br><br>1697–1773 |
| Sword cutler | Michael Reason<br>• From August 1557 to January 1558, supplied swords to the Drapers Company for defense of their settlement in Calais, France, during the French War. |  | 1550–1560 |
| Gun and<br>accoutrement<br>maker<br>Fine steel<br>worker<br>Steel hilt maker<br>Sword cutler | George S. Reddell<br>G. (George) S. Reddell<br>G. (George) S. Reddell<br>G. (George) S. Reddell<br>• Gun maker and sword cutler to His Royal Highness the Prince Regent, the Duke of Sussex, and the Duke of Cambridge. | 138 Jermyn St., Parish of St. James<br>47 Piccadilly<br>236 Piccadilly<br>28 Leadenhall St. | 1800–1810<br>1811–1813<br>1814–1819<br>1820–1821 |
| Sword cutler | Richard Rees<br>• Master of the Cutlers Company, 1826. |  | 1816–1831 |
| Sword cutler | Richard James Rees |  | 1832–1865 |
| Gun maker<br>Sword and<br>bayonet maker | Charles Reeves<br>• Successor to Isaac Hebberd (see Isaac Hebberd).<br>• Sword, cutlass, and bayonet maker.<br>• Became part of Wilkinson & Son in 1880 but kept the Reeves name.<br>(see Wilkinson & Son)<br>Charles Reeves & Co. Ltd.<br>Charles Reeves & Co. Ltd.<br>Charles Reeves & Co. Ltd.<br>(see Birmingham listing) | 8 Air St., Piccadilly<br><br><br><br><br>18 St. Martins St.<br>9 West St., Golden Square<br>9 Newburgh St., Regent St. | 1853–1900<br><br><br><br><br>1901–1909<br>1910–1937<br>1938–1955 |
| Cutler | Thomas Reid<br>• Master of the Cutler Company, 1765. |  | 1755–1770 |
| Rifle and pistol<br>maker<br>Goldsmith<br>Jeweler<br>Hilt maker<br>Sword maker | Joseph Charles Reilly<br>Joseph Charles Reilly<br>• Son Edward Michael Reilly began working for him, 1811.<br>Joseph Charles Reilly<br>Edward Michael Reilly<br>E. (Edward) M. (Michael) Reilly & Co.<br>• Began making rifles and pistols.<br>E. (Edward) M. (Michael) Reilly & Co.<br>E. (Edward) M. (Michael) Reilly & Co.<br>E. (Edward) M. (Michael) Reilly & Co.<br>E. (Edward) M. (Michael) Reilly & Co. | 12 Middle Row, Holborn<br>316 High Holborn<br><br>502 Oxford St.<br>502 Oxford St.<br>315 Oxford St.<br><br>16 New Oxford St.<br>277 New Oxford St.<br>277 & 295 New Oxford St.<br>295 New Oxford St. | 1816–1835<br>1836–1847<br><br>1848–1855<br>1856–1860<br>1861–1881<br><br>1882–1898<br>1899–1902<br>1903<br>1904–1908 |
| Mathematical<br>instrument<br>maker<br>Sword cutler | Isaac Rennoldson<br>Isaac Rennoldson | 23 Camden Row, Bethnal Green Rd.<br>Cambridge Place, Hackey Road | 1814–1825<br>1826–1847 |

| Role | Name | Address | Dates |
|---|---|---|---|
| Saddler<br>Army accoutrement maker<br>Sword cutler | Rentor & Frazer & Co.<br>• Sold naval cutlasses. | 29 Coventry St. & 48 Great Pulteney St. | 1811–1817 |
| Silversmith<br>Hilt maker<br>Sword cutler | John Reynolds<br>John Reynolds<br>John Reynolds | New Street, Fetter Lane<br>25 New St.<br>36 Little Old Bailey | 1768–1772<br>1773–1774<br>1775–1780 |
| Cutler | James Alfred Rhodes<br>• Master of the Cutlers Company, 1899. | | 1889–1904 |
| General cutler<br>Jeweler<br>Fine steel worker<br>Goldsmith<br>Silversmith<br>Hilt maker<br>Sword cutler | William Riccard<br>Riccard (William) & Littlefear (Samuel)<br>• A 1783 trade card read:<br>   *The neatest & newest fashion of all sorts of cutlery*<br>   *Curious steel mounted swords*<br>   *Silver mounted swords*<br>   *Shooting instruments of all sorts*<br>   *Steel, silver, plated & chain spurs* | Castle St., Leicester Fields<br>Castle St., Upper Mews Gate | 1774–1781<br>1782–1793 |
| Gun maker<br>Sword cutler | Thomas Richards<br>• Jeweler only.<br>• Retired in 1774 and died in 1779. | 114 Strand | 1771–1774 |
| | John Richards<br>John Richards<br>John Richards<br>John Richards<br>John Richards | 114 Strand<br>25 Strand<br>54 Strand & 4 Ball Alley, Lombard St.<br>54 Strand<br>55 Strand | 1775–1777<br>1778–1780<br>1781–1794<br>1795–1808<br>1809–D1821 |
| Gun maker<br>Sword maker<br>Gun and shot warehouse | Richards (Henry) & Hall (Samuel)<br>Henry Richards<br>Henry Richards<br>• Exported swords to Richards (Stephen) & Upson (George), a relative in New York, USA.<br>• Sold sword blades to John Salter at 35 Strand (see John Salter). | 39 Fish St. Hill<br>39 Fish St. Hill<br>125 Strand | 1807<br>1808–1809<br>1810–1816 |
| Goldsmith<br>Hilt maker<br>Buckle maker<br>Sword cutler | James Richardson<br>• Master of the Cutlers Company, 1763.<br>Thomas Richardson<br>• Probably son of James Richardson. | 3 Racquet Court, Fleet St.<br><br>447 Strand | 1753–D1778<br><br>1778–1800 |
| Sword cutler | Richardson & Co.<br>• Sold swords made by the Wilkinson Sword Company. | 26 King St., St. James | 1920–1930 |
| Gun maker<br>Sword maker | William Ridgeway<br>• Supplied swords to the Board of Ordnance, 1640–1658.<br>• Also repaired swords for the Board of Ordnance.<br>Elizabeth Ridgeway<br>• Widow of William Ridgeway. | Parish of St. Katherine<br><br><br>Parish of St. Katherine | 1630–D1661<br><br><br>1661–1665 |
| Sword maker | Henry Risby<br>Henry Risby | London<br>Hounslow Heath | c1612–1633<br>1633–1649 |

| | Henry Risby<br>(see Chapter 3) | London | 1649–1650 |
|---|---|---|---|
| Goldsmith<br>Hilt maker<br>Sword cutler | Robert Ritherdon<br>• Made the sword presented to Gen. Sir David Baird in 1806 for his victory at Seringapatem, India.<br>• Richard Teed decorated some of Ritherdon's presentation swords (see Richard Teed). | 3 Aldergate | 1800–1810 |
| Sword maker<br>Armourer | Robert of Ipswich | | 1301 |
| Silversmith<br>Hilt maker<br>Sword cutler | John Roberts | Threadneedle St. | 1716–1726 |
| Silversmith<br>Hilt maker<br>Sword cutler | Thomas Roberts<br>Thomas Roberts | Bread St.<br>Newgate St. | 1703–1719<br>1720–1730 |
| Sword cutler | John Robins | Parish of St. Mary at the Hill | 1598–D1615 |
| Sword cutler | Robinson & Green | Hanover Square | 1850–1861 |
| Sword cutler | Walter Robotham | Popes Head Alley | 1710–1720 |
| Sales agent<br>for a sword<br>maker | Rochussen & Co.<br><br>Rochussen & Co.<br><br>Rochussen & Co.<br>Heintzman & Rochussen<br>Heintzman & Rochussen<br>• Sales agents for sword dealer H.W. Lang of Solingen, Prussia, 1851–1869. | 13 Great St.,<br>Parish of St. Thomas Apostle<br>15 Cannon St., West End, & 13 Great St.,<br>Parish of St. Thomas Apostle<br>114 Fenchurch St.<br>9 Friday St.<br>23 Abchurch Lane | 1857–1858<br><br>1858–1861<br><br>1861–1863<br>1863–1867<br>1867–1869 |
| Bladesmith | Robert Roderam (Ruderam)<br>• Master of the Bladesmiths Company, 1376. | | 1370–1380 |
| Sword cutler | Davie Rogers | | 1606 |
| Sword cutler | John Rogers | Parish of St. Clements Danes | 1622–1637 |
| Cutler | Joseph Rogers<br>• Master of the Cutlers Company, 1636. | | 1626–1641 |
| Sword maker | Thomas Rogers<br>• Repaired swords for the Board of Ordnance, 1625–1641. | | 1620–1650 |
| Gold and silver<br>lacemen<br>Embroiderer<br>Army and navy<br>cap maker<br>Sword cutler | William & John Rogers<br>Hamburger, Rogers & Co.<br>• Successor to Hamburger, Harwood & Co. (see Hamburger & Co.)<br>• William and John Rogers were partners, 1840–1870.<br>• William and John Rogers became owners in 1870–1917 but kept the Hamburger, Rogers name.<br>• Appointed lacemen and embroiderers to His Majesty the King and His Royal Highness the Prince of Wales.<br>Rogers & Co. Ltd.<br>Rogers & Co. Ltd.<br>Rogers & Co. Ltd. | <br>30 King St., Covent Garden<br><br><br><br><br>8 New Burlington St.<br>57–58 Jermyn St.<br>57 Jermyn St. | 1820–1840<br>1840–1917<br><br><br><br><br>1918–1941<br>1942–1953<br>1954–1960 |

|  |  |  |  |
|---|---|---|---|
|  | Rogers & Co. Ltd. & John Jones & Co. | 33 Bruton St. | 1961–1962 |
|  | • At 33 Bruton Street, successors to John Jones & Co. Ltd. | | |
|  | • Succeeded by Robert Pulford & Son (see Pulford & Nicoll). | | |
| Jeweler | William Rogers | Bath | 1740–1769 |
| Goldsmith | • Had apprentice Philip Rundell, 1760–1765 (see Philip Rundell). | | |
| Hilt maker | William Rogers | Ludgate Hill | 1770–1780 |
| Sword cutler | | | |
| Sword cutler | Rogerson & Andrews | | 1822 |
| Sword cutler | Joseph Thomas Rolph | | 1807 |
|  | Edward Rolph | | |
|  | (see John Bennett Sr.) | | |
| Goldsmith | E. (Emick) Romer | Oslo, Norway | B1724–c1760 |
| Hilt maker | • Immigrated to London, c. 1760. | | |
| Sword cutler | E. (Emick) Romer | 123 High Holborn | c1760–1779 |
|  | • Sold silver work to Parker & Wakelin (see Parker & Wakelin). | | |
|  | • John Christopher Romer (son of Emick Romer) worked for Parker & Wakelin. | | |
| Goldsmith | Michael Rooke (Rook) | 15 Little New St., Shoe Lane, | 1796–1799 |
| Hilt maker | Rooke (Michael) & Co. | Savile Row | 1800–1810 |
| Sword cutler | | | |
| Sword cutler | Christopher Rooke (Rook) | Falcon St. | 1897–1900 |
|  | • Successor to Simpson & Rooke (see Simpson & Rooke). | | |
| Hatter | Thomas Rose | Ludgate Hill | 1703–D1728 |
| Sword cutler | | at the sign of Two Golden Eagles | |
|  | Robert Roskell | | |
|  | (see Hunt & Roskell) | | |
| Goldsmith | John Rowe | | Bc1727–1748 |
| Silversmith | • Apprenticed to Richard Bayley, 1741–1748 (see Richard Bayley). | | |
| Hilt maker | John Rowe | Gutter Lane, Royal Exchange | 1749–1773 |
| Sword cutler | John Rowe | Monkwell St. | 1774–1780 |
|  | • Mounted silver-hilted small swords. | | |
| Military | Owen Rowe (Row) | | 1642–D1661 |
| outfitter | • In 1645, was a colonel in the London militia. | | |
| Armour and | • On April 3, 1645, contracted for 1,000 suits of armour (breast plate, back plate, and | | |
| pike cutler | helmet) at 20 shillings a suit | | |
|  | • On April 9, 1645, contracted for 1,000 long pikes (16 feet long with steel heads) at 4 shillings each. | | |
|  | • The armour and pikes were contracted for the "New Model" army (parliamentary forces) under | | |
|  | Gen. Sir Thomas Fairfax during the English Civil War. | | |
|  | Anthony Rowed | | |
|  | (see Nicoll & Franklin) | | |
| Military | Charles Rowley & Co. | 13 Archer St., Haymarket | 1864–1865 |
| accoutrement | • Successor to John Lilly (see John Lilly). | | |
| maker | | | |
| Sword cutler | | | |

| Trade | Name | Location | Dates |
|---|---|---|---|
| Goldsmith<br>Hilt maker<br>Sword cutler | **James Rowse**<br>• Hilted small swords. | Bond St. | 1760–1770 |
| Military<br>outfitter<br>Sword cutler | **Richard Rumsey**<br>• On January 10, 1645, contracted for 200 carbine belts and 200 cartridge girdles.<br>• On July 17, 1645, contracted for approximately 700 bandoleers. On August 3, 1645, contracted for 500 swords (infantry) with belts and "Dutch" (Solingen, Prussia) blades at 4 shillings, 6 denier each.<br>• All contracts were for the "New Model" army (parliamentary forces) under Gen. Sir Thomas Fairfax. | Holborn | 1642–1649 |
| Goldsmith<br>Jeweler<br>Gun dealer<br>Hilt maker<br>Sword and<br>knife cutler | **Philip Rundell**<br>• Apprentice to William Rogers (goldsmith and jeweler of Bath), 1760–1766.<br>• Shopman to Thread (Henry) & Picket (William), 1767–1771 (see Thread & Picket). | Bath | B1743–1771 |
| | **Picket (William) & Rundell (Philip)**<br>• Successors to Thead & Picket (see Thead & Picket). | 32 Ludgate Hill<br>at the sign of the Golden Salmon | 1772–1774 |
| | **William Picket & Philip Rundell**<br>• Had apprentice Samuel Goldney of Bath, 1784–1791.<br>• Goldney had married Philip Rundell's sister, Eleanora. | 32 Ludgate Hill | 1775–1784 |
| | **Philip Rundell** | 32 Ludgate Hill | 1785–1787 |
| | **Philip Rundell & John Bridge** | 32 Ludgate Hill | 1788–1803 |
| | **Rundell (Phillip) & Bridge (John) & Co.**<br>• Appointed goldsmith and jeweler to His Majesty the King, 1797. | 32 Ludgate Hill | 1804 |
| | **Rundell (Philip), Bridge (John) &<br>Rundell (Edward Walker)**<br>• Edward Walker Rundell was Philip Rundell's nephew.<br>• Partner: William Thead.<br>• Paul Storr of Storr & Co. ran the manufacturing workshop for the company of 53 Dean Street, 1807–1819 (see Paul Storr).<br>• Paul Storr and William Thead left in 1819.<br>• James Morisset and Ray & Montaque mounted some of the swords sold by Rundell & Bridge and Rundell, Bridge & Rundell (see Ray & Montaque). | 32 Ludgate Hill | 1805–1818 |
| | **Rundell (Phillip), Bridge (John)<br>& Rundell (Edward Walker)**<br>• Their trade card read:<br>    *All sorts of Turkish ornaments in the newest taste after the most approved drawings and models from their agents in Costantinople Rich diamond and enamelled sabres, daggers, knives, snuff boxes, watches, pistols, etc.* | 32 Ludgate Hill &<br>76 Dean St. (Philip Rundell only) | 1819–1829 |
| | **Rundell (Philip), Bridge (John)<br>& Rundell (Edward Walker)**<br>• Appointed goldsmith and jeweler to His Majesty the King and the royal family.<br>• The location at 75–76 Dean Street was managed by Philip Rundell only.<br>• Philip Rundell retired in 1833 and died in 1837. | 32 Ludgate Hill &<br>75–76 Dean St. | 1830–1833 |
| | **Rundell (Edward), Bridge (John) & Co.** | 32 Ludgate Hill | 1834–1842 |
| Merchant<br>Military<br>importer<br>Sword and<br>sword blade<br>importer | **J. (Johann) J. (Justus) Runkel**<br>• Immigrated to London, England, in 1783. | Renysdorff Parish,<br>Neuweid, Prussia | Bc1751–1783 |
| | **J. (John) J. (Justus) Runkel**<br>• Huge importer of Solingen, Prussia, sword blades and swords, which he sold to many English cutlers (sword assemblers), goldsmiths, and silversmiths (sword mounters).<br>• Also sold many complete swords to English sword retailers.<br>• In 1786, a quality trial was arranged by the East India Company, demanded by English tool, sword blade, and sword maker Thomas Gill. | 8 Tookes Ct.<br>Castle St., Holborn | 1784–D1808 |

- Of 1,400 Runkel blades tested, 28 were rejected (see Thomas Gill).
- Became a naturalized British subject, 1796.
- Became a freeman of the Cutlers Company, 1790.
- Imported 500 swords and 100 blades from Solingen, Prussia, 1803.
  They were shipped from Enden, Saxony, a large seaport that handled products being exported from the Ruhr Valley, on the ship *Vrouw Maria*.
- The swords and blades were ordered for the use of the Prince of Wales' regiment of dragoons.
- In 1806, 1,650 swords and blades imported by Runkel were confiscated because they were undervalued when he tried to reduce his import tax levied by the British government.
- Runkel was required to pay a 5 pound per unit penalty.
- John Justus Runkel died in 1808.

| Trade | Name | Address | Dates |
|---|---|---|---|
| Sword and bayonet maker | **William Rush** | | 1637–1683 |
| | • Used "pomegranate" mark, 1637–1664. | | |
| | • Used "bunch of grapes with long stem" mark, 1664–1683. | | |
| | • Made plug bayonets. | | |
| Cutler | **Edmund Walter Rushworth** | | 1889–1904 |
| | • Master of the Cutlers Company, 1899. | | |
| Cutler | **Anthony Russell** | | 1710–1725 |
| | • Master of the Cutlers Company, 1720. | | |
| Gold and silver lacemen | **Russell & Allen** | 18 Old Bond St. | 1853–1872 |
| | • Successors to M. & I. Witton (see M. & I. Witton). | | |
| Silk mercer (dealer) | **Russell & Allen** | 18 & 19 Old Bond St. | 1873–1875 |
| Sword cutler | **Russell & Allen** | 17–18–19 Old Bond St. | 1876–1906 |
| Sword cutler | **Charles Frederick William Rust** | 129 London Wall | 1860–1874 |
| Sword cutler | **R. (Robert) Rutherdon** | | 1804–1810 |
| Bladesmith | **Nicholas Ryngwode (Rindwode)** | | 1435 |
| Bladesmith | **Richard Ryngwode (Ringwode)** | | 1410–1420 |
| | • Master of the Bladesmith Company, 1416. | | |
| Goldsmith Silversmith Jeweler | **John Salter** | | Bc1760–1800 |
| | • Worked for Joseph & Edward Greensill, c. 1780–1800. | | |
| Hilt maker Sword cutler | **John Salter** | 35 Strand | 1801–1824 |
| | • Successor to Joseph & Edward Greensill (see Joseph & Edward Greensill). | | |
| | • Sword cutler to the Duke of Sussex, Lord Admiral Horatio Nelson, and Lord Exmouth. | | |
| | • Sword cutler to the Patriotic Fund. | | |
| | • Large exporter of swords, including eaglehead pommel infantry and artillery officers swords, to United States sword dealers, including Lemuel Wells, A.W. Spies, and Richards, Upson, all of New York. | | |
| | • Bought sword blades from Henry Richards, Henry Osborne, and the Gills of Birmingham. | | |
| | **John Salter** | 73 Strand | 1825–1829 |
| | **Salter (John) & Co.** | 73 Strand | 1830–1831 |
| | **Salter (John), Widdowson (William) & Tate (Henry)** | 73 Strand | 1832–1834 |
| | • Succeeded by Widdowson & Veal (see Widdowson & Veal). | | |
| Army tailor | **Samuel Bros.** | 29 Ludgate Hill | 1830–1865 |

| | | | |
|---|---|---|---|
| Clothier | Samuel Bros. | 50 Ludgate Hill | 1866–1876 |
| Outfitter | Samuel Bros. | 65 & 67 Ludgate Hill | 1877 |
| Sword cutler | Samuel Bros. | 65 & 66 Ludgate Hill | 1878–1919 |
| | Samuel Bros. | 65 & 66 Ludgate Hill & 223 Oxford St. | 1920–1925 |
| | Samuel Bros. | 4 St. Pauls Churchyard | 1926 |
| | Samuel Bros. Ltd. | 4 St. Pauls Churchyard | 1927–1941 |
| | Samuel Bros. Ltd. | 1-2-3-4 St. Pauls Churchyard | 1942–1946 |
| | Samuel Bros. Ltd. | 4 St. Pauls Churchyard | 1947–1949 |
| | Samuel Bros. Ltd. | 11 & 12 Bartholomew Close | 1950–1957 |
| | Samuel Bros. Ltd. | 12 Bartholomew Close | 1958–1965 |
| Military tailor | Samuelson (Henry), Son & Linney (James) | 49 Maddox St. | 1913–1918 |
| Sword cutler | • Successors to Sibley, Linney, Samuelson & Son (see William Sibley). | | |
| | Samuelson (Henry), Son & Linney (James) | 49 & 51 Maddox St. | 1919–1933 |
| | Samuelson (Henry), Son & Linney (James) | 49 Maddox St. | 1934–1942 |
| Cutler | George Sanders | | 1640–1655 |
| | • Master of the Cutlers Company, 1650. | | |
| Armourer | William Sanders (Saunders) | | Bc1587–1611 |
| Sword maker | William Sanders (Saunders) | Tower Wharf | 1612–1640 |
| | • Repaired swords for the Board of Ordnance, 1613–1635. | | |
| Gun maker | John Sanders (Saunders) | | B1607–1626 |
| Sword maker | John Sanders (Saunders) | Tower Wharf | 1627–1640 |
| | • Son of William Sanders. | | |
| | • Repaired swords for the Board of Ordnance, 1628–1632. | | |
| Military tailor | Sandilands (John) & Nicoll (Henry) | 12 Conduit St. | 1840–1841 |
| Sword cutler | • Successors to Nicoll & Sandiland (see Nicoll & Franklin). Tailor to Queen Victoria, the Prince of Wales, the Duke of Edinburgh and Prince Arthur of Connaught. | | |
| | John Sandilands | 12 Conduit St. | 1842–1858 |
| | John Sandilands & Son | 12 Conduit St. | 1859–1895 |
| | John Sandilands & Son | 11 & 12 Conduit St. | 1896–1905 |
| | John Sandilands & Son | 12 Conduit St. | 1906–1941 |
| Gun, pistol and | Sargant (William) & Son (William Lucas) | 3 Coleman St. | 1835–1836 |
| rifle maker | Sargant (William) & Son (William Lucas) | 2 Coleman St. | 1837 |
| Sword maker | W.L. (William Lucas) & H. (Henry) Sargent | 2 Coleman St. | 1838–1839 |
| | Sargant Bros. (William Lucas & Henry) (see Birmingham listing) | 2 Coleman St. | 1840–1857 |
| Sword cutler | John Sartine | Parish of St. Thomas Apostle | 1620–1625 |
| Gun maker | Richard Savage | Parish of St. Mary Woolnoth Langbourne Ward | 1670–1699 |
| Sword cutler | • Gunmaker to the Board of Ordnance, 1686–1699. | | |
| | Jacob Sawbridge (see the Sword Blade Bank) | | |
| Cutler | Richard Saywell | | 1659–1674 |
| | • Master of the Cutlers Company, 1669. | | |

| | | | |
|---|---|---|---|
| Sales agent for a sword maker | **Schmitt & David**<br>• Sales agent for sword maker F. Horster of Solingen, Prussia, 1862–1863. | 102 Leadenhill St. | 1862–1863 |
| Armourer<br>Furbour | **John Scorfyn**<br>• Refurbished and refitted armour and edged weapons. | | 1333–1379 |
| Cutler | **Charles James Scott**<br>• Master of the Cutlers Company, 1906. | | 1896–1911 |
| Cutler | **James Scott**<br>• Master of the Cutlers Company, 1782. | | 1772–1787 |
| Cutler | **William Cumin Scott**<br>• Master of the Cutlers Company, 1898. | | 1888–1903 |
| Gun maker<br>Goldsmith<br>Sword cutler | **William Scott**<br>• Shop manager for Thomas Potts, 1839–1842.<br>**William Scott**<br>**William Scott** | <br><br>27 Leman St.<br>33 Leman St. | B1816–1842<br><br>1842–1849<br>1850–1853 |
| Hatter<br>Sword cutler | **Scott (James) & Co.**<br>• Succeeded by Christy & Co. Ltd., 1963. | 1 Old Bond St. | 1875–1962 |
| Silversmith<br>Hilt maker<br>Sword cutler | **George Scrivener**<br>• Had apprentice John Wilkins, c. 1681–1689.<br>**Dorah Scrivener**<br>• Widow of George Scrivener.<br><br>**Samuel Seagrove**<br>(see J. Gieve & Sons Ltd.) | Fetter Lane<br><br>White Alley<br>Chancery Lane | 1660–D1697<br><br>1697–1710 |
| Sword cutler | **Samuel Selby** | St. Martins Court | 1700–1715 |
| Cutler | **William Seton**<br>• Master of the Cutlers Company, 1470, 1471, 1478, 1479, 1484, 1485, 1496, and 1497. | | 1460–1502 |
| Silversmith<br>Goldsmith<br>Hilt maker<br>Sword cutler | **Robert Sharp**<br>• Apprentice to Gawen Nash, 1747.<br>• Apprentice to Thomas Gladwin, 1747–1757.<br>**Robert Sharp**<br>**Daniel Smith & Robert Sharp**<br>• Sold silver work to Parker & Wakelin (see Edward Wakelin).<br>• Had apprentice Robert Sharp (nephew), 1770.<br>**Carter (Richard), Smith**<br>**(Daniel) & Sharp (Robert)**<br>**Daniel Smith & Robert Sharp**<br><br>**Robert Sharp** | <br><br><br><br>50 Aldermanbury<br><br><br>14 Westmoreland Building,<br>Aldergate St.<br>14 Westmoreland Building,<br>Aldergate St.<br>14 Westmoreland Building,<br>Aldergate St. | Bc1733–1757<br><br><br>1758–1762<br>1763–1777<br><br><br>1778–1779<br><br>1780–1786<br><br>1787–D1803 |
| Silversmith<br>Hilt maker<br>Sword cutler | **George Shaw (Sr.)**<br><br><br>**George Shaw (Jr.)** | Parish of St. Andrews<br>Fullwoods Rents, Holborn<br>at the sign of the Flaming Sword<br>Parish of St. Andrews<br>Fullwoods Rents, Holborn<br>at the sign of the Flaming Sword | 1678–D1717<br><br><br>1717–1770 |

| | | | |
|---|---|---|---|
| Goldsmith<br>Silversmith<br>Hilt maker<br>Sword cutler | William Shaw<br>• Apprenticed to Edward Holliday, 1715–1723.<br>William Shaw | corner of Maxfield St. & Gerrard St., Soho | Bc1701–1723<br>1724 |
| | William Shaw | Gerrard St., Soho<br>at the sign of the Golden Ball | 1725–1744 |
| | William Shaw | Parish of St. Anne, Westminster | 1745 |
| Goldsmith<br>Silversmith<br>Hilt maker<br>Sword cutler | William Shaw<br>• Apprentice to John Swift, 1736–1748.<br>William Preist & William Shaw<br>William Preist & William Shaw<br>• Had apprentice John King, 1751–1756 (see John King).<br>William Shaw<br>William Shaw<br>(see William Preist Jr.) | Maiden Lane<br>corner of Lad Lane & Wood St.<br><br>22 Wood St., Cheapside<br>Bishopsgate | Bc1722–1748<br>1749–1750<br>1751–1762<br><br>1763–1768<br>1769–1773 |
| Silversmith<br>Goldsmith<br>Sword and<br>knife cutler | Daniel Shelmerdine<br>• Apprentice to Francis Williams, 1676–1681 (see Francis Williams).<br>Daniel Shelmerdine<br>Daniel Shelmerdine | Exchange Alley, Cornhill<br>New St. near Shoe Lane<br>Parish of St. Brides<br>at the sign of the Golden Dagger | Bc1662–1681<br>1682–1719<br>1720–1728 |
| | Daniel Shelmerdine<br>David Shelmerdine<br>• Son of Daniel Shelmerdine. | Noble St., near Foster Lane<br>Noble St., near Foster Lane | D1729<br>1729–1740 |
| Cutler | Joseph Shepard<br>• Master of the Cutlers Company, 1711. | | 1701–1716 |
| Army and navy<br>button maker<br>Accoutrement<br>maker<br>Gold and silver<br>lacemen<br>Embroiderer<br>Sword cutler | Thomas Sherlock<br>Thomas Sherlock & Co. | 15 King St., Covent Garden<br>15 King St., Covent Garden | 1837–1846<br>1847–1887 |
| Sword cutler | H. (Henry) W. Sherwood | | c1850 |
| Hilt maker<br>Sword cutler | James Shrapnell (Sr.)<br>James Shrapnell (Sr.)<br>Shrapnell (James Sr.) & Son (James Jr.)<br>James Shrapnell & Son (James (Jr.)<br>James Shrapnell (Jr.)<br>James Shrapnell (Jr.)<br>James Shrapnell (Jr.) | 36 Ludgate St.<br>60 Charing Cross<br>60 Charing Cross<br>60 Charing Cross<br>60 Charing Cross<br>61 Charing Cross<br>61 Charing Cross & 15 New St. Square | 1769–1771<br>1772–1799<br>1800<br>1801<br>1802–1804<br>1805–1826<br>1827–1838 |
| Military<br>tailor<br>Sword cutler | William Sibley<br>Sibley (William) & Snellgrave (David)<br>William Sibley<br>Sibley (William) & Linney (James)<br>Sibley (William), Linney (James), | 7 South Molton St.<br>46 Albemarle St.<br>46 Albemarle St.<br>46 Albemarle St.<br>49 Maddox St. | 1881–1882<br>1883<br>1884–1885<br>1886–1909<br>1910–1912 |

| | Samuelson (Henry) & Son | | |
|---|---|---|---|
| | • Succeeded by Samuelson, Son & Linney (see Samuelson, Son & Linney). | | |
| Cutler | **Henry Siddon** | | 1590–1606 |
| | • Master of the Cutlers Company, 1600–1601. | | |
| Gold and silver lacemen | **J. & J. Silver** | 28 Hatton Garden | 1800–1805 |
| | **J. & J. Silver** | 25 Bedford St., Covent Garden | 1806–1826 |
| Sword cutler | **J. & K. Silver** | 25 Bedford St., Covent Garden | 1827–1830 |
| Gold and silver lacemen Sword cutler | **James Silver** | 10 Camden St., Camdentown | 1811–1820 |
| Clothier | **Silver (Stephen M.) & Co.** | 9 Cornhill | 1825–1830 |
| Linen | • Successor to Arrow, Smith & Silver (see Arrow, Smith & Silver). | | |
| draper | **Silver (Stephen M.) & Co.** | 9 & 10 Cornhill | 1831–1838 |
| (maker) | **S. (Stephen) M. Silver & Co.** | 10 & 66 & 67 Cornhill | 1839–1844 |
| Army and navy | **Silver (Stephen Winkworth) & Co.** | 10 & 66 & 67 Cornhill & Bishopsgate | 1845–1856 |
| outfitter | **Silver (Stephen Winkworth) & Co.** | 66 & 67 Cornhill & 3 & 4 Bishopsgate | 1857–1875 |
| Sword cutler | **S. (Stephen) W. (Winkworth) Silver & Co.** | 66 & 67 Cornhill & 5 Bishopsgate | 1876–1898 |
| | • Partner: Benjamin Edgington. | | |
| | **S. (Stephen) W. (Winkworth) Silver & Benjamin Edgington** | 67 Cornhill | 1899–1913 |
| | **S. (Stephen) W. (Winkworth) Silver & Benjamin Edgington** | William King House 29 East Cheapside | 1914–1929 |
| | **Silver & Edgington Ltd.** | 29 East Cheapside,& 313 Regent St. | 1930–1940 |
| | **Silver & Edgington Ltd.** | 29 East Cheapside, & 69 Great Queen St. | 1941–1943 |
| | **Silver & Edgington Ltd.** | 29 East Cheapside & 29 Great Queen St. | 1944–1957 |
| | **Silver & Edgington Ltd.** | 29 East Cheapside & 144 Shaftesbury | 1958–1965 |
| | **Silver & Edgington Ltd.** | 29 East Cheapside | 1966–1967 |
| | (see Liverpool and Portsmouth listings) | | |
| Lacemen | **Simpson & Rooke** | | 1875–1896 |
| Sword cutler | • Successor to Joseph, Simpson & Simons (see Joseph, Simpson & Simons). | | |
| | • Succeeded by Christopher Rook (see Christopher Rook). | | |
| Silversmith | **Joseph Singleton (Shingleton)** | Foster Lane | 1677–1710 |
| Hilt maker | | | |
| Sword cutler | • Made silver-hilted small swords (some exported to USA). | | |
| Cutler | **George Skelton** | | 1795–1815 |
| | • Master of the Cutlers Company, 1605 and 1610. | | |
| Brass founder | **Abraham Skinner** | 25 Shoe Lane | 1785–1790 |
| Hilt maker | **A. (Abraham) Skinner** | Featherstone St. | 1791–1804 |
| Sword cutler | **Abraham Skinner** | 8 Dean St., Fetter Lane | 1805–1808 |
| Cutler | **Abraham Skinner** | | 1814–1829 |
| | • Master of the Cutlers Company, 1824. | | |
| Cutler | **Nicholas Skinner** | | 1607–1636 |
| | • Master of the Cutlers Company, 1617 and 1631. | | |
| | **Richard Skinner** (see Francis Thurkle) | | |

| | | | |
|---|---|---|---|
| Military tailor | William Skinner | 34 Bury St. | 1836–1843 |
| | William Skinner | 50 Jermyn St. | 1844–1864 |
| Sword cutler | W. (William) Skinner & Co. | 50 Jermyn St. | 1865 |
| | • Sold hunting swords. | | |
| | • Appointed sword cutler to His Royal Highness the Prince of Wales. | | |
| Gold and silver lacemen | Smith (Charles) & Trimnell (Henry) | 12 Piccadilly | 1820–1835 |
| | Charles Smith & Son | 12 Piccadilly | 1836–1859 |
| Army accoutrement maker | Charles Smith & Co. | 5 New Burlington St. | 1860–1878 |
| | Charles Smith & Son | 5 New Burlington St. | 1879–1914 |
| | • Appointed gold and silver lacemen to His Majesty the King. | | |
| | Charles Smith & Son | 6 New Burlington St. | 1915–1928 |
| Silversmith | Abraham Smith | Cheapside | 1690–D1749 |
| Goldsmith | • Master of the Cutlers Company, 1694. | | |
| Hilt maker | William Smith | at the sign of the Blackmoors Head opposite Gutter Lane, Cheapside, at the sign of the Blackmoors Head | 1749–1773 |
| Sword cutler | | | |
| | William Smith | 32 Cheapside | 1774–D1781 |
| | Wildman Smith | 32 Cheapside | 1781–1800 |
| | • Son of William Smith Sr. | | |
| Silversmith | Daniel Smith | | Bc1726–1753 |
| Goldsmith | • Apprenticed to Thomas Gladwin, 1740–1753. | | |
| Hilt maker | Daniel Smith | 50 Aldermanbury | 1754–1762 |
| Sword cutler | Daniel Smith & Robert Sharp | 50 Aldermanbury | 1763–1777 |
| | Carter (Richard), Smith (Daniel) & Sharp (Robert) | 14 Westmoreland Building Aldergate St. | 1778–1779 |
| | • Did some silver work for Parker & Wakelin (see Edward Wakelin). | | |
| | Daniel Smith & Robert Sharp | 14 Westmoreland Building Aldergate St. | 1780–1796 |
| | • Succeeded by Robert Sharp. (see Richard Carter & Robert Sharp) | | |
| Sword cutler | Edward Smith | 5 Boyle St. | 1870–1890 |
| Cutler | Frederick Smith | | 1859–1874 |
| | • Master of the Cutlers Company, 1869. | | |
| | John Smith (see George & William Almond) | | |
| Metal worker | Richard Smith | 21 Bowling Green Lane | 1810–1815 |
| Sword decorator | • Advertised as a fancy painter and embosser of swords. | | |
| | • All kinds of steel work. | | |
| | Thomas Smith (see Thomas Read) | | |
| Cutler | William Smith | | 1712–1727 |
| | • Master of the Cutlers Company, 1722. | | |
| Button maker | William Smith & Co. | 33 George St., Hanover Square | 1843 |
| Accoutrement maker | • Successor to King & Preston (see Thomas King). | | |
| Sword cutler | William Smith & Co. | 5 St. James St. | 1844–1845 |

| | | | |
|---|---|---|---|
| Sword cutler | **William Smithe** | Parish of St. Bride | 1597–1607 |
| Cutler | **John Smyth**<br>• Master of the Cutlers Company, 1550–1552. | | 1540–1557 |
| Cutler | **Samuel Smyth**<br>• Master of the Cutlers Company, 1743. | | 1733–1748 |
| Cutler | **Thomas Smyth**<br>• Master of the Cutlers Company, 1665. | | 1655–1670 |
| Sword cutler | **Walter Smyth** | | 1667 |
| | **David Snellgrave**<br>(see William Sibley) | | |
| Sword maker | **J. (James) Sorlay**<br>• Made naval cutlasses. | | 1790–1800 |
| Sword maker | **Robert South**<br>**Robert South**<br>• Freeman of the Cutlers Company, 1603.<br>• Master of the Cutlers Company, 1629.<br>• Royal cutler to King James I and King Charles I.<br>• Made richly decorated swords and daggers for the king's wardrobe and the Prince of Wales.<br>• On June 13, 1621, by order of the court of the Cutlers Company of London, master John Porter, past master Thomas Chesshire, and royal cutler Robert South were to inspect a sword blade mill, believed to be the mill of Thomas Murrey (see Thomas Murrey), and report their findings to the court at the next meeting.<br>• It is believed that Thomas Chesshire, who became freeman of the Cutlers Company in 1603 (same year as Robert South), was an associate of Robert South, since Chesshire, like South, provided richly decorated swords to the king's wardrobe, even though he was not the royal cutler.<br>• In 1631, the Cutlers Company advanced South 100 pounds to investigate the making of sword and rapier blades "for the good of the company and the kingdom."<br>• South used some blades made by Solingen sword and bladesmith Clemens Horn on his swords.<br>• In 1635, South was married at St. Leonards Church in the town of Heston in Hounslow Heath.<br>• Since South was born in Heston, it is believed that he suggested the Hounslow Heath area as a location for a sword and blade mill to Charles I in 1629 when Charles decided to bring some Solingen, Prussia, sword and bladesmiths to England to bolster English production during the Thirty Years War.<br>• On August 7, 1632, South registered the "Murrian head" mark with the Cutlers Company (to strike only his sword blades).<br>• In 1638, Benjamin Stone delivered 5,000 sword and rapier blades to the Board of Ordnance.<br>• Stone was contracted to hilt and make scabbards for 3,000 of them. South and William Cave were contracted to hilt and make scabbards for 1,000 each.<br>• In 1639, in a petition to the Council of War, South stated that he had provided 4,000 swords to the Office of Ordnance for the king's northern expedition and proposed that he provide at least 4,000 more.<br>• During the English Civil War (1642–1649), South provided many swords to the parliamentary forces. | Town of Heston, Hounslow Heath<br>London and Town of Heston, Hounslow Heath | Bc1580–1600<br>1600–1650 |
| Hardwareman (dealer)<br>Cutler<br>Sword cutler | **Robert Sparling** | 56 Haymarket | 1772–1811 |

| | | | |
|---|---|---|---|
| Sword cutler | Edward Spencer | | 1650 |
| Sword cutler | Major Spencer | | 1686–1687 |
| Goldsmith | Marshall Spink | 2 Gracechurch St. & 30 Barbican St. | 1803–1807 |
| Silversmith | M. (Marshall) Spink & Son | 2 Gracechurch St. | 1808–1881 |
| Jeweler | Spink & Son | 1 & 2 Gracechurch St. | 1882–1892 |
| Gun repository | Spink & Son Ltd. | 17 & 18 Piccadilly | 1893–1919 |
| Hilt maker | Spink & Son Ltd. | 6 King St., Covent Garden | 1920–1927 |
| Sword cutler | Spink & Son Ltd. | 5–6–7 King St., St. James St. | 1928–1994 |

| | | |
|---|---|---|
| Sword maker | Peter Spitser (Spitzer) (Sr.) | 1620–1664 |
| | • Granted "unicorn head and dagger" mark, 1621. | |
| | Peter Spitzer (Jr.) | 1664–1730 |
| | • Master of the Cutlers Company, 1725. | |
| | • Used "unicorn head and dagger mark" until 1698. | |
| | • Obtained the "heart, fleur-de-lis, and dagger" mark, 1698. | |

| | | | |
|---|---|---|---|
| Silversmith<br>Hilt maker<br>Sword cutler | Francis Springgall | New St. | 1693–1710 |

| | | |
|---|---|---|
| Sword cutler | Thomas Squire | 1760–1780 |
| | • Master of the Cutlers Company, 1774. | |
| | • Used "star and half moon" mark. | |
| Cutler | William Squire | 1815–1844 |
| | • Master of the Cutlers Company, 1825 and 1839. | |

William Stalker
(see Welch & Stalker)

John Stamper
(see Edward Aldridge Sr.)

| | | |
|---|---|---|
| Silversmith<br>Sword hilter<br>Sword cutler | Richard Stanford | 1661 |
| | • Stanford signed an advertisement warning sword purchasers about counterfeit cast silver hilts. | |

| | | | |
|---|---|---|---|
| Bladesmith | John Stannard | Fleet St., Parish of St. Bride | 1400 |

| | | |
|---|---|---|
| Silversmith<br>Hilt maker<br>Sword cutler | John Stanton | 1661 |
| | • In 1661, signed an advertisement warning sword purchasers about counterfeit cast silver hilts. | |

| | | | |
|---|---|---|---|
| Gold and silver<br>laceman<br>Embroiderer<br>Accoutrement<br>maker<br>Sword cutler | J. (Joseph) Starkey & Co. | 1 Spur St., Leicester Square | 1835–1836 |
| | Joseph Starkey | 5 Old Bond St. | 1837–1855 |
| | • Gold and silver lacemen for the army and navy, diplomatic service, and consular service. | | |
| | Joseph Starkey | 4 Old Bond St. | 1856–1858 |
| | Joseph Starkey | 23 Conduit St. | 1859–1914 |
| | Joseph Starkey | 45 Conduit St. | 1915–1918 |
| | Joseph Starkey | 21 George St., Hanover Square | 1919–1934 |
| | Joseph Starkey | 16 & 18 Beak St., Regent St. | 1935–1951 |
| | Joseph Starkey | 135 Rye Lane | 1952–1964 |
| | Joseph Starkey | 86 Bellandon Road | 1965–1969 |

| | | | |
|---|---|---|---|
| Silversmith<br>Hilt maker<br>Sword cutler | Robert Stebbing | | 1690–1720 |
| Gun maker<br>Sword maker | Francis Stedman<br>• Repaired swords for the Board of Ordnance, 1666–1668. | Parish of St. Katherine | 1660–1670 |
| Gun maker<br>Sword maker | John Stedman<br>• Repaired swords for the Board of Ordnance, 1637–1642. | Parish of St. Katherine | 1630–1645 |
| Gun maker<br>Sword maker | Robert Stedman<br>• Supplied swords to the Board of Ordnance, 1627–1663.<br>• Also repaired swords for the Board of Ordnance.<br>• Was furbisher of rich weapons at the Tower of London, 1654–1664.<br>• In 1663, with Joseph Hudley and Samuel Low, repaired and cleaned<br>  500 short swords and bayonets for the Board of Ordnance. | Parish of St. Katherine | 1620–Dc1664 |
| Hatter<br>Sword cutler | Abraham Stevens<br>Abraham Stevens<br>Abraham Stevens<br>• Sold small swords with chiseled and pierced steel hilts. | Fleet St.<br>6 Temple Gate<br>17 Fleet St. | 1755–1759<br>1760<br>1761–1777 |
| Wire drawer<br>Gold and silver<br>lacemen<br>Accoutrement<br>maker<br>Sword cutler | Edward Swift Stillwell<br>• Succeeded by his son Edward Stillwell (see Edward Stillwell).<br>Stillwell (Edward) & Co.<br>• Successors to Atherly & Stillwell (see Atherly & Stillwell).<br>• Successors to Edward Swift Stillwell (see Edward Swift Stillwell).<br>Edward Stillwell & Son<br>Edward Stillwell & Son<br>Stillwell (Edward), Son & Ledger (John)<br>• In 1865, David Lewis Tappolet at 6 Little Britain Street merged with Stillwell.<br>Edward Stillwell & Son<br>Edward Stillwell & Son<br>Edward Stillwell & Son<br>Edward Stillwell & Son<br><br>• Webb & Co. at 29 Sackville Row merged with Stillwell, 1895.<br>Edward Stillwell & Son<br><br>• Gold and silver lacemen by royal appointment to Queen Victoria.<br>Edward Stillwell & Son Ltd.<br><br>Edward Stillwell & Son<br><br>Edward Stillwell & Son<br><br>Edward Stillwell & Son<br>• Gold and silver lacemen to the army, navy, and air force. | 4 Bridgewater Square<br><br>3 & 4 Bridgewater Square<br><br><br>3 Bridgewater Square & 25 Barbican St.<br>25 Barbican St.<br>25 Barbican St.<br><br>25 Barbican St. & 6 Little Britain<br>26 & 27 Barbican St. & 6 Little Britain<br>25 to 27 Barbican St. & 6 Little Britain<br>25 to 27 Barbican St. & 6 Little Britain<br>& 22 Sackville St., Piccadilly<br><br>25 to 27 Barbican St. & 6 Little Britain,<br>29 Savile Row<br><br>25 to 27 Barbican St. & 6 Little Britain<br>29 Savile Row<br>25 to 27 Barbican St., & 6 Little Britain<br>29 Savile Row, 18 Ramillies Place<br>25 to 27 Barbican St., & 6 Little Britain<br>18 Ramillies Place<br>6 Little Britain, 18 Ramillies Place | 1838–D1851<br><br>1852<br><br><br>1853–1859<br>1860<br>1861–1864<br><br>1865–1867<br>1868–1874<br>1875–1887<br>1888–1894<br><br><br>1895–1907<br><br><br>1908–1924<br><br>1925–1932<br><br>1933–1934<br><br>1935–1957 |
| Sword cutler | William Stockley | Parish of St. Botolph, Altergate | 1605–1621 |
| | Benjamin Stone<br>(see Chapter 3) | | |
| Sword maker | Guy Stone (Sr.) | | 1695–1720 |

- Freeman of the Cutlers Company, 1695.
- Master of the Cutlers Company, 1719.
- Supplied swords to the Board of Ordnance, 1704–1715.
- On March 15, 1715, appointed sword cutler to the Board of Ordnance.
- Supplied plug bayonets to the Board of Ordnance, 1708–1711.

| | | | |
|---|---|---|---|
| Sword maker | **Guy Stone (Jr.)** | | 1719–1730 |
| | • Freeman of the Cutlers Company, 1719. | | |

| | | | |
|---|---|---|---|
| Goldsmith | **Paul Storr** | | B1771–1790 |
| Silversmith | • Apprenticed to William Rock, 1784–1791. | | |
| Hilt maker | **Paul Storr** | Tothill Fields, Westchester | 1791 |
| Sword cutler | **Frisbee (William) & Storr (Paul)** | 5 Cock Lane, Snowhill | 1792 |
| | **Paul Storr** | 30 Church St., Soho | 1793–1795 |
| | **Paul Storr** | 20–23 Air St., Parish of St. James | 1796–1806 |
| | **Storr (Paul) & Co.** | 53 Dean St., Soho | 1807–1817 |

- Partner: Peter Bogerts.
- In 1807, Storr became the manufacturing workshop for Rundell, Bridge & Rundell. He was partner in that company from 1811–1817 (see Rundell, Bridge & Rundell).

| | | |
|---|---|---|
| **Paul Storr** | 17 & 18 Harrison St., Grays Inn Road | 1817–1838 |

- New workshop and retail store.
- In 1822, Storr bought out the shop of Meyer (Edward Simeon) & Mortimer (John) at 13 New Bond Street as an additional separate shop to retail his goods.

| | | |
|---|---|---|
| **Storr (Paul) & Mortimer (John)** | 13 New Bond St. | 1822–1838 |

- A separate retail store.
- Successors to Meyer & Mortimer (see Meyer & Mortimer).
- French goldsmith Antoine Vechte worked for them, 1822–1838.

| | | |
|---|---|---|
| **Storr (Paul) & Mortimer (John)** | 156 New Bond St. | 1838 |

- Paul Storr retired in 1838 and died in 1844.
- Succeeded by Mortimer & Hunt (see Mortimer & Hunt).

| | | | |
|---|---|---|---|
| Sword cutler | **Samuel Stovel** | New Bond St. | 1845 |

| | | | |
|---|---|---|---|
| Sword cutler | **William Stower** | | 1550–1560 |

- On January 22, 1558, Stower supplied some swords to the Drapers (cloth maker) Company for the defense of its settlement in Calais, France, during the French War.

| | | | |
|---|---|---|---|
| Goldsmith | **Samuel Strange** | | 1682–D1716 |
| Silversmith | | | |
| Hilt maker | | | |
| Sword cutler | | | |

| | | | |
|---|---|---|---|
| Goldsmith | **William Strange** | | Bc1702–1724 |
| Silversmith | • Son of Samuel Strange. | | |
| Hilt maker | • Apprenticed to goldsmith John Carmen, 1716–1724 (see John Carmen). | | |
| Sword cutler | **William Strange** | New Street Square, Parish of St. Bride | 1725–1743 |

**Henry Stratton**
(see John Garden)

| | | | |
|---|---|---|---|
| Gold and silver laceman | **Thomas Street** | next door to Somerset House, Strand, at the sign of the Peacock | 1724–1766 |
| Sword cutler | • Succeeded by Newham & Binham (see Newham & Binham). | | |

| | | | |
|---|---|---|---|
| Sword cutler | **Strickland & Sons** | Savile Row | 1899–1909 |
| Engraver | **John Strongitharm** | 19 Pall Mall | 1799–1819 |
| Seal maker | **John Strongitharm** | 1 Waterloo Place, Pall Mall | 1820–1856 |
| Sword decorater | • Seal engraver to the Prince of Wales.<br>• Succeeded by John & Richard Longman and John Strongitharm.<br>(see John & Richard Longman, John Strongitharm) | | |
| Silversmith | **Isaac Stuart** | | Bc1697–1720 |
| Hilt maker | • Apprentice and journeyman to Thomas Bass, c. 1712–1720 (see Thomas Bass). | | |
| Sword cutler | **Isaac Stuart** | New St., Shoe Lane | 1721–1730 |
| Sword cutler | **John Stubbs** | | 1606 |
| Sword hilt, pommel cap, and chape maker | **Lawrence Subelane** | Blackfriars | 1625–1635 |
| | • On March 19, 1630, the Cutlers Company seized 14 hilts, pommels, and chapes from Subelane because he was a Frenchman illegally working in London.<br>• He was not a member of the Cutlers Company. | | |
| Military tailor | **George Sully** | 40 Castle St., Holborn | 1856–1857 |
| | **George Sully** | 14 Bishopsgate within | 1858–1869 |
| Draper (cloth maker) Sword cutler | • Successor to Wisker & Sully (see Wisker & Sully). | | |
| Bladesmith | **Thomas Sutton** | | 1500–1510 |
| | • In 1506, moved from the Armourers Company to the Bladesmiths Company. | | |
| Sword cutler | **John Swen** | | 1608 |
| Silversmith Hilt maker Sword cutler | **Thomas Swetman (Sweetman)** | | 1661 |
| | • In 1661, signed an advertisement warning sword purchasers about counterfeit cast silver hilts. | | |
| Sword cutler | **C. P. Swinburn & Son** | | 1863–1865 |
| | • Sold naval cutlasses. | | |
| Sword and sword blade maker | **The Sword Blade Bank**<br>(see Chapter 4) | Birchin Lane | 1704–1724 |
| Goldsmith Jeweler | **J. & H. Sydenham** | 126 New Bond St. | 1797–1809 |
| | • Successor to Nodes & Sydenham (see Nodes & Sydenham). | | |
| Hilt maker | **J. Sydenham** | 126 New Bond St. | 1810–1812 |
| Sword cutler | **J. Sydenham** | 133 New Bond St. | 1813–1823 |
| Hatter | **Daniel Symes** | 7 Bartholomew Lane | 1805–1808 |
| Sword cutler | **Daniel Symes** | 7 Whalebone Ct., Lothbury | 1809–1826 |
| Cutler | **Robert Taber** | | 1798–1813 |
| | • Master of the Cutlers Company, 1808. | | |
| Sword cutler | **James Tackley** | | 1610–1621 |
| Pole arm maker | **James Tanner** | | 1555–1560 |
| | • Supplied pikes and bills to the Drapers Company for the defense of their settlement in Calais, France, during the French War. | | |

| Sword cutler | Henry Tapp | Strand | 1740 |

| Gold and silver laceman | David Lewis Tappolet | 6 Little Britain | 1839–1843 |
| | David Lewis Tappolet & Co. | 6 Little Britain & 44 Lombard St. | 1844–1855 |
| Sword cutler | David Lewis Tappolet & Co. | 6 Little Britain | 1856–1864 |

• Merged with Edward Stillwell & Son (see Edward Stillwell & Son).

| Modeler | James Tassie | Pollokshaws, Scotland | B1735–1763 |

Medallion maker • Pollokshshaws was a suburb of Glasgow.
Cameo maker • Immigrated to Dublin, 1763.

| Sword hilt and scabbard decorator | James Tassie | Dublin, Ireland | 1763–1767 |

• Immigrated to London, 1767.

| | James Tassie | Great Newport St. | 1767–1771 |
| | James Tassie | Compton St., Soho | 1772–1777 |
| | James Tassie | 20 Leicester Fields | 1778–D1799 |

• Designed and made scabbard and sword hilt medallions and decorations for patriotic swords made by Richard Teed.

| | William Tassie | 20 Leicester Fields | 1799–1835 |

• Nephew of and successor to James Tassie.

| Gun maker | Henry Tatham (Sr.) | | Bc1770–c1790 |
| Army cap, hat accoutrement maker | Henry Tatham (Sr.) | 60 Frith St. | c1790–1797 |
| | Henry Tatham (Sr.) | 35 Charing Cross, near the Admiralty | 1798–1799 |
| | Henry Tatham (Sr.) | 37 Charing Cross, near the Admiralty | 1800–1801 |
| Sword belt maker | Tatham (Henry Sr.) & Egg (Joseph) | 37 Charing Cross, near the Admiralty | 1802–1816 |

• Appointed gun maker and sword cutler to King George III.
• Army sword maker.
• Made naval cutlasses.
(see Joseph Egg)

| | Henry Tatham (Sr.) & Son (Henry Jr.) | 37 Charing Cross, near the Admiralty | 1817–1834 |

• Appointed sword cutler to King George III.
• Tatham's sons Charles (B1805–D1846) and Henry Jr. worked with him.
• Henry Tatham Sr. died in 1834.

| | Henry Tatham (Jr.) | 37 Charing Cross, near the Admiralty | B1804 1834–D1860 |

• Became army and cap accoutrement maker.
• John Prosser Jr. was a tenant of Tatham at 37 Charing Cross, 1853–1860 (see John Prosser Jr.).
• Tatham sold cavalry officer swords and presentation swords.
• Henry Tatham (Jr.) died in 1860.

| Silversmith Jeweler Hilt maker | James Taylor | | Bc1773 |

• Son of wire drawer John Taylor.
• Apprentice to James Kirk, 1787–1794.

| Sword cutler | James Taylor | 121 Cheapside | 1795–1796 |
| | James Taylor & Thomas Hobbs | 12 St. Ann's Lane, Aldergate | 1797–1806 |

| Silversmith | John Taylor | 98 Strand | 1783–1784 |
| Hilt maker Sword cutler | John Taylor | 31 Iron Monger Row, Old Street | 1785–1790 |

| Wire drawer Hilt wire maker | John Taylor | Waterlane, Blackfriars | c1753–1793 |

• Father of James Taylor Silversmith.

| Military tailor | John Walker Taylor | 126 New Bond St. | 1862–1863 |

• Successor to Preston & Taylor (see Preston & Taylor).

| | | | |
|---|---|---|---|
| Sword cutler | • Succeeded by Henry Huntsman (see Henry Huntsman). | | |
| Cutler | **Thomas Taylor**<br>• Master of the Cutlers Company, 1635. | | 1625–1640 |
| Sword cutler | **Christopher Tedcastle** | | 1584–D1597 |
| Antique jewelry dealer | **Richard Teed** | | B1765–1784 |
| Jeweler | **Richard Teed** | Lancaster Court, Strand | 1785–1798 |
| Goldsmith<br>Silversmith<br>Hilt maker<br>Sword cutler | **R. (Richard) Teed**<br>• Mounted most of the Patriotic Fund swords.<br>• Used scabbard and sword hilt decorations and medallions made or designed by James Tassie and Edward Burch on some of his Patriotic Fund swords. (see James Tassie, Edward Burch)<br>• Did sword decorations for some of John Prosser's and Robert Ritherdon's presentation swords.<br>• Some swords engraved "Richard Teed, dress sword maker to the Patriotic Fund." | 3 Lancaster Court, Strand | 1799–D1816 |
| Bladesmith | **William Temple** | St. Clements Lane, Candlewick Ward | 1370–1377 |
| Steel worker<br>Pike maker | **John Thacker**<br>• Freeman of the Cutlers Company, 1623. | Fenchurch St. | 1623–D1645 |
| | **Elizabeth Thacker**<br>• Widow of John Thacker.<br>• On April 19, 1645, during the English Civil War, had a contract for 500 long pikes (16 feet long with steel heads at 4 shillings, 2 denier each) for the "New Model" army (parliamentary forces) under Gen. Sir Thomas Fairfax. | Fenchurch St. | 1645 |
| | **John Thacker**<br>• Son of Henry and Elizabeth Thacker.<br>• In 1645, had contracts for 600 long pikes for the "New Model" army (parliamentary forces) under Gen. Sir Thomas Fairfax.<br>• Had a contract dated July 5, 1645, for 200 long pikes.<br>• Had a contract dated December 23, 1645, for 400 long pikes to be delivered to the Tower Armoury (100 at a time) by the last day of December 1645 and January, February, and March 1646 (3 shillings, 10 denier each). | Fenchurch St. | 1645–1649 |
| Armourer<br>Sword maker | **Adam de Thaxted (Taxted)** | | 1307 |
| Hafter (sword hilt maker) | **Richard de Thaxted (Taxted)** | Broad St. | 1287 |
| Goldsmith<br>Jeweler<br>Hilt maker<br>Sword cutler | **Thead (Henry) & Pickett (William)**<br>• Phillip Rundell was shopman, 1767–1771.<br>• Successor to Henry Hurt (see Henry Hurt).<br>• Succeeded by Pickett & Rundell (see Pickett & Rundell). | 32 Ludgate Hill<br>at the sign of the Golden Salmon | 1758–1772 |
| Goldsmith<br>Jeweler<br>Sword cutler | **William Thead**<br>• Worked for Rundell, Bridge & Rundell, 1811–1819 (see Rundell, Bridge & Rundell). | | 1819–1830 |
| | **Henry Third**<br>(see Charles Marr) | | |
| Cutler | **Alfred James Thomas**<br>• Master of the Cutlers Company, 1915. | | 1905–1920 |

| | | | |
|---|---|---|---|
| Military tailor | Thomas Jeremy Thomas & Ellis Jones | 91 Watling St. | 1872–1875 |
| Army clothier | Thomas Jeremy Thomas & Ellis Jones | 138 Queen Victoria St. | 1876–1882 |
| Uniform maker | | & 20 & 23 Knightrider St. | |
| Sword cutler | Thomas Jeremy Thomas & Son | 1 Clark St., Commercial Rd. | 1883–1885 |
| | Thomas Jeremy Thomas, Son & Co. | 1a Fore St., 29 & 30 Jewin St., 33 & 34 Redcross St. | 1886–1892 |
| | Thomas Jeremy Thomas, Son & Co. | 42 to 46 Whitecross St. | 1893 |
| Cutler | Charles Thompson • Master of the Cutlers Company, 1721. | | 1711–1726 |
| Gun maker Sword cutler | James Thompson | | 1810–1835 |
| Silversmith Hilt maker Sword cutler | Francis Thompson | Abchurch Lane | 1699–1725 |
| Silk mercer (dealer) | Francis Thompson • Successor to William Clementson (see William Clementson). | 344 Oxford St. | 1820–1830 |
| Woollen draper (maker) | Francis Thompson | 11 Conduit St. | 1831–1849 |
| | Francis Thompson & Son | 11 Conduit St. | 1850–1867 |
| Military tailor Sword cutler | Thompson (Francis) & Son • Succeeded by Austin & Oaker (see Austin & Oaker). | 11 Conduit St. | 1868–1880 |
| Sword cutler | John Thompson | 5 Air St. | 1815–1820 |
| Military tailor | Peter Thompson (Sr.) | 3 Queen St. | 1802–1804 |
| Army uniform | Peter Thompson (Sr.) | 11 Frith St., Soho | 1805–1810 |
| maker | Peter Thompson (Sr.) & Son (Peter Jr.) | 11 Frith St., Soho | 1811–1820 |
| Army clothier | Peter Thompson & Son (Peter Jr.) | 12 Frith St., Soho | 1821–1824 |
| Sword cutler | Peter Thompson (Jr.) | 12 Frith St., Soho | 1825–1839 |
| | Peter Thompson (Jr.) & Sons • Appointed army clothier to Her Majesty the Queen, 1840. | 12 Frith St., Soho | 1840–1843 |
| | Peter Thompson (Jr.) & Co. | 12 Frith St., Soho | 1844–1862 |
| Sword cutler | William Thompson | Dover St. | 1910–1915 |
| Goldsmith | William Thompson | 29 Grafton St., Soho | 1802–1803 |
| Ornament maker | William Thompson | 35 Piccadilly | 1804 |
| Army | William Thompson | 54 Jermyn St., Parish of St. James | 1805–1817 |
| accoutrement maker Sword cutler | William Frost & William Thompson (see William Frost) | 11 Air St., Piccadilly | 1818–1826 |
| Cutler | Joseph Thorne • Master of the Cutlers Company, 1882. | | 1872–1887 |
| Sword maker | W. (William) Thornhill & Co. | 144 New Bond St. | 1850–1876 |
| Hosier Shirt maker Military outfitter | Richard Thresher • Successor to Newham & Thresher (see Newham & Thresher). • Appointed hosier to Her Majesty the Queen. | 152 Strand | 1780–1819 |
| | John Thresher | 152 Strand | 1820–1833 |
| Sword cutler | Thresher (John) & Son & Glenny (David) | 152 Strand | 1834–1841 |

|  |  |  |  |
|---|---|---|---|
|  | Thresher (John) & Glenny (David) | 152 Strand | 1842–1898 |
|  | Thresher (John) & Glenny (David) | 152 & 153 Strand | 1899–1914 |
|  | Thresher (John) & Glenny (David) | 152 & 153 & 166 Strand & 5 Conduit St. | 1915–1917 |
|  | Thresher (John) & Glenny (David) | 152 & 153 Strand | 1918–1919 |
|  | Thresher (John) & Glenny (David) | 152 & 153 Strand & 19 Clifford St. | 1920–1931 |
|  | Thresher (John) & Glenny (David) | 152 & 153 Strand | 1932–1936 |
|  | Thresher (John) & Glenny (David) | 152 & 153 Strand & 85 Gracechurch St. | 1937–1964 |
|  | Thresher (John) & Glenny (David) | 122 Drury Lane & 85 Gracechurch St. | 1965 |

| | | | |
|---|---|---|---|
| Silversmith | Francis Thurkle | | Bc1740–1760 |
| Hilt maker | Francis Thurkle (Sr.) | 15 Great New Street square | 1760–1789 |
| Sword maker | • Master of the Cutlers Company, 1766. | | |
| Sword belt | • Employee William Henry Whawell was a designer, etcher, and gilder. | | |
| maker | Thurkle (Francis) (Sr.) & Son (Frances Jr.) | 15 Great New Street square | 1790 |
| Sword cutler | Francis Thurkle (Jr.) | 15 Great New Street square | 1791–1801 |

• Freeman of the Cutlers Company, 1779.
• Master of the Cutlers Company, 1795.
• Made silver and brass hilts with his mark "F.T." on them.
• Exported swords, including infantry and artillery officers swords with eaglehead pommels, to the USA.

| | | | |
|---|---|---|---|
| Silversmith | George M. (Moses) Thurkle | 15 Great New St. square | 1789–1805 |
| Hilt maker | • Son of Francis Thurkle Sr. | | |
| Sword maker | • Apprenticed to Francis Thurkle Sr., 1782–1788 (see Francis Thurkle Sr.). | | |
| Sword belt | Thurkle (George Moses) & Skynner (Richard) | 15 Great New St. square | 1806–1812 |
| maker | George Moses Thurkle | 15 Great New St. square | 1813–1820 |
| Sword cutler | • Master of the Cutlers Company, 1814–1826. | | |
| | George Moses Thurkle | 5 New St. square | 1821–1825 |
| | Abraham Thurkle | 5 New St. square | 1826 |
| | • Son of George Moses Thurkle. | | |
| | • Apprenticed to George Moses Thurkle, 1818–1825 (see George Moses Thurkle). | | |
| | Abraham Thurkle | 115–116 New St. square & Dean St., Fetterlane | 1827–1833 |
| | Benjamin Thurkle | 238 High Holborn | 1834–1835 |
| | Benjamin Thurkle | 104 High Holborn | 1836–1863 |
| | Edward Thurkle | 104 & 200 High Holborn | 1864 |
| | Edward Thurkle | 104 High Holborn | 1865–1875 |
| | Edward Thurkle | 5 Denmark St., Soho | 1876–1899 |

• Edward Thurkle died in 1899.
• J.R. Gaunt succeeded Edward Thurkle in 1900 at the 5 Denmark St. address. (see J.R. Gaunt)

| | | | |
|---|---|---|---|
| | George Thurkle | 23 Lisle St. | 1900–1902 |
| | George Thurkle | 4 High St., Bloomsbury | 1903–1906 |
| | George Thurkle | 21 Denmark Place, Charing Cross | 1907–1912 |
| | George Thurkle | 149 High Holborn | 1913–1919 |

• Wilkinson Sword Company absorbed the George Thurkle Company, 1920.

| | | | |
|---|---|---|---|
| Army tailor | John Timewell | 3 Great Vine St. | 1842 |
| Sword cutler | John Timewell | 3 Upper Vine St., Regents | 1843–1844 |
| | John Timewell | 24 Duke St., St. James | 1845–1864 |
| | Henry Borne Timewell | 2 Sackville St., Piccadilly | 1865–1869 |

|  |  |  |  |
|---|---|---|---|
|  | Henry Borne Timewell | 8 Sackville St., Piccadilly | 1870–1882 |
|  | Timewell & Fulcher | 8 Sackville St., Piccadilly | 1883–1888 |
|  | Timewell & Co. | 8 Sackville St., Piccadilly | 1889–1913 |
|  | Timewell, Hooper & Co. | 8 Sackville St., Piccadilly | 1914–1936 |
|  | Timewell, Hooper & Co. | 38 Sackville St., Piccadilly | 1937–1941 |
|  | Timewell, Hooper & Co. | 20 Sackville St., Piccadilly | 1942–1943 |
| Sword cutler | **John Tipping**<br>• Successor to Thomas Blayney.<br>(see Thomas Blayney) | Exchange Alley,<br>Parish of St. Mary Woolnoth | 1705–1723 |
| Cutler | **Clement Tookey**<br>• Master of the Cutlers Company, 1700. |  | 1690–1705 |
| Silversmith<br>Spoon maker | **James Tookey**<br>• Apprenticed to Henry Green, 1733–1740. |  | Bc1719–1749 |
| Hilt maker | **James Tookey** | Noble St., Foster Lane | 1750–1751 |
| Sword cutler | **James Tookey**<br>• Master of the Cutlers Company, 1755. | 12 Silver St. | 1752–D1768 |
|  | **Elizabeth Tookey**<br>• Widow of James Tookey. | 12 Silver St. | 1768 |
|  | **Thomas Tookey**<br>• Son of James and Elizabeth Tookey.<br>• Apprenticed to father James Tookey, 1766–1768. | 12 Silver St. | B1751, 1769–D1791 |
| Silversmith | **Thomas Tookey** | 45 New Bond St. | 1773–1782 |
| Jeweler | **William Constable & Co.** | 45 New Bond St. | 1783–1784 |
| Toy (hardware)<br>maker | • Partner: Thomas Tookey (see William Constable & Co.). |  |  |
| Hilt maker | **Constable (William) & Tookey (Thomas)**<br>• Thomas Tookey died in 1791. | 45 New Bond St. | 1785–1791 |
| Sword cutler | **Ann Tookey**<br>• Widow of Thomas Tookey. | 49 New Bond St. | 1791–1808 |
|  | **William Tookey**<br>• Son of Thomas Tookey. | 49 New Bond St. | 1809–1815 |
| Sword cutler | **William Tooley**<br>• Master of the Cutlers Company, 1645. | Parish of St. Mary at the Hill | 1612–1655 |
| Silversmith<br>Hilt maker<br>Sword cutler | **William Toone**<br>• Apprenticed to Joseph Smith, 1718–1723.<br>• Apprenticed to Charles Jackson, 1723–1725. |  | Bc1704–1725 |
|  | **William Toone** | Moor Lane, near Cripplegate | 1725–1736 |
| Sword cutler | **Tooting Cutlery Works** |  | 1900 |
| Pike maker | **Edward Touch**<br>• On December 22, 1645, Touch contracted for 600 long pikes (16 feet long with steel heads) for the "New Model" army (parliamentary forces) under Gen. Sir Thomas Fairfax. | Houndsditch | 1642–1649 |
| Silversmith<br>Goldsmith | **Louis Toussiant** | Greek St., Parish of St. Anne<br>Soho, Westminster | 1751–1758 |
| Jeweler<br>Enamelers | • James Morisset (Toussiant's brother-in-law) probably apprenticed<br>and worked for him, 1752–1763. |  |  |
| Hilt maker<br>Sword cutler | **Louis Toussiant** | 11 Denmark St.<br>Parish of St. Giles in the Fields, Holborn | 1759–1766 |

| | James Morisset & Louis Toussiant | 11 Denmark St. | 1767–1775 |
| | | Parish of St. Giles in the Fields, Holborn | |

- Did enamel work for Parker & Wakelin (see Parker & Wakelin).
- Toussiant retired in 1775 and died in 1785.

(see James Morisset)

| | | | |
|---|---|---|---|
| Gun maker<br>Sword cutler | John Tow | 10 New Bond St. | 1883–1893 |

- Successor to Griffen & Tow (see Griffen & Tow).

| | | | |
|---|---|---|---|
| Furrier | Richard Townsend | 18 Fenchurch St. | 1794–1808 |
| Hat maker | Richard Townsend & Son | 18 Fenchurch St. | 1809–1812 |
| Army | Richard Townsend & Son | 16 & 18 Fenchurch St. | 1813–1819 |
| contractor | William & Thomas Townsend | 16 & 18 Fenchurch St. | 1820–1836 |
| Sword cutler | William Townsend & Co. | 16 & 18 Fenchurch St. | 1837–1838 |
| | Thomas Townsend | 16 & 18 Fenchurch St. | 1839–1855 |
| | Thomas Townsend & Co. | 16 & 18 Fenchurch St.,<br>14 & 34 Fenchurch St. | 1856 1965 |
| | Thomas Townsend & Co. | 32 Great Pulteney St. | 1966–1969 |

| | | | |
|---|---|---|---|
| Hilt maker | John Tredwell (Treadwell) | | 1644 |

| | | | |
|---|---|---|---|
| Sword cutler<br>Gunsmith | Richard Tredwell | Parish of St. Brides | 1640–1670 |

- On April 3, 1645, during the English Civil War, contracted for 1,500 swords (infantry) with belts at 4 shillings, 8 denier each for the "New Model" army (parliamentary forces) under Gen. Sir Thomas Fairfax.
- After the Civil War, supplied swords for the Board of Ordnance.
- Master of the Cutlers Company, 1664.

| | | | |
|---|---|---|---|
| Sword cutler | William Tresham | | 1645–1655 |

- Supplied swords to the Board of Ordnance, 1650–1651.

| | | | |
|---|---|---|---|
| Cutler | Richard Tress | | 1863–1878 |

- Master of the Cutlers Company, 1873.

Henry Trimnell
(see Charles Smith)

| | | | |
|---|---|---|---|
| Sword cutler | Thomas Trotter | | 1814–1820 |

| | | | |
|---|---|---|---|
| Gun maker<br>Sword blade<br>maker | Francis Troulett | | 1680–1696 |

- In 1681, Troulett made a hollow ground small sword blade in the "old way" (hollows and flutes beat into the blade by hand) for the petitioners to King James II (Hollow Sword Blade Company). The blade was shown to Lord Dartmouth, Master of Ordnance, and compared to the petitioners' small sword blade made in the "new way" (hollows and flutes ground into the blade with small grind stones). (See Chapter 4.)

| | | | |
|---|---|---|---|
| Cutler | Thomas Tuck | | 1617–1632 |

- Master of the Cutlers Company, 1627.

| | | | |
|---|---|---|---|
| Sword and<br>pike maker | Robert Tucker | | 1619–1630 |

| | | | |
|---|---|---|---|
| Cutler | Thomas Tunn | | 1680–1695 |

- Master of the Cutlers Company, 1690.

| | | | |
|---|---|---|---|
| Sword cutler | **Thomas Turke** | | 1625–1630 |
| | • Supplied swords to the Board of Ordnance, 1627. | | |
| Gold and silver lacemen | **Walter Turner** | 3 Strand at the sign of the Whiteheart | 1740–1742 |
| Sword cutler | • Succeeded by Turner, Hill & Pitter (see Richard Hill). | | |
| Gun maker | **Edward Turvey** | next to Furnivalls Inn, Holborn | 1692–D1714 |
| Sword cutler | **William Turvey** | next to Furnivalls Inn, Holborn | 1714–D1744 |
| Gold and silver lacemen | **Tyler (E.) & Rogers (R.)** | 26 Bedford St., Covent Garden | 1846–1847 |
| | • Successor to Lonsdale & Tyler (see Lonsdale & Tyler). | | |
| Sword cutler | **E. Tyler & Co.** | 26 Bedford St., Covent Garden | 1848–1854 |
| | • Succeeded by Herbert & Boys (see Herbert & Boys). | | |
| Sword cutler | **George Tyler** | | 1801 |
| | **Henry Ulton** (see Linney & Ulton) | | |
| Sword cutler | **Underhill & Cooper** | | 1803 |
| Cutler | **Henry Thomas Underwood** | | 1827–1842 |
| | • Master of the Cutlers Company, 1837. | | |
| Cutler | **Yeeling Underwood** | | 1809–1824 |
| | • Master of the Cutlers Company, 1819. | | |
| Sword, gun and button exporter | **Henry Upson** | | 1800–1820 |
| | • Probably a partner in the company of Samuel Williams (a London-based International trader). | | |
| | • Ordered 1,000 gilt and plated officer and noncomissioned officer swords from Osborn & Gunby, London and Birmingham, in July 1812. | | |
| | • By August 1812, 200 swords had been shipped to Upson by Osborn & Gunby. | | |
| | • Upson also bought swords from Samuel Harvey and Samuel Walker of Birmingham. | | |
| | • Relative of George Upson of the New York, NY, military outfitter Richards (Stephen) & Upson (George), 1808–1816. | | |
| | • Exported swords to Richards, Upson & Co., some during the War of 1812, even though there was a U.S. embargo on all British goods. | | |
| Cutler | **Zachariah Uwins** | | 1828–1843 |
| | • Master of the Cutlers Company, 1838. | | |
| Cutler | **Richard Vale** | | 1576–1892 |
| | • Master of the Cutlers Company, 1586–1587. | | |
| Cutler | **William Vale** | | 1470–1496 |
| | • Master of the Cutler Company, 1480, 1481, 1486, 1487, 1490, and 1491. | | |
| Army tailor Sword cutler | **Robert Valentine** | 9 Sackville Row | 1951–1966 |
| | • Successor to Guthrie & Valentine (see Thomas Ansley Guthrie). | | |
| Sword cutler Gun maker | **George Vandebaise** | | 1700 |
| | • Made a sword-pistol. | | |

| | John Varnan<br>(see Hicks & Varnan) | | |
|---|---|---|---|
| Cutler | **Sir Walter Vaughn**<br>• Master of the Cutlers Company, 1911. | | 1901–1916 |
| | **Robert Veale**<br>(see William Widdowson) | | |
| Goldsmith<br>Gun and sword<br>decorator | **Antoine Vechte**<br>• In 1822, immigrated to London and worked for Storr & Mortimer, 1822–1838.<br>• Designer and artist in metal, decorated firearms and swords.<br>**Antoine Vechte** | Paris, France | Bc1800–1822<br><br><br>1839–D1868 |
| Military<br>outfitter | **Alexander Vener**<br>• On May 15, 1645, during English Civil War, contracted for ensigns, drums, partisans, and halberds for the "New Model" army (parliamentary forces under Gen. Sir Thomas Fairfax. | | 1640–1652 |
| | **John Henry Vere**<br>(see Edward Aldridge Jr.) | | |
| Cutler | **Henry Verinder**<br>• Master of the Cutlers Company, 1836 and 1844. | | 1826–1849 |
| Woollen draper<br>(maker) and<br>mercer (dealer)<br>Army clothier<br>Military<br>warehouse<br>Sword cutler | **William Vernon**<br>**William Vernon & Son**<br>**William Vernon & Son**<br>**William Vernon & Son**<br>**William Vernon & Son**<br>**W. (William) H. Vernon & Co.**<br>**William & John Vernon**<br>**William & John Vernon & Co.**<br>**Vernon (William & John) & Co**<br>**William Vernon** | 4 Charing Cross<br>4 Charing Cross<br>3 & 4 Charing Cross<br>4 Charing Cross<br>4 Charing Cross<br>4 Charing Cross<br>4 Charing Cross<br>4 Charing Cross<br>4 Charing Cross<br>4 Charing Cross | 1790–1815<br>1816–1817<br>1818–1819<br>1820–1837<br>1838<br>1839<br>1840–1841<br>1842<br>1843<br>1844–1845 |
| | • Succeeded by Thomas Edwards at the 4 Charing Cross address (see Thomas Edwards).<br>**William Vernon**<br>**William Vernon** | 1 Bloomsbury Square<br>4 Harpur St., Theobald's Road | 1846<br>1847 |
| Silversmith<br>Hilt maker<br>Sword cutler | **Thomas Vicaridge (Vickeridge)**<br>• Apprenticed to Joseph Jones (c. 1677–1682).<br>**Thomas Vicaridge**<br>• Had apprentice Elias Hosier, c. 1694–1701 (see Elias Hosier).<br>• Had apprentice John Carmen Sr., c. 1708–1715 (see John Carmen Sr.).<br>• Made fine silver-hilted small swords.<br>• John Carmen Sr. succeeded Thomas Vicaridge in 1716. | New St., Chancery Lane | Bc1663–1682<br><br>1683–D1716 |
| Silversmith<br>Knife and<br>sword hilt<br>maker<br>Knife and sword<br>cutler | **John Vincent** | Angle Court on Snowhill | 1730–1740 |
| Silversmith | **Thomas Waddington** | | 1661 |

ENGLISH SWORD MAKERS, CUTLERS, DEALERS, AND CRAFTSMEN WHO MOUNTED SWORDS

| | | | |
|---|---|---|---|
| Hilt maker<br>Sword cutler | • Signed an advertisement warning sword purchasers against counterfeit cast silver hilts. | | |
| Goldsmith<br>Silversmith<br>Jeweler<br>Hilt maker<br>Sword cutler | **Edward Wakelin**<br>• Apprentice and journeyman to John Hugh LeSage, 1730–1746 (see John Hugh LeSage). | | Bc1716–1746 |
| | **George Wickes & Edward Wakelin**<br>• George Wickes died in 1761.<br>• Successors to George Wickes<br>(see George Wickes). | Panton St.<br>two doors from the Haymarket<br>Parish of St. James | 1747–1761 |
| | **John Parker & Edward Wakelin** | Panton St.<br>two doors from the Haymarket<br>Parish of St. James, at the sign of the Kings Arms | 1762–1777 |
| | • Silversmiths and goldsmiths John Carter; Carter, Smith & Sharp; Smith & Sharp; and Emick Romer all did work for John Parker & Edward Wakelin. (see John Carter; Carter, Smith & Sharp; Smith & Sharp; Emick Romer)<br>• Employee: James Morisset, 1762–1765.<br>• Employee: John Christopher Romer (son of Emick Romer), 1762–1777 (see John Christopher Romer).<br>• Employee: Stephen Gilbert, 1762–1777<br>• Son John Wakelin (Bc1752) apprenticed and worked for him, 1766–1777.<br>• Edward Wakelin retired in 1777 and died in 1784.<br>(see John Parker) | | |
| | **John Wakelin & William Taylor**<br>• Employee: Robert Garrard, 1780–1791. | Panton St.<br>two doors from the Haymarket,<br>Parish of St. James | 1777–1791 |
| | **John Wakelin & Robert Garrard**<br>• Wakelin retired in 1801.<br>• Succeeded by Robert Garrard (see Robert Garrard).<br>(see William Garrard) | 31 Panton St. | 1792–1801 |
| Cutler | **John Wakeman**<br>• Master of the Cutlers Company, 1463–1464. | | 1453–1469 |
| Hafter<br>(hilt maker) | **Richard Waldron** | | 1840–1850 |
| | **Henry Walford**<br>(see Robert Makepeace) | | |
| Sword cutler | **W. Walker**<br>**Walker (W.) & Lark (S.)**<br>**W. Walker & Sons** | | 1710–1724<br>1725–1735<br>1736–1796 |
| | **William Walker**<br>(see Kemp Brydes) | | |
| | **William Walker**<br>(see Chapter 3) | | |
| Cutler | **James Beaumont Waller**<br>• Master of the Cutlers Company, 1905. | | 1895–1910 |
| Gun maker<br>Sword cutler | **John Wallis**<br>• Son of W.R. Wallis (owner of the Manton Co.).<br>• London agent for Manton & Co.<br>• Successor to Samuel Nock. | 116 Jermyn St. | 1859–1864 |

| | | | |
|---|---|---|---|
| Cutler | **Adam Ward** | | 1622–1631 |
| | • Master of the Cutlers Company, 1632. | | |
| Cutler | **Geoffrey Ward** | | 1616–1631 |
| | • Master of the Cutlers Company, 1626. | | |
| Silversmith<br>Hilt maker<br>Sword cutler | **Thomas Ward**<br>• Signed an advertisement warning sword purchasers about counterfeit cast silver hilts. | | 1661 |
| | **William Ward**<br>(see James Lawson) | | |
| Sword cutler | **William Ward** | York | 1382 |
| Sword cutler | **Edward Warnett** | | 1651 |
| Sword belt maker<br>Sword cutler | **Richard Warren** | | 1608–1610 |
| Sword belt maker | **Thomas Warren**<br>• Successor to John David Cripps & Co. (see John David Cripps & Co.). | 23 Cursitor St., Chancery Lane | 1846–D1861 |
| Sword cutler | **Mary Warren**<br>• Widow of Thomas. | 23 Cursitor St., Chancery Lane | 1861–1871 |
| Goldsmith<br>Silversmith<br>Jeweler | **Samuel Wastell**<br>• Apprenticed to Benjamin Braford, 1694–1699.<br>• Apprenticed to John Fawdery, 1699–1700. | | Bc1680–1700 |
| Hilt maker | **Samuel Wastell** | Finch Lane | 1701–1704 |
| Sword cutler | **Samuel Wastell** | Leadenhall St. | 1705–1720 |
| | • Had apprentice George Wickes, 1712–1719.<br>• Succeeded by George Wickes (see George Wickes). | | |
| Sword cutler | **William J. Waterer** | 9 Pugh Lane, Carnaby St.<br>Parish of St. Regent | 1832–1930 |
| Silversmith<br>Hilt maker<br>Sword cutler | **John Watkins**<br>• Had apprentice George Harrison, 1751–1759 (see George Harrison). | | 1730–1760 |
| Sword cutler | **William Watkins** | | 1606 |
| Sword cutler | **Edward Watkins** | | 1830–1840 |
| Silversmith<br>Hilt maker<br>Sword cutler | **George Watson**<br>• Mounted silver-hilted dirks. | | 1766–1791 |
| Pole arm maker | **William Watson**<br>• In 1557–1558, supplied infantry pikes to the Drapers Company for the defense of their settlement in Calais, France, during the French War.<br>• In 1562, made two ceremonial halberds for the city of London. | | 1557–1560 |
| Sword maker | **H. Watts**<br>• Made M1856 pioneer swords. | | 1850–1876 |

| | William Watts<br>(see W. Locke) | | |
|---|---|---|---|
| Sword cutler | J. Waycott & Sons | | 1830 |
| Hatter | Henry Wayte | 11 Panton St. | 1788–1809 |
| Hosier | Henry & Charles Wayte | 11 Panton St. | 1810–1815 |
| Sword cutler | Henry Wayte | 10 Panton St. | 1816–1820 |
| Cutler | Simon Weaver | | 1668–1683 |
| | • Master of the Cutlers Company, 1678. | | |
| Gold and silver lacemen | Charles Webb | 57 Piccadilly | 1804–1820 |
| | Charles Webb | 48 Piccadilly | 1821–1826 |
| Sword cutler | Charles Webb & Co. | 48 Piccadilly | 1827–1833 |
| | Webb (Charles) & Co. | 41 Piccadilly | 1834–1837 |
| | William Webb | 48 Piccadilly | 1838–1850 |
| Gold and silver lacemen | C. (Charles Sr.) Webb & Co. | 20 Old Bond St. | 1824–1829 |
| | Webb (Charles Sr.) & Co. | 23 Old Bond St. | 1830–1836 |
| Embroiderer | Charles Webb (Jr.) | 23 Old Bond St. | 1837–1857 |
| Sash maker | Charles Webb (Jr.) & Co. | 23 Old Bond St. | 1858–1868 |
| Military warehouse | Webb (Charles Jr.) & Bonella (John) | 23 Old Bond St. | 1869–1887 |
| | • Appointed sash and sword manufacturers to His Majesty the King and His Royal Highness the Prince of Wales. | | |
| Sword cutler | Webb (Charles Jr.) & Bonell (John) | 4 Savile Row | 1888–1891 |
| | Webb (Charles Jr.) & Co. | 29 Savile Row | 1892–1894 |
| | • Merged with Edward Stillwell & Son, 1895. | | |
| Sword cutler | Nicholas Webb | | 1650–1660 |
| Pike maker | Frances Webber | | 1642–1649 |
| | • On July 26, 1645, during the English Civil War, contracted for 200 long pikes (16 feet long with steel heads) at 4 shillings, 2 denier each for the "New Model" army (parliamentary forces) under Gen. Sir Thomas Fairfax. | | |
| Pike maker | Anthony Webster | Milk St. | 1642–1649 |
| | • On December 22, 1645, during the English Civil War, contracted for 1,250 long pikes (16 feet long with steel heads) at 3 shillings, 10 denier each for the "New Model" army (parliamentary forces) under Gen. Sir Thomas Fairfax. | | |
| Jeweler | Henry Webster | 8 Plough Court, Fetter Lane | 1805–1810 |
| Goldsmith | Henry Webster | 10 Rolls Building, Fetter Lane | 1811–1836 |
| Hilt maker<br>Sword cutler | Henry Webster | 21 Bedford St., Strand | 1837–1840 |
| Cutler | Charles Welch | | 1887–1926 |
| | • Master of the Cutlers Company, 1907. | | |
| Outfitter | Welch (James) & Stalker (William) | 134 Leadenhall St. | 1795–1806 |
| Agent for | • Successor to Evans & Welch (see Evans & Welch). | | |
| the Honorable | Welch (James) & Welch (Arthur D.) | 134 Leadenhall St. | 1806–1809 |
| East India | Stalker (William) & Welch (Arthur D.) | 134 Leadenhall St. | 1810–1814 |
| Company | Stalker (William), Welch (Arthur D.) & Hilburn (Henry) | 134 Leadenhall St. | 1815–1818 |
| Sword cutler | Stalker (William), Welch (Arthur D.) | 134 Leadenhall St. | 1819 |

|  |  |  |  |
|---|---|---|---|
|  | Arthur D. Welch | 134 Leadenhall St. | 1820–D1836 |
|  | James D. Welch | 134 Leadenhall St. | 1836–D1841 |
|  | • Son of Arthur D. Welch. | | |
| Cutler | Samuel Welch | | 1908–1923 |
|  | • Master of the Cutlers Company, 1918. | | |
| Silversmith<br>Goldsmith<br>Hilt maker<br>Sword cutler | John Welles<br>• Had apprentice Benjamin Corbett, 1718–1722 (see Benjamin Corbett). | | 1690–D1722 |
| Cutler | Richard Wellom | | 1406–1422 |
|  | • Master of the Cutlers Company, 1416–1417. | | |
| Hatter<br>Sword cutler | Edmund Wells | Fenchurch St. | 1761–1763 |
| Chiseller<br>Hilt maker<br>Sword cutler | Joseph Wells<br>• Made pierced steel sword hilts. | St. Johns Square, Clerkenwell | 1760–1770 |
|  | J. (Joseph) B. Wells | 57 Conduit St. | 1770–1780 |
| Goldsmith<br>Silversmith<br>Jeweler<br>Hilt maker<br>Sword cutler | J. (Joseph) & T. (Thomas) Wells<br>J. (Joseph) & T. (Thomas) Wells<br>J. (Joseph) & T. (Thomas) Wells<br>T. (Thomas) Wells<br>T. (Thomas) Wells & Co. | 3 Chettenham, Lambeth Place<br>101 Strand<br>76 Holborn Bridge<br>76 Holborn Bridge<br>76 Holborn Bridge | 1797<br>1798<br>1799–1800<br>1801–1803<br>1804–1818 |
|  | Henry Westall<br>(see Samuel Firming) | | |
| Sword cutler | John Whaley | Parish of St. Sepulchre<br>without Newgate | 1585–1589 |
| Sword hilt<br>designer<br>etcher and<br>gilder | William Henry Whawell<br>• Worked for Edward Thurkle Sr., 1746–1785 (see Edward Thurkle Sr.). | | 1725–1785 |
| Sword<br>ornamentor | William Whawell | 13 Dean St., Fetter Lane | 1830–1833 |
| Sword cutler | Wheeler & Robinson | Hanover Square | 1827–1830 |
| Cutler | Amos White | | 1723–1738 |
|  | • Master of the Cutlers Company, 1733. | | |
|  | Henry Thomas White<br>(see Thomas Hawkes & Co.) | | |
| Hatter<br>Sword cutler | Jonathan White | 187 Strand, Arandel St. | 1779–1795 |

| | | | |
|---|---|---|---|
| Sword maker<br>Armourer | J. (John) White | Aldershot | 1818–1820 |
| Jeweler<br>Embroiderers<br>Gold and silver<br>lace maker<br>Sword cutler | Robert White<br>Robert White<br>Robert White & Sons<br>Robert White & Sons<br>Robert White & Sons<br>Robert White & Sons | 8 Gibbons St., Waterloo Lane<br>30 Bow St., Covent Garden<br>30 Bow St., Covent Garden<br>43 Drury Lane<br>19 Stakeley St.<br>57 & 59 Neal St. | 1850–1855<br>1856–1895<br>1896–1902<br>1903–1940<br>1941–1946<br>1947–1969 |
| Pike maker | Robert White<br>Susanna White<br>• Widow of Robert White.<br>• On December 22, 1645, contracted for 500 long pikes (16 feet long with steel heads) at 3 shillings, 10 denier each for the "New Model" army (parliamentary forces) under Gen. Sir Thomas Fairfax. | Milk St.<br>Milk St. | 1630–D1642<br>1642–1649 |
| Sword cutler | Thomas White | Aldershot | 1714 |
| Silversmith<br>Goldsmith<br>Hilt maker<br>Sword cutler | White & Campbell | New Bond St. | 1870–1880 |
| Goldsmith<br>Hilt maker<br>Buckle maker<br>Sword cutler | Samuel Whitford (Sr.)<br>• Partners: Thomas Whitford (son), and Mary Whitford (wife).<br>• Samuel Whitford Sr. died in 1777.<br>Mary Whitford & William Ballantine<br><br>Samuel Whitford (Jr.)<br>• Son of Samuel & Mary Whitford.<br>Samuel (Jr.) & George Whitford<br>Samuel Whitford (Jr.)<br>Samuel Whitford (Jr.) & George Pizey<br>Samuel Whitford (Jr.)<br>Samuel Whitford (Jr.)<br>• Samuel Whitford Jr. died in 1855.<br>(see William Ballantine) | 6 Kings Head Court<br>Parish of St. Martins Le Grand<br><br><br>6 Kings Head Court<br>Parish of St. Martins Le Grand<br>1 Smithfield Bars<br><br>15 Denmark Court, Strand<br>15 Denmark Court, Strand<br>15 Denmark Court, Strand<br>25 Grafton St., Soho<br>4 Porter St., Newport Market | 1760–D1777<br><br><br><br>1777–1787<br><br>1788–1801<br><br>1802–1806<br>1807–1809<br>1810–1811<br>1812–1816<br>1817–D1855 |
| Cutler | Henry Whittaker<br>• Master of the Cutlers Company, 1673. | | 1663–1678 |
| Goldsmith<br>Silversmith<br>Jeweler<br>Hilt maker<br>Sword cutler | George Wickes<br>• Apprenticed to Samuel Wastell, 1712–1719 (see Samuel Wastell).<br>George Wickes<br>• Successor to Samuel Wastell.<br>George Wickes<br>George Wickes & John Craig<br>• Successors to silversmith John Craig.<br>• Had apprentice David Craig (son of John), 1731–1735.<br>• John Craig became ill in 1735 and died in 1736.<br>George Wickes | <br><br>Leadenhall St.<br><br>Threadneedle St.<br>corner of Norris St.<br>and the Haymarket, Parish of St. James<br><br><br>Panton St., two doors from the Haymarket<br>Parish of St. James<br>at the sign of the Kings Arms and Feathers | Bc1698–1719<br><br>1720–1722<br><br>1722–1730<br>1730–1735<br><br><br><br>1735–1747 |

|   |   |   |   |
|---|---|---|---|
|  | **George Wickes & Edward Wakelin**<br>• Silversmith only. | Panton St., two doors from the Haymarket<br>Parish of St. James | 1747–1761 |
|  | **George Wickes & Samuel Netherton**<br>• Goldsmith and jeweler only.<br>• George Wickes had apprentice and journeyman John Parker, 1751–1761.<br>• George Wickes had a large business with many nobles as his clients.<br>• Wickes became silversmith to the Prince of Wales in 1735.<br>• Other clients included the Dukes of Devonshire, Chandos, Kingston, Roxburgh, Montrose, and Bridgewater; the Earls of Scarborough and Kildare; Admiral Vernon; the Duchess of Norfolk; James Fitzgerald (Duke of Leinster); and the Marquess of Caernarron.<br>• George Wickes died in 1761.<br>• Succeeded by Parker & Wakelin.<br>(see Edward Wakelin and John Parker) | Panton St., two doors from the Haymarket<br>Parish of St. James | 1747–1761 |
| Goldsmith<br>Pistol dealer<br>Sword cutler | **Wicks (John) & Bishop (William)**<br>• Succeeded by William Bishop (see William Bishop). | 170 New Bond St. | 1823–1825 |
| Goldsmith<br>Jeweler<br>Hilt maker<br>Sword cutler | **Widdowson (William) & Veale (Robert)**<br>• Successors to Salter, Widdowson & Tate (see Salter, Widdowson & Tate).<br>• Sword cutler to Prince Albert.<br>• William, Henry & Louis Dee decorated some of their presentation swords.<br>• Merged with Longman & Strongitharm in 1877.<br>(see John & Richard Longman) | 73 Strand | 1835–1876 |
| Sword maker | **John Widmore**<br>• Apprenticed to Richard Blayney, 1678–1683 (see Richard Blayney).<br>**John Widmore** |  | Bc1663–1683<br><br>1684–1710 |
| Silversmith<br>Goldsmith<br>Hilt maker<br>Sword cutler | **Edward Wigan**<br>• Apprentice and journeyman to James Stamp, 1772–1785.<br>**Goodbehere (Samuel) & Wigan (Edward) & Co.**<br>• Partner: James Bult.<br>**Goodbehere (Samuel), Wigan (Edward) & Bult (James)**<br>(see Samuel Goodbehere, James Bult) | <br><br>86 Cheapside<br><br><br>86 Cheapside | Bc1798–1775<br><br>1786–1799<br><br><br>1800–1818 |
| Gun maker<br>Sword cutler | **George Wilbraham**<br>**George Wilbraham**<br><br>• Succeeded by John Wilbraham at 26 Goulston Square.<br>**George Wilbraham** | 26 Goulston Square<br>26 Goulston Square<br>& 123 Leadenhall St.<br><br>123 Leadenhall St. | 1819–1825<br>1826<br><br><br>1827–1836 |
| Gun maker<br>Sword maker | **Joseph Wilbraham**<br>• Successor to William Childe (see William Childe).<br>**Joseph Wilbraham**<br>• Succeeded by George Fuller at 280 Strand (see George Fuller).<br>**Joseph Wilbraham** | 280 Strand<br><br>280 Strand & 404 Strand<br><br>404 Strand | 1851–1853<br><br>1854<br><br>1855–1856 |
| Gun maker<br>Sword cutler | **John Wilbraham**<br>• Successor to George Wilbraham.<br>**John Wilbraham**<br>**John Wilbraham** | 26 Goulston Square<br><br>17 White Row, Spital Fields<br>14 Black Church Lane | 1827–1831<br><br>1832–1840<br>1841–1852 |

| | | | |
|---|---|---|---|
| | John Wilbraham | 4 Pavilion Terrace, Battersea Fields | 1853–1868 |
| Silversmith<br>Hilt maker<br>Sword cutler | Nathaniel Wilby | Churchyard, Bloomsbury | 1778–1788 |
| Silversmith<br>Hilt maker<br>Sword cutler | George Wilcox (Willcocks)<br>• Had apprentice Thomas North, c. 1716–1723 (see Thomas North).<br>• Had apprentice William Wilson, 1719–1724 (see William Wilson). | Wine Office Court, Fleet St. | 1715–1730 |
| Cutler | John Wilcox (Willcocks)<br>• Master of the Cutlers Company, 1695. | | 1685–1700 |
| Bladesmith | Richard Wilcox (Willcocks)<br>• Freeman of the Bladesmiths Company, 1465.<br>• Master of the Bladesmiths Company, 1489. | | 1465–1500 |
| Sword cutler | Wilcox & Son | Argyle St. | 1850 |
| Gold and silver<br>lacemen | Samuel Wilding | 186 Strand | 1789–1804 |
| Sword cutler | Wilding (Samuel) & Childe (John)<br>• Succeeded by John Childe (see John Childe). | 186 Strand | 1805–1826 |
| Gun maker<br>Sword cutler | James Wilkes<br>James Wilkes | 1 Coventry Ct.<br>21 St. James St. | 1796–1798<br>1799–1810 |
| Silversmith<br>Hilt maker<br>Sword cutler | John Wilkins<br>• Apprenticed to George Scrivener, c. 1682–1689 (see George Scrivener).<br>John Wilkins | <br><br>Exchange Alley | Bc1668–1689<br><br>1690–1700 |
| Gun maker<br>Sword maker | James Wilkinson<br>• Apprenticed to gun maker Henry Nock, c. 1778–1784.<br>• Wilkenson married Nock's daughter.<br>Henry Nock & James Wilkinson<br>• Nock died in 1804 and willed his company to Wilkinson.<br>James Wilkenson<br>• Appointed gun maker to His Majesty the King.<br>James Wilkenson<br>James Wilkenson & Son (Henry)<br>• James Wilkenson wrote the following books: *Engines of War* (1841), *Observations on Firearms*, and *Observations on Swords*.<br>• In 1844 he invented a machine for testing sword blades.<br>• James Wilkenson retired in 1824 and died in 1848.<br>Henry Wilkinson<br>• Appointed gun maker to Her Majesty the Queen.<br>Henry Wilkinson<br>  next to the Board of Ordnance<br>• John Latham (B1820–D1880) joined the company in 1847 as a mercantile clerk. He later became a partner and owner.<br>• In 1846, Wilkinson began to make swords and eventually became England's largest sword maker. Latham expanded the sword production.<br>• They made army and navy officers swords, enlisted men's swords, presentation swords, and custom-order swords.<br>Henry Wilkinson | <br><br><br>10 Ludgate St.<br><br>10 Ludgate St.<br><br>12 Ludgate St.<br>12 Ludgate St.<br><br><br><br>17 Pall Mall<br><br>27 Pall Mall<br><br><br><br><br><br>27 Pall Mall & 18 St. Marys Axe | B1759–1784<br><br><br>1785–1804<br><br>1805–1806<br><br>1807–1817<br>1818–1824<br><br><br><br>1825–1828<br><br>1829–1848<br><br><br><br><br><br>1849–1852 |

**Henry Wilkinson & Son**　　　　27 Pall Mall & 18 St. Marys Axe　　1853–1854
- Started to number their swords in 1854.

**Wilkinson (Henry) & Son**　　　　27 Pall Mall　　　　1855–c1882
- Appointed gun, sword, and rifle maker to Her Majesty the Queen, His Royal Highness the Prince of Wales, the Duke of Edinburgh, and the Duke of Connaught.
- In 1858, Wilkinson (in ill health) drew up an agreement stating that his partner, John Latham, could purchase the Wilkinson & Son company when he died.
- Henry Wilkinson died in 1861 and Latham became the owner of Wilkinson & Son.
- John Latham was an accomplished swordsman.
- In 1871, John Francis Latham (John's son, B1856–D1889) began working with father.
- In 1877, Henry Wilkinson Latham (John's son, B1860–D1904) joined the company.
- Wilkinson, under Latham, expanded into accoutrements such as uniforms and became a large army contractor.
- Gun making became secondary to sword and bayonet making.
- John Latham wrote two theses: *The Shape of Sword Blades* and *Notes on Swords in the International Exposition of 1862*.
- John Latham died in 1880 and left Wilkinson & Son to his eldest son, John Francis Latham.
- In 1880, Latham acquired the Charles Reeves Company.
- Latham used the Reeves factory (8 Air Street) to proof swords.
- John Francis Latham decided to expand production. In c. 1882, he leased a steam-powered factory on Baldwins Place off Grays Inn Road.

**Wilkinson & Son**　　　　27 Pall Mall　　　　c1882–1887
- Workshops and offices located at 27 Pall Mall.
- Factory located at Baldwins Place off Grays Inn Road.
- Henry Wilkinson Latham (younger brother of Francis Latham) worked with him to modernize the company. New machinery was added.
- At the Baldwins Place factory, blades were forged and proofed, scabbards were made, and hilts and metal parts were cast and polished.
- At the basement workshop at 27 Pall Mall, sword blades were mounted to hilts, the blades were etched, and the swords finished.
- In 1884, John Francis Latham was a member of the War Office Committee investigating the purchasing of swords and bayonets when complaints of flaws surfaced after the Gordon expedition.
- The War Office decided to concentrate production of swords and bayonets in England rather than purchase abroad.
- In 1887, Latham obtained a contract for 150,000 1886 pattern bayonets.
- His factories could not produce such quantities, and he needed expert help to set up production. Latham traveled to Solingen, Germany, and formed a partnership with Rudolph Kirschbaum of the Weyersburg & Kirschbaum Company.
- He purchased a sawmill on Kings Road, Chelsea, and converted it to a sword and bayonet factory. It was called the Oakley Works.
- He outfitted the factory with English sword- and bayonet-making machinery approved by Kirschbaum.
- Kirschbaum brought German sword makers and craftsmen to the factory to train the English workers.
- His brother, Henry Wilkinson Latham, a superb engineer, helped lay out the factory.

**Wilkinson & Son**　　　　27 Pall Mall　　　　1887–1889
- Factory on Kings Road (Oakley Works), Chelsea.
- Appointed sword cutler to Queen Victoria.
- The Oakley Works employed 300 workmen.
- It could make 6,000 bayonets and 6,000 swords and lances annually.
- In 1889, the partnership with Rudolph Kirschbaum was dissolved. A joint stock company was formed called Wilkinson Sword Company Limited.
- John Francis Latham was chairman and managing director.

**Wilkinson Sword Company Ltd.**　　　　27 Pall Mall　　　　1889–1901
- Factory on Kings Road (Oakley Works), Chealsea.
- John Francis Latham died in 1898. His brother, Henry Wilkinson Latham, took over the company.

**Wilkinson Sword Company Ltd.**　　　　27 Pall Mall　　　　1901–1909
- Factory located on Southfield Road, Acton.
- Began making other products such as safety razors (patented in 1903).

- Eventually made typewriters, bicycles, motorcycles, motor cars, and sporting and hunting equipment.
- Henry Wilkinson Latham died in 1904. His son, Enderby Latham, took over the company.
- Enderby Latham took over sword production until his death in 1969.
- In 1907, John Wilkinson Latham Sr. (B1891–D1952, son of John Francis Latham) joined the company.
- In 1915, he became managing director.

**Wilkinson Sword Company Ltd.**  53 Pall Mall  1910–1949
- Factory located on Southfield Road, Acton.
- Made more than two million bayonets during World War I.
- In 1920, began making garden tools.

**Wilkinson Sword Company Ltd.**  53a Pall Mall  1950–1957
- Factory located on Southfield Road, Acton.
- John Wilkinson Latham (Sr.) died in 1952.
- In 1956, John Wilkinson Latham Jr. (B1920–D1999) joined the company.

**Wilkinson Sword Company Ltd.**  16 Pall Mall  1958–1969
- Factory, called the Southfield Road Sword Works, located in Acton.
- Still made razor blades.
- Now produced scissors and kitchen knives.
- In 1969, Enderby Latham died and John Wilkinson Latham Jr. took over the company.
- He took over sword production.

**Wilkinson Sword Ltd.**  19–21 Brunel Road, Acton  1969–2003
- John Wilkinson Latham retired in 1976 and died in 1999.
- The Wilkinson company absorbed the following companies:
  - Charles Reeves, London (1880)
  - Robert Mole & Son, Birmingham (1920)
  - George Thurkle, London (1920)
  - George Pillin, London (1922)
- Over the years, the Wilkinson company sold swords to many sword cutlers, including Carr & Son, J. Dece & Sons, Harrods Ltd., W.C. Cater, Flights Ltd., Hawkes & Co., William Cater, Gieves Ltd., and Conway Williams.
- Wilkinson Sword Ltd. has offices in Solingen, Germany.
- Wilkinson Sword Ltd. still makes the following products:
  - British Army officer sword (infantry, cavalry, artillery)
  - Royal Navy officer swords
  - Royal Air Force officer swords
  - U.S. military swords
  - Miniature swords
  - Fraternal swords
  - Personal and commercial presentation swords
  - Limited edition swords
  - Sword belts, knots, frogs, bags
  - Sword desk sets
  - Commando and survival knives
  - Embroidered badges

| | | | |
|---|---|---|---|
| Sword cutler | **A. Williams** | Exeter Change, Strand | 1796 |
| Military outfitter | **Conway Williams**<br>• Sold Wilkinson Sword Company swords. | 48 Brock St. | 1915 |
| Cutler | **Ebenezer Williams**<br>• Master of the Cutlers Company, 1893. | | 1883–1898 |
| Cutler | **Edward Williams**<br>• Master of the Cutlers Company, 1716. | | 1706–1721 |

| | | | |
|---|---|---|---|
| Cutler | **Edwin Hadley Williams** | | 1912–1927 |
| | • Master of the Cutlers Company, 1922. | | |
| Silversmith<br>Hilt maker<br>Sword cutler | **Francis Williams** | | 1656–1690 |
| | • Had apprentice Daniel Shelmerdine, c. 1676–1681 (see Daniel Shelmerdine). | | |
| Cutler | **Francis William Williams** | | 1900–1915 |
| | • Master of the Cutlers Company, 1910. | | |
| Gun maker<br>Sword cutler | **Joseph Williams** | 67 Threadneedle St. | 1834–1850 |
| | • Successors to Gameson & Co. (see Gameson & Co.). | | |
| Button maker<br>Accoutrement<br>maker<br>Sword cutler | **William & John Williams** | 103 St. Martins Lane | 1783–1788 |
| | **William Williams & Son** | 103 St. Martins Lane | 1789–1816 |
| | **Walter Williams** | 103 St. Martins Lane | 1817–1828 |
| | • Succeeded by Phipson, Doughty & Co. (see Phipson, Doughty & Co.). | | |
| Sword cutler | **Gregory Williamson** | Fleet St., Parish of St. Bride | 1567–1568 |
| Sword cutler | **Jodocum Williamson** | White Chapel, Middlesex | 1583 |
| Hatter<br>Hosier<br>Sword cutler | **Williamson (John Sr.) & Son (John Jr.)** | 26 High Holborn | 1770–1771 |
| | **John Williamson Jr.** | 41 High Holborn | 1772–1779 |
| | **J. (John Jr.) & W. (William) Williamson** | 41 High Holborn | 1780–1783 |
| | **Williamson (John Jr.) & Hollier (David)** | 41 High Holborn | 1784–1790 |
| Knife maker | **Thomas Wills** | | 1560–D1607 |
| | • In 1607, his "thistle" mark was obtained by John Jenks (see John Jenks). | | |
| Silversmith<br>Hilt maker<br>Sword maker | **Daniel Wilson** | | 1645–1700 |
| | • Had apprentice Joseph Reason, c. 1690–1696. | | |
| | • Master of the Goldsmiths Company, 1688. | | |
| | • In 1661, he signed an advertisement warning sword purchasers about counterfeit cast silver hilts. | | |
| | **John Wilson**<br>(see Charles Henry Gilks) | | |
| Gold and silver<br>laceman<br>Sword cutler | **Thomas Wilson** | 31 Lombard St. | 1858–1865 |
| Silversmith<br>Hilt maker<br>Sword cutler | **William Wilson** | | Bc1705–1725 |
| | • Apprenticed to George Wilcox, 1719–1724 (see George Wilcox). | | |
| | • Apprenticed to Thomas North, 1724–1725 (see Thomas North). | | |
| | **William Wilson** | New Street | 1726–1730 |
| Military tailor<br>Sword cutler | **Wilson & Willman** | 18 Old Bond St. | 1833–1838 |
| | • Successor to Allen & Wilson (see Allen & Wilson). | | |
| | **Wilson, Willman & Wilson** | 18 Old Bond St. | 1839–1844 |
| | **M. & I. Wilson** | 18 Old Bond St. | 1845–1852 |
| | • Succeeded by Russell & Allan (see Russell & Allan). | | |
| Goldsmith<br>Silversmith | **Thomas Wiltshire** | 46 Lombard St. | 1794–1817 |
| | **Wiltshire (Thomas) & Sons** | 46 Lombard St. | 1818–1819 |

| Trade | Name | Address | Dates |
|---|---|---|---|
| Hilt maker<br>Sword cutler | **Wiltshire (Thomas) & Sons**<br>• Succeeded by Barber & Smith (see Barber & Smith). | 36 Cornhill | 1820–1843 |
| Cutler | **Henry Winchester**<br>• Master of the Cutlers Company, 1829. | | 1819–1834 |
| Brazier<br>Iron monger (dealer)<br>Horn maker<br>Silversmith<br>Hilt maker<br>Sword cutler | **Nicholas Winkins**<br>• Horn maker to the royal hunt (1763). | Red Lion St., Holborn | 1751–D1782 |
| Jeweler<br>Enameler<br>Goldsmith<br>Buckle maker<br>Toy (hardware) maker | **Gabriel Wirgman (Sr.)** | 14 Red Lion Square, Clerkenwell | 1769–1775 |
| | **Morisset (James) & Wirgman (Gabriel Sr.)**<br>(see James Morisset) | 11 Denmark St., Parish of St. Giles in the Field, Soho | 1776–1778 |
| Hilt maker<br>Sword cutler | **Gabriel Wirgman (Sr.)** | 11 Denmark St., Parish of St. Giles in the Field, Soho | 1779–1791 |
| | • Partner and son Peter Wirgman ran their store at 69 St. James Street from 1779–1791.<br>• Gabriel Wirgman Sr. died in 1791. | | |
| | **Wirgman (Gabriel Jr.) & Son (George) & Colibert (Henry)** | 11 Denmark St., Parish of St. Giles in the Field, Soho | 1791–1798 |
| | **G. (Gabriel Jr.) & G. (George) Wirgman** | 11 Castle St., Holborn | 1799–1803 |
| | **G. (Gabriel Jr.) & G. (George) Wirgman** | 31 Castle St., Holborn | 1804 |
| | **Gabriel Wirgman Jr.** | 31 Castle St., Holborn | 1805–1808 |
| Cutler | **Richard Wise**<br>• Master of the Cutlers Company, 1697.) | | 1687–1702 |
| Sword cutler | **M. Wiseman** | Sackville St. | 1850 |
| Military tailor<br>Draper (cloth maker)<br>Sword cutler | **Wisker, Butler & Bodkin** | 7 Staple Inn, Holborn | 1846–1849 |
| | **Wisker, Butler & Bodkin** | 40 Castle St., Holborn | 1850 |
| | **Wisker & Sully** | 40 Castle St., Holborn | 1851–1855 |
| | **Wisker, Bodham & Butler** | 44 Castle St., Holborn | 1856–1866 |
| | **Wisker & Fricker** | 44 Castle St., Holborn | 1867–1870 |
| Cutler | **Henry Withers**<br>• Master of the Cutlers Company, 1628. | | 1618–1633 |
| Gun maker<br>Sword cutler | **D. (David) W. (William) Witton** | 2 Crosby Square, 67 Threadneedle St. | 1806–1814 |
| | **Lacy (J.D.) & Witton (David William)** | opposite North Gate<br>Royal Exchange & 63 Fenchurch St. | 1815–1826 |
| | • Witton worked for the J.D. Lacy Co., 1826–1854. | | |
| | **David William Witton** | 21 Great St., Dunnings Alley<br>Bishopsgate, Parish of St. Helens | 1854–1856 |
| | **Witton Bros.** | 21 Great St., Dunnings Alley<br>Bishopsgate, Parish of St. Helens | 1857–1863 |
| | • Partners: David William Witton and Thomas William Witton. | | |
| | **Witton Bros.** | 20 Great Helens, Dunnings Alley<br>Bishopgate, Parish of St. Helens | 1864–1869 |

| | | | |
|---|---|---|---|
| Gun maker | John Sergeant Witton | 82 Old Bond St. | 1842–1850 |
| Sword cutler | Witton (John Sergeant), Daw (George Henry) & Co. | 57 Threadneedle St. | 1851–1856 |
| | Witton (John Sergeant) & Daw (George Henry) | 57 Threadneedle St. | 1857–1860 |
| | • Succeeded by George Henry Daw at 57 Threadneedle Street. | | |
| Sword cutler | W. (William) Wolmershusen | 48 Curzon St., Mayfair | 1850–1860 |
| Sword cutler | E. (Edmund) Wood | Bow St. | 1822–1836 |
| | Edmund Wood | 10 Great Wild St., Lincohn's Inn Fields | 1837–1840 |
| Sword cutler | John Woodcraft | | 1680–1720 |
| | • Appointed sword cutler to the Board of Ordnance, November 6, 1693. | | |
| | • Master of the Cutlers Company, 1702. | | |
| | • Supplied swords to the Board of Ordnance, 1693–1720. | | |
| | • Also repaired swords for the Board of Ordnance. | | |
| | • Supplied plug bayonets to the Board of Ordnance, 1708–1711. | | |
| Pole axe maker | John Woodley | | 1660–1680 |
| | • Granted "pole axe" mark, 1664. | | |
| Silversmith Hilt maker Sword cutler | Edward Woodward | | 1661 |
| | • Signed an advertisement warning sword purchasers about counterfeit cast silver hilts. | | |
| Sword maker Bayonet and knife maker Silversmith Hilt maker Gun ramrod maker | Woolley (James) & Deakin (Thomas) | 138 Leadenhill St. | 1798–1802 |
| | Wooley (James) & Deakin (Thomas) | 10 Bush Lane, Cannon St. | 1802–1805 |
| | Wooley (James), Deakin (Thomas) & Co. | 10 Bush Lane, Cannon St. | 1805–1806 |
| | • Partners: John and Joseph Dutton. | | |
| | Woolley (James), Deakin (Thomas), Dutton (John & Joseph) & Johnston (Richard) | 10 Bush Lane, Cannon St. | 1806–1811 |
| | Woolley (James), Deakin (Thomas), Dutton (John & Joseph) | 10 Bush Lane, Cannon St. | 1811–1815 |
| | • Also made edged tools, frying pans, and agricultural tools. | | |
| | • Succeeded by Woolley & Sargant, Birmingham. | | |
| | (see Richard Johnston and John Dutton of London and William Sargant, Thomas Deakin, James Woolley of Birmingham) | | |
| Sword maker | Humphrey Woolrich (Woolridge) | near the Tower, Parish of St. Katherines | 1720–1740 |
| Goldsmith Jeweler Spectacle maker Hilt maker Sword cutler | Arthur Worboys (Sr.) | Vinson Court | 1758–1766 |
| | Arthur Worboys (Sr.) | 4 Wine Office Court, Fleet St. | 1767–1773 |
| | Arthur Worboys (Sr.) | 94 Fleet St. near Bride Lane | 1774–D1787 |
| | • Son John apprenticed to him, 1771–1778. | | |
| | • Son Thomas apprenticed to him, 1776–1783. | | |
| | • Son Arthur (Jr.) apprenticed to him, 1780–1787. | | |
| | Arthur Worboys (Jr.) | 9 Bells Building, Salisbury Square | 1787–1800 |
| Cutler | Job Worrall | | 1705–1720 |
| | • Master of the Cutlers Company, 1715. | | |
| Cutler | Sir Thomas Worsfold | | 1913–1928 |
| | • Master of the Cutlers Company, 1923. | | |

| | | | |
|---|---|---|---|
| Cutler | **John Wort** | | 1824–1840 |
| | • Master of the Cutlers Company, 1834–1835. | | |
| Pike maker | **Bartholomew Wray (Rayn)** | Parish of St. John Zachary | 1640–1649 |
| | • In 1645, during the English Civil War, Wray contracted for 550 long pikes (16 feet long with steel heads) for the "New Model" army (parliamentary forces) under Gen. Sir Thomas Fairfax: | | |
| |     100 pikes at 4 shillings, 2 denier each (May 15, 1645) | | |
| |     200 pikes at 4 shillings, 2 denier each (July 5, 1645) | | |
| |     250 pikes at 4 shillings, 2 denier each (July 26, 1645) | | |
| Gun maker<br>Sword cutler | **Charles Wright & Co.** | 1 Fenchurch St. | 1850–1857 |
| Gun maker<br>Sword cutler<br>Sword belt maker | **John Wright**<br>• Appointed royal cutler to His Royal Highness the Duke of Kent. | 72 St. James St. | 1799–1802 |
| Bladesmith | **John Wright** | | 1471 |
| | **Henry Yardley**<br>(see Charles Hill) | | |
| Gun maker | **James Yardley** | 33 Penton Place, Pentonville | 1799–1811 |
| Sword cutler | **James Yardley** | 14 Dean St., Holborn | 1812–1836 |
| Goldsmith<br>Hilt maker<br>Buckle maker<br>Sword cutler | **John Yardley** | Hatton Garden | 1760–1779 |
| | **Joel Jacobson & John Yardley** | 37 Charles St., Hatton Garden | 1780–1786 |
| | **John Yardley** | 37 Charles St., Hatton Garden | 1787–1805 |
| | **John Yardley** | 18 Plum Tree St. | 1806–1812 |
| | **John Yardley** | 29 Queen St., Bloombury | 1813–1814 |
| | • Succeeded by John Ash (see John Ash).<br>(see Joel Jacobson) | | |
| Gun maker | **Ralph Yardley** | 15 Field Gate St., Whitechapel | 1799–1822 |
| Sword cutler | **Ralph Yardley** | Mile End Rd. | 1823–1829 |
| Goldsmith<br>Hilt maker<br>Buckle maker<br>Sword cutler | **William Yardley** | 23 Thorney St., Bloombury | 1760–1779 |
| | **Mary Beedall & William Yardley** | 23 Thorney St., Bloombury | 1780 |
| | **William Yardley** | 5 Thorney St., Bloombury | 1781–1794 |
| | **William Yardley** | 5 Thorney St., Bloombury & Vine St. | 1795–1798 |
| | **J. (James) & W. (William) Yardley** | 5 Thorney St., Bloombury | 1799–1810 |
| | **W. (William) Yardley** | 5 Thorney St., Bloombury | 1811–1813 |
| | **William Yardley & Son**<br>(see John Beedall) | 5 Thorney St., Bloombury | 1814–1819 |
| Military tailor | **John Yates** | 28 King St., Parish of St. James | 1845–1847 |
| Sword cutler | **John Yates** | 66 Jermyn St., Parish of St. James | 1848–D1857 |
| | **Sarah Yates** | 66 Jermyn St., Parish of St. James | 1858 |
| | • Widow of John Yates. | | |
| Cutler | **Thomas Yelloley** | | 1730–1745 |
| | • Master of the Cutlers Company, 1740. | | |

| | | | |
|---|---|---|---|
| Gun maker | James Yoemans (Sr.) | 46 Chamber St. | 1802–1833 |
| Sword maker | James Yoemans (Sr.) | 68 Chamber St. | 1834–1840 |
| | • James Yoemans (Jr.) worked for his father. | | |
| | James Yoemans (Jr.) | 4 Magdalen Row, Great Prescott St. | 1841–1849 |
| | James Yoemans (Jr.) & Son (Horace) | 67 Chamber St., Goodmans Fields | 1850–D1852 |
| | Elizabeth Yoemans | Poultry St. & Tenter St. West, Goodmans Fields | 1852–1854 |
| | • James Yoemans Jr.'s widow. | | |
| | Elizabeth Yoemans & Son (Horace) | Poultry St. & Tenter St. West, Goodmans Fields | 1855–1863 |
| | Elizabeth Yoemans & Son (Horace) | 7 St. Mildreds Court & Poultry St. & Tenter St. West, Goodmans Fields | 1864 |
| | • In 1864, Yoemans was bought by John Alkin Blake & Co. (see John Alkin Blake & Co.). | | |
| | Horace Yoemans & Co. | 42 Great Tower St. | 1865 |
| | Horace Yoemans & Co. | 35 Upper East St., Smithfield | 1866–1870 |
| Swordsmith | Edward Yonger (Younger) | Parish of St. Clement Dane | 1640–D1679 |

- On March 19, 1660, King Charles II issued Royal Letters stating a patent had been issued granting Edward Yonger the Royal Appointment as royal sword Damascener, sword furbisher, sword slipper (English takeoff on German *schliefer*—grinder and polisher of blades), and sword cutler.
- Edward Yonger was a member of the Armourers Company. His family had immigrated to London from Solingen, Prussia, in the early 1600s.
- In 1660, King Charles II petitioned the Duke (Duchy) of Berg (a German state) for permits to allow four swordsmiths from Berg to immigrate to London and work under Edward Yonger.

| | | | |
|---|---|---|---|
| Goldsmith | Thomas Young | | Bc1766–1780 |
| Silversmith | • Apprentice to Cornelius Bland, 1773–1780 (see Cornelius Bland). | | |
| Hilt maker | Thomas Young | Aldergate St. | 1781–1800 |
| Sword cutler | • Had apprentice James Huell Bland, 1787–1793. | | |

## BIRMINGHAM

| | | | |
|---|---|---|---|
| Metal button | Edward Armfield | Holloway Rd. | 1763–1788 |
| Maker | Edward Armfield | Newhall St. | 1789–1790 |
| Sword cutler | Edward Armfield | St. Pauls Square | 1891–1910 |
| | Edward Armfield & Co. Ltd. | St. Pauls Square | 1911–1940 |
| | Edward Armfield & Co. Ltd. | 51 Carpenter Rd. | 1941–1968 |
| Steel sword | Thomas Armfield | Lower Gosty Green | 1767–1773 |
| hilt maker | Thomas Armfield | Great Bow St. | 1774–1776 |
| Toy (hardware) | Thomas Armfield | 12 Duke St. | 1777–1789 |
| maker | Thomas Armfield | St. Pauls Square | 1790–D1819 |
| Steel snuff box maker | | | |
| Bayonet maker | R. (Richard) & W. (William) Aston | 8 Upper Priory | 1850–1854 |
| Gun and pistol | R. (Richard) & W. (William) Aston | 27 Edmund St. | 1855–1859 |
| furniture | R. (Richard) & W. (William) Aston | 26 Edmund St. | 1860–1864 |
| maker | R. (Richard) & W. (William) Aston | Townhall Works & 26 Edmund St. | 1865–1869 |
| Sword maker | R. (Richard) & W. (William) Aston | 26–28 Edmund St. | 1870–1871 |
| | Richard Aston | 8 Shadwell St. | 1872–1873 |
| Sword cutler | Henry Atkins | 50 1/2 & 58 1/2 Price St. & 6 Lichfield St. | 1810–1872 |
| | Henry Atkins | back of 50 & 58 1/2 Price St. | 1873 |

|  |  |  |  |
|---|---|---|---|
|  | Henry Atkins | back of 50 Price St. | 1874–1882 |
| Bladesmith | George Atkins | 11 Court, Staniforth St. | 1890–1895 |
| Hilt maker | William Bacchus | Hill St. | 1780–1784 |
| Snuff box maker | William Bacchus | Summer St. | 1785 |
| Sword cutler | John Bagley | 20 Park St. | 1800–1818 |
| Army accoutrement maker | Frederick Barnes & Co. | 27 Whittall St | 1831–1832 |
|  | Frederick Barnes & Co. | 14 Sand St. | 1833–1840 |
|  | Frederick Barnes & Co. | 182 Livery St. | 1841–1855 |
| Sword cutler | Frederick Barnes & Co. | 57 Livery St. | 1856–1871 |
|  | Frederick Barnes & Co. | 57 Livery St. & 15 Lionel St. | 1872–1902 |
|  | Frederick Barnes & Co. | 15 Lionel St. | 1903 |
|  | (see London and Sheffield listings) |  |  |

| | | | |
|---|---|---|---|
| Gun maker | Thomas Bate | | Bc1773–1806 |
| Silversmith | • Gun maker and silversmith only; made silver sword mountings. | | |
| Toy (hardware) maker | Reddell (John) & Bate (Thomas) | Carey's Court, Dale End | 1807–1813 |
| | Thomas Bate | Aston St. | 1814–1815 |
| Sword maker | • Had a contract with the Board of Ordnance for 560 rifleman's swords, 1815. | | |
| Hilt maker | Thomas Bate | Lancaster St., Summer Lane | 1816–1820 |
| | • Exported swords to the United States, including eaglehead pommel officers swords. (see John Reddell) | | |

| | | | |
|---|---|---|---|
| Silversmith | Richard Bickley | 8 Smallbrook St. | 1767–D1792 |
| Dirk and sword hilt maker | • Made silver sword hilts and gun furniture. | | |
| | William Bickley & Samson Tomlinson | Moor St. | 1792–1800 |
| Silver caster maker | | | |

| | | | |
|---|---|---|---|
| Gold and silver Laceman | V. & R. Blakemore | Charlotte St. | 1866–1906 |
| | V. & R. Blakemore | 38 Charlotte St. | 1907–1916 |
| Accoutrement maker | (see London listing) | | |
| Sword cutler | | | |

| | | | |
|---|---|---|---|
| Sword cutler | Alexander Bloomer | 28 Adderly St. | 1870–1872 |
| Agent for sword maker | Frederick D. Blyth | 72 Bath St. | 1865–1875 |
| | • Agent for Solingen, Prussia, sword maker A. (August) & A. (Albert) Schnitzler, 1865–1875. | | |
| Gun maker | R. (Richard) Bolton (Boulton) & Co. | Newhall St. | 1805–1811 |
| Sword cutler | • Sold silver-hilted pillow pommel infantry officers swords. | | |
| | • Exported swords, including eaglehead pommel infantry and artillery officers swords, to the United States. | | |

| | | | |
|---|---|---|---|
| Steel hardware maker | Matthew Boulton (Jr.) | 38 Snow Hill, County of Warwick | B1728, 1745–1762 |
| Silver, gold and Sheffield plated ware maker | • Began working with father Matthew Boulton Sr. (c. 1700–D1759) as steel toy (hardware) maker. | | |
| | • As the years passed, Boulton expanded into many steel products. | | |
| | • Made gun furniture and assembled guns. | | |
| | • Bought gun making machinery and parts from Thomas Gill. | | |
| Steam engine maker | • Made steam engines (machinery). | | |
| | • Made polished steel products (buttons, buckles, jewelry). | | |

Merchant
Industrialist
Steel
sword hilt
maker

- Made polished steel sword hilts, which he sold to sword cutlers.
- Developed a sword blade testing machine (striking and bending blades).
- Made medals and coins (government copper coins).
- Became one of the largest merchants and industrialists of his time.
- He was a good friend of gun and sword maker William Henry Jr. of Philadelphia, Pennsylvania, USA.
- Henry bought gun making machinery from Boulton.
- After Matthew Boulton died in 1809, Henry named his new gun factory in Bushkill, Pennsylvania, after Boulton. Called Boulton Works, it was built in 1812.
- In early 1761, Boulton purchased a land lease from Mr. Edward Ruston for 1,000 pounds, where he built a polished steel products manufactory. It was located in the Parish of Handsworth, County of Statford, near the road from Birmingham to Wolverhampton (3/4 mile from his Snowhill shop in Birmingham).
- The land contained an old water-powered rolling mill on Hockley Brook. Boulton tore it down and built a new one.
- In July 1761, he began building houses for himself and his workmen.
- Warehouses and shops were also built.
- He called it the Soho Foundry and Manufactory, and it was completed in 1762.
- Between 1761–1775, he spent 4,000 pounds on the manufactory

**Soho Foundry & Manufactory**            **Parish of Handsworth,**            **1762–1809**
                                          **County of Statford**

- John Fothergill became partner from June 24, 1762 until 1782.
- Fathergill handled product sales and foreign marketing.
- The Snowhill shop was still kept as a warehouse and dwelling.
- Boulton produced a large quantity of mostly polished steel products at his new factory.
- Boulton's pattern book had around 1,475 designs in it.
- Boulton's polished steel products included:
    Buttons
    Buckles
    Buckle chapes
    Chatelaines
    Watch guards
    Cut steel sword hilts (highly burnished and faceted) for small swords
    Sword scabbard mounts
    Jewelry
- As time went on, Boulton began making "Ormolu" (gilt bronze) products such as vace mounts.
- He also made gold plated ware.
- Boulton became heavily involved with silver-plated products (started silver plating in 1766).
- He was instrumental in having a silver assay office opened in Birmingham.
- The Birmingham silver mark was an anchor.
- When the assay office opened in 1773, Boulton had 841 ounces of silverware assayed.
- Boulton had a silver sword hilt assayed in 1773.
- Boulton also mounted some silver-hilted small swords, some with gold inlays.
- Boulton employed goldsmith and designer Francis Edginton (Eginton), 1781–1800, and engraver and chaser John Edginton (Eginton), 1781–D1796.
- Johann Andreas Kern of London did some gold enlay work for Boulton.
- Boulton bought large amounts of silver from Samuel Garbett, Birmingham refiner.
- In 1770, Boulton had over 200 workman at his manufactory.
- John Fothergill died in 1782 and the partnership ended.
- Matthew Boulton died in 1809.

**Matthew Boulton & Plate Co.**                                               **1765–1809**
- Made silver-plated buttons and tableware.
- Made Sheffield plated (silver plate over copper plate) buttons and tableware,

**Boulton (Matthew) & Watt (James) & Sons    Smethwick**                      **1795–1800**
- Partners: James Watt (1736–1819), his sons James Jr. and Gregory, and Matthew Boulton's son Matthew Robinson Boulton.

|  |  |  |  |
|---|---|---|---|
|  | • Large producer of steam engines (machinery), including steam-powered machines for making gun parts and machines for making coins and medals. <br> • Large iron foundry. |  |  |
|  | **Boulton (Matthew & Matthew Robinson), Watt (James Jr. & Gregory) & Co.** <br> • Steam engine maker and iron foundry. | Smethwick | 1800–1809 |
|  | **Matthew Boulton & Button Co.** <br> • Made buttons and buckles. |  | 1782–1809 |
|  | **Boulton (Matthew) & Scale (John)** <br> • Made medals and copper coins. |  | 1795–1796 |
|  | **Boulton (Matthew) & Wedgwood (Josiah)** <br> • Made cameos. <br> • Some cameos were used on Boulton's small sword hilts. | Burslem | c1769–c1795 |
| Sword cutler | James Boydell | Kingswinford, Dudley, Oak Farm County | 1845–1848 |
| Gun maker <br> Sword maker | Edward Brooks & Son <br> Edward Brooks & Son <br> Brooks (Edward) & Co. <br> • Also made bayonets. <br> (see London listing) | 22 Russell St. <br> 35 Whittall St. <br> 34 & 35 Whittall St. | 1847–1854 <br> 1855 <br> 1856–1865 |
| Bayonet maker | George Brown <br> • Made spring bayonets. | Woodcock St. | 1818–1820 |
| Sword and bayonet maker | Stephen Brown | 19 Phillip St. | 1800–1810 |
| Sword cutler | John Bryan & Co. <br> Bryan Brothers & Co. <br> • Partners and brothers: John and Henry. | Willow Walk, Bermondsey <br> 31 St. Pauls Square | 1856–1869 <br> 1869–1894 |
| Gun maker <br> Silversmith <br> Silver sword hilt maker <br> Silver gun furniture maker | Joseph Bunney <br> • Made silver sword hilts, pistols, and gun furniture. <br> • Also supplied silver gun mountings for John Bennett of London. <br> Bunney (Joseph) & Son (Joshua) <br> • Advertised in the 1818 city directory as maker of airguns, crossbows, rifleguns, fowling pieces, and hair trigger pistols. | 25 Snow Hill <br><br><br> 25 Snow Hill | 1770–c1790 <br><br><br> c1790–1812 |
| Scabbard maker | George Butler <br> George Butler <br> George Butler | Back of 16 Jennings Row <br> 84 a Belmont Row <br> 84 a Colehill St. | 1882 <br> 1883 <br> 1884–1887 |
| Sword mounter | William Butler | 8 Court, Loveday St. | 1878–1882 |
| Sword mounter | William Henry Butler | 220 Bradford St. | 1915–1939 |
| Pattern maker <br> Sword maker <br> Lance, axe and boarding pike maker | John Byworth <br> John Byworth <br> John Byworth <br> John Byworth & Son <br> William Byworth | 35 1/2 Constitution Hill <br> 81 1/2 Constitution Hill <br> 127 1/2 Constitution Hill <br> 127 1/2 Constitution Hill <br> 127 1/2 Constitution Hill | 1870–1885 <br> 1886–1889 <br> 1890–1898 <br> 1899–1900 <br> 1901 |

| | | | |
|---|---|---|---|
| Silversmith<br>Hilt maker | W. Cartwright & S. Horton<br>• Made silver sword hilts. | High St. | 1801–1815 |
| Gas apparatus maker<br>Sword and bayonet maker | Samuel Chambers<br>• Sword and bayonet maker for the Board of Ordnance.<br>S. (Samuel) Chambers & Son | Aston St., Camphill<br><br>Aston St., Camphill | 1810–1830<br><br>1830–1835 |
| Sword cutler | Horace Charasse & Co.<br>Horace Charasse & Co. | 63 Oxford St., Alma St., Aston Newtown<br>Alma St. Works, Alma St., Summerlane | 1860–1864<br>1865–1868 |
| Sword cutler | William Cheshire<br>William Cheshire | Lichfield St.<br>Woodcock St. | 1815–1817<br>1818–1821 |
| Hilt maker | Thomas Clive<br>• Made cast iron hilts. | | 1840–1850 |
| Merchants<br>Sword cutler<br>Banker | Robert Coales<br>Robert Coales<br>Robert Coales<br>Coales (Robert)<br>Woolley (James) & Co.<br>• Robert Coales died in 1804.<br>(see James Woolley) | Bartholomew Row<br>68 Chapel Row & Bartholomew Row<br>Bartholomew Row<br>Bartholomew Row | 1767–1773<br>1774–1779<br>1780–1799<br>1800–1804 |
| Silversmith<br>Hilt maker | James Cooke<br>Elizabeth Cooke<br>• Widow of James Cooke.<br>• Made silver hilts, sword furniture, and scabbard mountings. | 29 High St.<br>29 High St. | 1780–D1791<br>1791–1800 |
| Sword cutler | Cooper & Banks | | 1790–1815 |
| Sword cutler | Cooper & Goodman | | 1860–1876 |
| Sword cutler | John Cooper | | 1649–1691 |
| Sword maker<br>Fireplace fender maker | John Cooper<br>Craven (Thomas) & Cooper (John)<br>• Made bayonets for Board of Ordnance, 1809.<br>Cooper (John) & Craven (Thomas)<br>• Had a Board of Ordnance contract for 560 rifleman's swords, 1815.<br>John Cooper | <br>96 Moor St.<br><br>96 Moor St.<br><br>Bartholomew St. | 1782–1802<br>1803–1814<br><br>1815–1817<br><br>1818–1820 |
| Silversmith<br>Hilt maker | Robert B. Cooper<br>• Made silver sword hilts and furniture. | Temple Row | 1775–1800 |
| Leather worker<br>Leather sword and bayonet scabbard maker | Jonathon Cope<br>Cross (John) & Cope (Jonathon) | Masshouse Lane<br>Masshouse Lane | 1815–1818<br>1819–1825 |
| | George Crane<br>(see James Woolley) | | |

English Sword Makers, Cutlers, Dealers, and Craftsmen Who Mounted Swords

| | | | |
|---|---|---|---|
| Sword maker | Thomas Craven | | Bc1777–1796 |
| Fireplace fender maker | Craven (Thomas) & Bradbury (George) | 96 Moor St. | 1797–1798 |
| | • Had a Board of Ordnance contract for 1,750 cultlasses, c. 1799. | | |
| | Thomas Craven | 96 Moor St. | 1799–1802 |
| | Craven (Thomas) & Cooper (John) | 96 Moor St. | 1803–1814 |
| | • Records show Thomas Craven supplied 200 cutlasses in 1798 and 200 cutlasses in 1804. | | |
| | Cooper (John) & Craven (Thomas) | 96 Moor St. | 1815–1817 |
| | • Had a Board of Ordnance contract for 560 rifleman's swords, 1815. | | |
| | Thomas Craven | Moland St. | 1818–1820 |
| Sword cutler | Henry Creasy | Wellington Works, Floodgate St. | 1880–1883 |
| Scabbard maker | Thomas Cross | | 1810–1818 |
| | John Cross (see Jonathon Cope) | | |
| Silversmith Hilt maker Sword cutler | J. (John) Crum | | 1755–1780 |
| Sword and bayonet maker | John Dawes | High St. | 1767–1769 |
| | John Dawes & Son (Samuel) | 25 High St. | 1770–1773 |
| | Dawes (John) & Son (Samuel) | 25 High St. | 1774–1776 |
| | Dawes (John) & Son (Samuel) | 25 Cannon St. | 1777–D1785 |
| Steel toy (hardware) maker | Samuel Dawes (Sr.) | 18 Bull Ring | 1774–1776 |
| | Samuel Dawes (Sr.) | 26 Cannon St. | 1777–1795 |
| | Samuel Dawes (Sr.) & Sons | Snowhill | 1796–1802 |
| Button maker Army accoutrement maker | • Made M1796 light and heavy cavalry sabers. | | |
| | • Made naval cutlasses and bayonets. | | |
| | • Bought sword blades from Thomas Gill. | | |
| | • Partners and sons William and Samuel Dawes Jr. | | |
| Gun maker Sword and bayonet maker | W. (William) & S. (Samuel) Dawes (Jr.) | Snowhill | 1803–1813 |
| | • William died in 1813. | | |
| | • Had a Board of Ordnance contract for 560 rifleman's swords, 1815. | | |
| | S. (Samuel Jr.) & J. (John) Dawes | Livery St. | 1813–1823 |
| | • John was Samuel Jr.'s son. | | |
| | Samuel (Jr.) & John Dawes | 6 Livery St. | 1824–1830 |
| | Samuel (Jr.) & John Dawes | 8 Livery St. | 1831–1835 |
| Sword maker | John Dawkes | | 1800–1810 |
| Sword and bayonet maker | Francis Deakin | Suffolk St. | 1811–1839 |
| | • Made bayonets for the Board of Ordnance, 1812–1813. | | |
| | • Made M1796 light and heavy cavalry sabers. | | |
| | F. (Francis) S. Deakin | Suffolk St. | 1840–1850 |
| Sword maker | Thomas Deakin | | 1855–1898 |
| Bayonet and knife maker | Woolley (James) & Deakin (Thomas) | 74 Edmund St. | 1898–1805 |
| Silversmith | Woolley (James) & Deakin (Thomas) & Co. | 74 Edmund St. | 1805–1806 |
| Steel and silver Hilt maker | • Partners: John and Joseph Dutton. | | |
| | • Had 29 silver sword hilts assayed in 1805 and 1806. | | |
| | • Made silver scabbard mounts. | | |
| Steel and silver | Woolley (James), Deakin (Thomas), | 74 Edmund St. | 1806–1811 |

| | | | |
|---|---|---|---|
| toy (hardware) maker | **Dutton (John & Joseph) & Johnston (Richard)** | | |
| | • Exported cavalry sabers to the United States. | | |
| Gun ramrod maker | **Woolley (James), Deakin (Thomas) & Dutton (John & Joseph)** | 74 Edmund St. | 1811–1815 |
| | • Also made edged tools, frying pans, and agricultural tools. | | |
| | • Succeeded by Woolley & Sargant (see Woolley & Sargant). | | |
| | **J. (James Sr.) Deakin** | Great Brook St. | 1815–1817 |
| | **William Deakin** | Small Brook St. | 1817–1835 |
| | **William Deakin** | 70 Navigation St. | 1835–D1845 |
| | **William Deakin & Sons** | Hazelwell Mill | 1845–1865 |
| | • Owners: William Harvey Deakin and James Deacon Jr. (sons of William). | | |
| | • Made M1821 light cavalry sabers and M1855 sword bayonets. | | |
| | **W. (William) Deakin & Sons** | Kings Norton | 1865–1870 |
| | • Owners: William Harvey Deakin and James Deakin Jr. | | |
| | • Made sword bayonets. | | |
| | (see John Dutton, London) | | |
| | (see Richard Johnston, London) | | |
| | (see James Wooley, Birmingham and London) | | |
| Sword cutler | **Sampson Dean** | | 1800–1810 |
| Sword cutler | **William Dunn** | | 1800–1810 |
| Sword cutler | **Joshua Eddels & Co.** | | 1800–1810 |
| Engraver | **John Eginton** | Soho | 1760–1780 |
| Chaser | **Matthew Boulton** | Soho | 1781–1796 |
| Sword hilter | • Employee: John Eginton, 1781–D1796. | | |
| | (see Matthew Boulton) | | |
| Goldsmith | **Francis Eginton** | Soho | 1760–1780 |
| Designer | **Matthew Boulton** | Soho | 1781–1800 |
| Sword hilter | Employee: Francis Eginton, 1781–1800. | | |
| | (see Matthew Boulton) | | |
| | **Francis Eginton** | Soho | 1801–1823 |
| | **John Fathergill** | | |
| | (see Matthew Boulton) | | |
| Silversmith | **Joseph Fendall** | Dale End | 1780–1796 |
| Hilt maker | • Made silver sword hilts. | | |
| Sword cutler | **Thomas Field** | | 1800–1810 |
| Steel button maker | **Firmin & Sons Ltd.** | 31 & 32 St. Pauls Square | 1880–1892 |
| | **Firmin & Sons Ltd.** | Globe Works, Villa St. | 1893–1959 |
| Badge maker | **Firmin & Sons Ltd.** | Globe Works, Portland St. | 1960–c1980 |
| Gold lacemen | **Firmin & Sons PLG** | Globe Works, 82–86 New Town Row | c1980–2003 |
| Embroiderer | (see London listing) | | |
| Trimming warehouse | • Current maker of: | | |
| |     Officers swords | | |
| Sword cutler |     Sword belts | | |
| |     Braid | | |
| |     Tassels | | |
| |     Pennants | | |

|  |  |  |  |
|---|---|---|---|
| | Buttons | | |
| | Regalia badges | | |
| | Awards | | |
| | Metalware | | |
| | Commemoratives | | |
| | Trophies | | |
| | Plaques | | |
| Sword cutler | **Thomas Fitter** | | 1800–1810 |
| Silversmith | **Thomas Frattorini** | **Yorkshire** | 1827–c1910 |
| Goldsmith | • Had retail stores in Skipton, Harrogate, and Bradford. | | |
| Jeweler | **Thomas Frattorini** | **Birmingham, Regent Street Works** | c1910–2003 |
| Sword cutler | • Have offices at Skipton Castle, Skipton, and North Yorkshire. | | |
| | • Have London offices at 189 Regent Street. | | |
| | • Sword production started c. 1978. | | |
| | • Current maker of: | | |
| |     Officers regulation swords | | |
| |     Presentation swords | | |
| |     Miniature swords | | |
| |     Replica swords | | |
| |     Badges | | |
| |     Medals | | |
| |     Insignia | | |
| |     Trophies | | |
| |     Commemoratives | | |
| |     Silverware | | |
| |     Small metalware | | |
| | • Current officers: | | |
| |     Managing Director: Thomas P. Fattorini | | |
| |     Director: H.T. Fattorini | | |
| |     Director: A.F. Fattorini | | |
| |     Director: T.R.B. Fattorini | | |
| |     Director: G.T.B. Fattorini | | |
| Brass founder | **Charles Freeth** | | B1731–1755 |
| Silversmith | • Would have apprenticed to a founder (brass and silver) since he used Sheffield plate, c.1745–1755. | | |
| Dirk and sword | **Charles Freeth** | **65 Park St.** | 1756–1773 |
| hilt maker | • Casted brass and silver gun furniture. | | |
| Gun furniture | **Charles Freeth** | **Great Charles St.** | 1774–D1804 |
| maker | • Became silver casters. | | |
| | • Made silver hilts and gun furniture. | | |
| | • Between 1788 and 1803, had 667 silver hilts assayed (some sterling silver). | | |
| | • Made silver-mounted dirk hilts for Henry Osborne. | | |
| | • Made cast silver products. | | |
| | • Made silver-plated pistol guards, forks, pistol furniture, handles, gun furniture, dirk hilts, seals, sword hilts, medals, dirk and sword scabbard furniture, shoe buckles, and lions. | | |
| | **John Freeth** | **Great Charles St.** | 1804–D1807 |
| | **Freeth (Ann & Elizabeth) & Jones (William)** | **Livery St.** | 1808–1820 |
| | • Ann and Elizabeth Freeth were wife and daughter of John Freeth. | | |
| | **Freeth (Ann & Elizabeth) & Jennings (William)** | **Livery St.** | 1821–1834 |
| | **William Jennings** | **Livery St.** | 1835–1846 |
| Sword maker | **J.R. Gaunt & Son Ltd.** | **Warstone Parade** | 1900–1969 |
| Sword belt | • Stopped sword production in 1942. | | |
| maker | (see London listing) | | |

Badge maker
Military
ornament maker

## THE GILL FAMILY OF SWORD MAKERS

| | | | |
|---|---|---|---|
| Saw, spur file, tool maker | **James Gill** • Son Thomas Gill (Sr.) apprenticed and worked at his factory. | Prescott, Lancashire | Bc1730–Dc1773 |

Watch and clock maker

**Thomas Gill (Sr.)**      Prescott, Lancashire      Bc1750–1773
- Apprenticed and worked for his father James Gill.

Saw, file and tool maker

**Thomas Gill (Sr.)**      95 Dale End      1774–1776
**Thomas Gill (Sr.)**      92 Dale End      1777–1782

Steel toy (hardware) maker
Gun maker
Sword blade maker
Sword maker
Silversmith
Machine maker

**Thomas Gill (Sr.)**      Jennons Row, near St. Bartholomew Chapel      1783–D1801

- Began sword and sword blade production in 1783.
- Made silver sword hilts, sword furniture, and scabbard furniture.
- Gill was a machinery maker. He made cotton spinning machines, gun making machines, and sword testing machines.
- Gill's sons, Thomas Jr., James, and John, worked in his factory with him.
- In 1783, the Earl of Surrey (a member of the Treasury Board) sent a letter to a Mr. Eyre of the town council of Sheffield, informing him of a petition by the London sword sellers (cutlers) under consideration to allow German sword blades into England duty free. (The reason was that English blade makers could not provide quality sword blades at a reasonable price.) The earl wanted sword blade quality and price information from Eyre on Sheffield makers.
- Since few sword blades were being made at Sheffield, Mr. Eyre sent a copy to Thomas Gill of Birmingham. Gill sent a letter to the Treasury Board, saying he could make sword blades of equal quality to German blades. He sent a letter to the Board of Ordnance requesting a test of his blades. The Board of Ordnance agreed and requested the Treasury Board to send some German blades and other English blades to the Tower of London for testing. The test, however, never took place.
- In 1786, the East India Company announced an order for 10,000 cavalry (horseman's) swords (the sword contracts were to be distributed to English and German makers).
- The only cavalry sword makers in England were Thomas Gill, James Woolley, and Samuel Harvey of Birmingham and J.J. Runkel of London, an importer of German swords and blades.
- Gill appealed to the East India Company for a quality test of his swords against his current competitors.
- All four sword makers sent their swords to the East India Company warehouse. Colonel Windus, Inspector of Arms for the East India Company, tested the swords using Matthew Boulton's method of striking and bending the blades. The test results were:

| Maker | Swords Sent | Rejected | Scabbard Serviceability |
|---|---|---|---|
| Gill | 2,650 | 4 | Serviceable |
| Woolley | 1,000 | 19 | Not Serviceable |
| Harvey | 1,700 | 42 | Not Serviceable |
| Runkel | 1,400 | 28 | Serviceable |

- Gill suggested that sword blades be tested by striking them flatways on a cast iron plate and edgeways on a cast iron cylinder. He developed a spring-activated machine that allowed swords to be tested in this way. It was eventually used by the Ordnance Department in the Tower of London armoury.
- In 1790, unhappy with the number of government and private orders for his cavalry swords, Gill published a pamphlet entitled "The Superiority of English Swords of Mr. Gills Manufacture to those of Germany or any other Nations (asserted and maintained), humbly submitted to the serious and candid consideration of the officers of the Army, Navy, etc."
- It expounded the quality of his swords and blades, listed the officers who recommend them, and showed the results of the 1786 East India Company test.
- Gill made cavalry sabers, including M1788 light and heavy cavalry swords and M1796 light and heavy cavalry swords.

- Gill made cavalry sabers for the 1st Life Guards in 1789.
- Gill made cavalry sabers for a regiment of dragoons sent to the United States during the U.S. Revolutionary War (1775–1783).
- Gill's early swords are marked:
    WARRANTEED NEVER TO FAIL
- Later swords are marked:
    WARRANTED TO CUT IRON OR GILL'S WARRANTEED
- Gill sold huge amounts of royal artificers tools and equipment and iron monger stores to the British Office of Ordnance.
- Gill sold sword blades to many English sword cutlers (assemblers), including Samuel Dawes of Birmingham and John Salter of London.
- Gill opened a warehouse at Charing Cross, London, to supply blades and swords to London sword cutlers (both assemblers and retailers).
- Gill made gun barrels and assembled muskets, carbines, and pistols for the Board of Ordnance.
- Gill made gun-making machinery (some were sold to Matthew Boulton).
- Gill also made bayonets for the Board of Ordnance.
- Henry Palmer of St. James Street, London, was an agent (exclusive retailer) of Gill's swords in London (see Henry Palmer).
- Thomas Gill (Sr.) died in 1801. His sons succeeded him.

| | | |
|---|---|---|
| **Thomas (Jr.), James (Jr.) & John Gill** | Jennons Row, near St. Bartholomew Chapel | 1801–1802 |

- Sons of Thomas Gill Sr.
- John Gill opened his own sword shop in 1803.

| | | |
|---|---|---|
| **Thomas (Jr.) & James (Jr.) Gill** | Jennons Row, near St. Bartholomew Chapel | 1803–1808 |

- James Gill (Jr.) died in 1808.

| | | |
|---|---|---|
| **Thomas Gill (Jr.)** | Jennons Row, near St. Bartholomew Chapel | 1809–D1826 |

- Had a Board of Ordnance contract for 560 rifleman's swords in 1815.
- Many swords with Gill or "G" marked on the blades were exported to the United States.

| | | | |
|---|---|---|---|
| Sword cutler | **Gill (Thomas Jr.) & Parkes (Richard)** | | 1809–1815 |

- A separate company.

| | | | |
|---|---|---|---|
| Sword maker | **John Gill** | 20 Masshouse Lane | 1803–1814 |

- Made brass-mounted band swords.

| | | |
|---|---|---|
| **John Gill** | 20 Masshouse Lane & St. Bartholomew Square | 1815 |
| **John Gill** | 20 Masshouse Lane | 1816–D1826 |

- John Gill died in 1826.

| | | |
|---|---|---|
| **Elizabeth Gill** | 20 Masshouse Lane | 1826–1837 |

- John's widow.
- Many swords with Gill or "G" marked on the blades were exported to the United States.

| | | | |
|---|---|---|---|
| Sword cutler | **George Goodwin** | | 1800–1810 |

| | | | |
|---|---|---|---|
| Steel bead maker | **Joseph Greaves** | 11 Bartholomew Row | 1800–1817 |
| Steel hilt maker | **Joseph Greaves** | Tanter St. | 1818–1828 |
| Steel toy (hardware) maker | • Also a victualler (army food supplier). | | |
| Gun barrel maker | • Succeeded by Reeves & Greaves. (see Charles Reeves) | | |

**John Gunby**
(see Henry Osborn)

| | | | |
|---|---|---|---|
| Gun maker<br>Sword maker | **Thomas Hadley** | 47 Bull St. | 1770–1820 |
| | • Made naval cutlasses and short swords.<br>• In 1815, contracted with the Board of Ordnance for 560 rifleman's swords. | | |
| Goldsmith<br>Jeweler<br>Sword cutler | **Thomas Hadley**<br>**Thomas Hadley & Sons** | 21 Newhall St.<br>21 Newhall St. | 1780–1792<br>1793–1805 |
| Steel button<br>maker<br>Steel buckle<br>maker<br>Steel toy<br>(hardware)<br>maker<br>Steel hilt<br>maker<br>Sword maker | **Joseph Harvey**<br>**Joseph Harvey**<br>• Successor to George Harvey at 21 Park Street.<br>• Exported eaglehead pommel officers swords to the United States.<br>(see S. & G. Harvey) | 16 Upper Priory<br>21 Park St. | 1800–1810<br>1811–1815 |
| Sword maker | **S. (Samuel Sr.) & G. (George) Harvey** | 21 Park St. | 1767–1778 |
| | • A separate company from Samuel Harvey (Sr.).<br>• Made bayonets for the Board of Ordnance, 1778.<br>• Samuel Harvey died in 1778. | | |
| | **George Harvey** | 21 Park St. | 1779–1810 |
| | • Succeeded by Joseph Harvey (see Joseph Harvey). | | |
| Sword maker<br>Silversmith | **Samuel Harvey (Sr.)**<br>**Samuel Harvey (Sr.)**<br>**Samuel Harvey (Sr.)**<br>(see S.&G. Harvey) | <br>Moor St.<br>74 High St. | B1698–1717<br>1718–1766<br>1767–D1778 |
| | • Made many basket hilt broadswords.<br>• Made naval cutlasses and infantry hangers.<br>• Made many cavalry (horseman's) sabers.<br>• His blades are marked S.H. or S.H. in a running fox.<br>• Made silver sword hilts.<br>• Samuel Harrey Sr. died in 1778. | | |
| | **Samuel Harvey (Jr.)** | 74 High St. | 1778–1789 |
| | • Son of Samuel Harvey Sr. | | |
| | **Samuel Harvey (Jr.)** | 4 Cannon St. | 1790–D1795 |
| | • Samuel Harvey Jr. died in 1795. | | |
| | **Samuel Harvey III** | St. Pauls Square | 1795–D1810 |
| | • Samuel Harvey III died in 1810.<br>• The Harveys made many silver-hilted swords.<br>• The Harveys sold sword blades to Dru Drury of London (see Dru Drury).<br>• Benjamin May made silver hilts for Samuel Harvey I and II (see Benjamin May). | | |
| Sword and<br>blade maker<br>Steel toy<br>(hardware)<br>maker | **William Harvey**<br>**William Harvey**<br>**William Harvey**<br>**William Harvey** | 6 High St.<br>Coventry Rd.<br>Bordesley St.<br>High St., Deritend | 1816–1834<br>1835–1836<br>1837–1838<br>1839–D1846 |
| | • Sold sword blades to the Board of Ordnance and many sword cutlers.<br>• Also made bayonets<br>• William Harvey died in 1846. | | |
| | **Mary Harvey** | High St., Deritend | 1846 |
| | • Widow of William Harvey. | | |

|  |  |  |  |
|---|---|---|---|
|  | John Harvey | 40 High St., Deritend | 1847–1854 |
|  | • Son of William and Mary Harvey. | | |
|  | John Harvey | Albert Works, Glover St. | 1855–1859 |
|  | John Harvey | 27 Adderley St. | 1860–1864 |
|  | John Harvey | 123 Steelhorse Lane | 1865–D1882 |
|  | Mrs. John Harvey | 113 Coleshill St. | 1882–1897 |
|  | • Widow of John Harvey. | | |
| Steel toy (hardware) maker  Sword cutler | James Heeley & Sons | 147 Great Charles St. | 1825–1840 |
| Bayonet maker | John Heighington | | 1847–1849 |
| Axe and hatchet maker | John Heighington | Bordesley Mills, Liverpool St. | 1850 |
| Sword cutler | John Heighington | Bordesley Rolling Mills, Liverpool St. | 1851–1852 |
|  | Heighington (John) & Lawrence (Thomas) | Albion Works; Liverpool St., Deritend | 1853–1858 |
|  | • Made M1855 sword bayonets for sappers for the East India Service. | | |
|  | • Made bayonets for the Board of Ordnance. | | |
|  | Thomas Lawrence & Co. | Albion Works; Liverpool St., Deritend | 1859–1877 |
|  | • Partner: John Heighington. | | |
|  | Thomas Lawrence & Co. | 50 Price St. | 1878–1879 |
|  | • Partner: John Heighington. | | |
|  | • Made bayonets. | | |
| Tin plate worker  Bayonet maker | John Hill | 35–36 Pritchett St. | 1812–1829 |
|  | • Made bayonets for Board of Ordnance, 1812. | | |
| Sword blade maker | Stephen Hill | 35–36 Pritchett St. | 1830–1848 |
|  | • Made M1855 sword bayonets for sappers for the East India Service. | | |
| Bayonet maker | S. (Stephen) Hill & Sons | 35 Pritchett St. | 1849–1855 |
| Ax and hatchet maker  Sword blade maker | Samuel Hill | Snow Hill | 1835–1845 |
| Sword cutler | William Hinton | Adderley St., 23 Court  High St., Bordesley | 1847–1848 |
| Gun maker | Soloman Jackson | Lionel St. | 1803–1814 |
| Gun lock maker | Soloman Jackson | Summer Lane | 1815 |
|  | • Exported swords to the United States, including eaglehead pommel officers swords. | | |
| Sword cutler | Jackson (Soloman) & Yanier (Robert) | Summer Lane | 1816–1817 |
|  | Thomas Jackson | Frazeley St. | 1818–1821 |
| Sword Blade maker | Richard Jarvice | Smallbrook St. | 1767–1770 |
| Scabbard maker | George Jennings | Back of 40 Potter St. | 1897–1906 |
|  | Samuel Jennings  (see Charles Freeth) | | |

| | | | |
|---|---|---|---|
| Steel worker<br>Sword hilt maker<br>Steel plate worker | **James Johnson**<br>• Steel sword hilt maker. | New Inkleys | 1815–1820 |
| Steel toy (hardware) maker<br>Steel sword hilt maker<br>Sword maker | **George Johnson**<br>**George Johnson** | 13 Digby St.<br>39 Digby St. | 1847–1854<br>1855 |
| Steel toy (hardware) maker<br>Steel sword hilt maker<br>Sword maker | **Benjamin Johnson**<br>**Benjamin Johnson**<br>**Benjamin Johnson**<br>• Also made bayonets. | 17 Court, Bartholomew St.<br>17 Court, Bartholomew St.<br>River St. | 1849–1855<br>1856–1860<br>1860–1862 |
| Gun maker<br>Sword cutler | **John Jones & Co.**<br>• Partners and sons: William Jones and Joseph Jones.<br>• Made flintlock pistols.<br><br>**Henry Jones**<br>(see Charles Freeth) | | 1800–1815 |
| Silversmith<br>Jeweler<br>Masonic and theatrical sword cutler | **Richard Kennedy**<br>**John Kennedy**<br>**John Kennedy & Co.**<br>**John Kennedy & Co.**<br>**John Ligatt Kennedy & Co.** | 53 Loveday St.<br>53 Loveday St.<br>46 Loveday St.<br>36 Bath St.<br>140 Steelhouse Lane | 1810–1864<br>1865–1867<br>1868–1869<br>1870–1871<br>1872–1890 |
| Sword cutler | **Thomas Kenning** | Edgebaston St. | 1835–1845 |
| Gun maker<br>Sword cutler | **William Ketland Sr.**<br>• Son and partner Thomas Ketland Sr.<br>• William Ketland retired in 1767 and died in 1780.<br>**Thomas Ketland Sr.**<br>• Partner and sons: William Ketland Jr., Thomas Ketland Jr., and John Ketland.<br>• Sent his sons Thomas Jr. and John to Philadelphia, PA, USA, from 1789 to 1800 (they remained partners in their father's company). They returned to England in 1800.<br>• Exported muskets and musket locks to his sons in Philadelphia (may have exported swords also).<br>• On November 15, 1797, his sons in Philadelphia obtained a Pennsylvania contract for 1,000 Charleville type muskets. The muskets were to be provided by his father, but few were shipped because of the British firearm exportation ban.<br>• Thomas Ketland (Sr.) died in 1800.<br>**William Ketland (Jr.) & Co.**<br>• Partners and brothers: Thomas Ketland Jr. and John Ketland.<br>• Partner and brother-in-law: Thomas Izon.<br>• William Ketland (Jr.) died in 1804.<br>**Ketland (Thomas Jr. & John) & Izon (Thomas)**<br>• Made bayonets for the Board of Ordnance.<br>**Ketland & Co.**<br>• Thomas Izon served as president. Partners: Thomas Ketland Jr. and John Ketland. | | Bc1710<br>1730–1767<br><br>Bc17301767–D1800<br><br><br><br><br><br><br><br><br><br>1800–D1804<br><br><br><br>1804–1805<br><br>1805–1871 |

| | | | |
|---|---|---|---|
| | • Between 1804–1812, they exported many eaglehead pommel infantry and artillery officers swords to dealers in the United States such as George Upson and Richards, Upson & Co. of New York, NY. | | |
| | • They also exported a complete line of pistols, muskets, and fowling pieces to the United States, including some to Richards, Upson & Co. of New York, NY. | | |
| Merchant Sword cutler | C.H. Lauderberg & Co. • Sold military and domestic products to the East India Company. | 2 King Alfred Place, Broad St. | 1884–1895 |
| | Thomas Lawrence (see John Heighington) | | |
| Chain maker Button maker Sword cutler | William Leonard & Co. Leonard (William) & Co. William Leonard & Son Leonard (William) & Son | Aston St. Aston St. Aston St. Aston St. | 1800–1814 1815 1816–1820 1821 |
| Shovel and tool maker Sword cutler | Joseph Lyndon W. (Walter) A. (Adam) Lyndon Walter Adam Lyndon | Minerva Works, Fazeley St. Fazeley St. | 1810–1836 1837–1849 1850–1853 |
| Sword maker Ax and hatchet maker | Ralph Martindale & Co. Ralph Martindale & Co. • Ax and hatchet maker only from 1949. | Alma St., Aston Newtown Crocodile Works, Alma St. Aston Newtown | 1870–1948 1949–1970 |
| Sword cutler | Richard Mason | 2 Castle St. | 1777–1780 |
| Sword cutler | William Mason | 87 Steelhouse St. | 1835–1837 |
| Silversmith Toy (hardware) maker Hilt maker Silver plater | Benjamin May • Made silver sword hilts for Samuel Harvey I and II. • Made silver-plated products, including sword hilts, bits, shoe buckles, sword hilt parts, coach furniture, snuff boxes, and scabbard furniture. • Assayed 216 silver hilts, 1776–1787. • Also made steel frying pans and candle snuffers. | 55 New St. | 1776–D1781 |
| Sword cutler | Samuel Mayo Samuel Mayo Samuel Mayo | Hill St. High St. 73 Worcester St. | 1815–1817 1818–1834 1835 |
| Sword cutler | Richard Middemore | New St. | 1816–1820 |
| Agent for a sword maker | Nathaniel Mills & Co. • Agent for Clemen & Jung, sword maker and bayonet maker of Solingen, Prussia. | 25 Mary Ann St. | 1865–1867 |

### THE MOLE FAMILY
### (MOHLL, MOLL)

| | | | |
|---|---|---|---|
| Sword and sword blade maker | Abraham Mohll • Master swordsmith. • Immigrated to Shotley Bridge, England, 1687. Abraham Mohll | Solingen, Prussia Shotley Bridge | Bc1640–1687 1687–1690 |
| | Herman Mohll • Master swordsmith. | Solingen, Prussia | Bc1660–1687 |

- Son of Abraham Mohll.
- Immigrated to Shotley Bridge, England, 1687.

**Herman Mohll**  Shotley Bridge  1687–D1716
- Operated and later owned a sword blade mill (see Chapter 4).
- Herman Mohll died in 1716.
- William Mohll the Elder (Herman's son) took over the mill.

**William Mohll the Elder**  Shotley Bridge  Bc1690
1716–Dc1740
- In 1724, William Mohll the Elder sold his mill to Robert Ohlig the Elder.

**John Moll the Elder**  Shotley Bridge  Bc1710–Dc1770
- Son of William Mohll the Elder.
- Changed his name to Moll.

**William Moll the Younger**  Shotley Bridge  Bc1740–Dc1790
- Son of John Moll the Elder.

**John Moll the Younger**  Shotley Bridge  Bc1780–Dc1832
- Son of William Moll the Younger.
- John Moll moved to Birmingham in 1832.

**John Moll the Younger (Jr.)**  Broad St., Islington  1832–1834
**J. (John Jr.) & R. (Robert Sr.) Mole**  Broad St., Islington  1835–1837
- Robert Mole Sr. was the son of John Moll Jr.
- They changed their name to Mole.

**John (Jr.) & Robert (Sr.) Mole**  171 Broad St., Islington  1837–1846
- John Mole Jr. died in 1846.

**Robert Mole (Sr.)**  171 Broad St., Islington  1847–1855
**Robert Mole (Sr.) & Son**  171–172 Broad St., Islington  1856–1874
- Son and partner: Robert Mole (Jr.).
- They became a large sword maker, selling many swords to the British Ordnance Department.
- Sold swords to both sides during the American Civil War.
- Sold brass-hilted M1853 cavalry sabers to the confederates and M1821 cavalry sabers to Tiffany & Co., which sold the swords to the Union government.
- Swords were marked R.M.S.B., Robert Mole & Son, Birmingham.

**Robert Mole (Sr.) & Son**  238–239 Broad St., Islington  1875–1879
**Robert Mole (Sr.) & Son**  93 to 99 Granville St., Broad St.  1880
**Robert Mole (Jr.)**  24 to 34 Granville St.  1881–1894
- Also made bayonets.
- In 1884, Mole allowed Wilkinson (Henry) & Son to borrow bladesmith Tom Beasley, sword maker Ernie Johnson, and sword maker Walter Johnson to help with a special project.

**Robert Mole (Jr.) & Son**  24 to 34 Granville St.  1895–1920
- In 1920, Wilkinson Sword Co. Ltd. purchased the Robert Mole & Co. sword division and Mole moved to Aston Newtown.

**Robert Mole & Son**  Alma St., Aston Newtown  1921–1967
- At Aston, Mole made axes, hatchets, edged tools, and agricultural implements.

Sword cutler  **Thomas Moor**  1800

Gun retailer  **Thomas Moxham (Sr.)**  37 Moland St & 19 New John St.  1820–1867
Sword cutler
- Sword cutler to King William IV, the Duke of York, and the Board of Ordnance.
- Also made bayonets.

**Thomas Moxham (Sr. & Jr.)**  37 Moland St.  1868–1878
**Thomas Moxham (Sr. & Jr.)**  37 Moland St. & 67 New John St.  1879–1880

| | **John Oakes** | | |
|---|---|---|---|
| | (see Terry & Oakes) | | |

| | | | |
|---|---|---|---|
| Sword cutler | **W. (William) Orton** | 2 Court St., Mount St. | 1856–1860 |

| | | | |
|---|---|---|---|
| Gun maker | **Henry Osborne (Osborn)** | | Bc1765–1784 |
| Silversmith | **Henry Osborne** | Brookhouse | 1785–1799 |
| Accoutrement maker | • Appointed sword cutler and accoutrement maker to His Majesty the King and His Royal Highness the Prince of Wales. | | |
| Hilt maker | • Osborne helped design the Model 1796 light and heavy cavalry swords. | | |
| Sword and dirk cutler | • Henry Osborne, James Woolley, William Sargant, and Thomas Ketland were founding members of the Loyal Birmingham Light Horse Volunteers (1797–1802). | | |
| | • Osborne made M1796 cavalry officers swords for the unit. | | |
| | **Henry Osborne** | Brookhouse, Bordesly Street Mills | 1800–1805 |
| | • Charles Freeth sold Osborne silver dirk hilts. | | |
| | • Osborne had 72 silver sword hilts assayed, 1799–1803. | | |
| | • He made silver gorgets, sword hilts, sword hilt parts, sword furniture, naval cutlasses, army swords, and gun furniture. | | |
| | **Osborne (Henry) & Gunby (John)** | Brookhouse, Bordesley Street Mills | 1806–1820 |
| | • Had a retail shop at 82 Pall Mall, London (see London listing). | | |
| | • In 1806, became sword cutler and accoutrement maker to the king and Prince of Wales. | | |
| | • Made silver sword furniture, sword hilt parts, and sword hilts. | | |
| | • Had 19 silver sword hilts assayed, 1805–1807. | | |
| | • Exported swords (including eaglehead pommel infantry and artillery officers swords) to the United States. | | |
| | • In 1809, became sword cutler and gun maker to the Prince Regent and the Honorable East India Company. | | |
| | • In 1815, had Board of Ordnance contracts for 640 rifleman's swords. | | |
| | • Became a large gun maker. | | |
| | **Henry Osborne** | Brookhouse, Bordesley Street Mills | 1821–1838 |
| | • Henry Osborne died in 1827. | | |
| | • His widow Hannah and son Thomas took over the company but kept the name. | | |
| | (see London listing) | | |
| | **Thomas Osborne** | Brookhouse, Bordesley Street Mills | 1839–1846 |
| | • He advertised as follows: | | |
| |     *Sword cutler, gun manufacturer* | | |
| |     *Accoutrement maker to the Honorable East India Company &* | | |
| |     *The right Honorable Board of Ordnance* | | |
| |     *Merchants supplied on short notice* | | |
| | • Osborne also made bayonets. | | |
| | **Thomas Osborne** | Liverpool St. & 85 Weaman St. | 1847–1848 |
| | **Thomas Osborne** | Upper Trinity St. | 1849 |

| | | | |
|---|---|---|---|
| Sword maker | **John Park** | Leadenhall Market | 1626–1630 |

| | | | |
|---|---|---|---|
| Sword maker | **Alfred Grey Parker** | 69 Icknield St. | 1900 |
| Sword repairer, | **Alfred Grey Parker** | 264 Icknield St. | 1901–1912 |
| polisher and | **Alfred Grey Parker** | 6 1/2 Whithall St. | 1912–1913 |
| plater | • Made sergeants and officers swords. | | |

| | | | |
|---|---|---|---|
| Sword cutler | **William Parr** | | 1800–1820 |

| | | | |
|---|---|---|---|
| Silversmith | **Thomas Parsons** | High St. | 1785–1799 |
| Hilt maker | • Made silver sword hilts, hilt parts, and sword furniture. | | |
| | • Had 10 silver hilts assayed, 1798. | | |

| | | | |
|---|---|---|---|
| Goldsmith | **Thomas Penberton** | Snow Hill | 1807–1820 |

| | | | |
|---|---|---|---|
| Hilt maker<br>Sword cutler | • Entered his mark as a goldworker, 1807. | | |
| Silversmith<br>Hilt maker | **John Piercy**<br>• Made silver sword grips, snuff boxes, and thimbles. | Snow Hill | 1775–1820 |
| Sword blade maker<br>Sword maker | **Robert Porter (Sr.)**<br>• In 1643, sold 15,000 swords to the parliamentary forces during the English Civil War (1642–1649).<br>• The swords were ordered by Prince Rupert.<br>• The royalists destroyed his factory in retaliation.<br>**Robert Porter (Jr.)**<br>• In 1686, offered his blades to the Cutlers Company of London for resale by them.<br>**Samuel Porter**<br>• Son of Robert Jr.<br>**Robert Porter**<br>• Son of Samuel. | | 1629–1649<br><br><br><br><br>1650–1686<br><br>1687–1698<br><br>1699–1729 |
| Wooden scabbard maker | **George Prince** | 6 Court Tower St. | 1888–1890 |
| Silversmith<br>Hilt maker | **William Pugh**<br>• Made silver sword hilts and scabbards. | Suffolk St. | 1800–1810 |
| Hilt maker<br>Sword cutler | **Charles Purser**<br>**Charles Purser** | 23 Court High St., Bordesley Court<br>corner of Moseley St. & Sherlock | 1835–1859<br>1860 |
| Sword cutler | **John Rawson** | | 1800–1820 |
| Sword cutler | **Benjamin Reddell**<br>**Benjamin Reddell**<br>**Benjamin Reddell** | Jamaica Row<br>Bradford St.<br>Aston Rd. | 1816–1817<br>1818–1820<br>1821 |
| Silversmith<br>Hilt maker<br>Toy (hardware) maker<br>Sword cutler | **John Reddell**<br><br>**Reddell (John) & Bate (Thomas)**<br>• Thomas Bate was a silversmith (see Thomas Bate).<br>• Had a Board of Ordnance contract for 560 rifleman's swords, 1815.<br>• Exported swords (including eaglehead pommel infantry and artillery officers swords) to the United States.<br>**J. (John) H. Reddell & Co.** | Carey's Court,<br>Dale End<br>Carey's Court, Dale End<br><br><br><br><br>Carey's Court, Dale End | 1796–1806<br><br>1807–1815<br><br><br><br><br>1816–1821 |
| Gun maker<br>Bayonet maker<br>Sword maker | **Charles Reeves (Sr.)**<br>**Charles Reeves (Sr.)**<br>• Gun maker only.<br>**Reeves (Charles Sr.) & Greaves (Joseph)**<br>• Sword, cutlass, and bayonet maker.<br>**Reeves (Charles (Sr.)) & Greaves (Joseph)**<br>**Reeves (Charles (Sr.), Greaves (Joseph) & Reeves (John)**<br>**Reeves (Charles (Sr.), Greaves (Joseph) & Reeves (John)**<br>**Reeves (Charles (Sr.), Greaves (Joseph) & Reeves (John)**<br>• In 1851, exhibited swords and field weapons at the Great Exhibition at the London Crystal Palace. | Weaman St.<br>62 Bartholomew St.<br><br>6 Frazeley St.<br><br>12 Fazeley St.<br>12 Fazeley St.<br><br>28 Bartholomew St.<br><br>Baggol St. Mills &<br>28 Bartholomew St. | 1815–1818<br>1819–1828<br><br>1829–1834<br><br>1835–1839<br>1840–1844<br><br>1845–1849<br><br>1850–1851 |

|  |  |  |  |
|---|---|---|---|
|  | Reeves (Charles (Sr.), Greaves (Joseph) & Co. | Charlotte St. | 1852–1854 |
|  | Charles Reeves (Jr.) | 28 Charlotte St., Toledo Works | 1855 |
|  | Charles Reeves (Jr.) | 29 Charlotte St. | 1856–1859 |
|  | Charles Reeves & Co. Ltd | 29 Charlotte St. | 1860–1865 |
|  | • In 1862, exhibited swords and guns at the International Exposition London. | | |
|  | Charles Reeves & Co. Ltd. | 21 George St., Parade | 1866–1872 |
|  | Charles Reeves & Co. Ltd. | 13 St. Mary's Row | 1873–1880 |
| Sword cutler | T. Reeves | | 1885–1892 |
| Sword cutler | William Reynolds | 101 Rea St. | 1847–1850 |
| Goldsmith Hilt maker Sword cutler | Theophilus Richards & Son | Colmore St. | 1815–1820 |
| Sword cutler | George Robinson & Co. | 35 Livery St. | 1860–1864 |
|  | George Robinson & Co. | 2 Coleshill St. | 1865–1877 |
|  | George Robinson & Co. | 42 Loveday St. | 1878 |
|  | George Robinson & Co. | 14 Steel House Lane | 1879–1880 |
| Sword maker | Robinson & Watts | | 1850 |
|  | • Made naval cutlasses. | | |
| Silversmith Hilt maker Gilt toy (hardware) maker | William Russell • Made silver sword hilts. | Great Hampton St. | 1815–1820 |
| Sword cutler | E. Sabell | 35–36 Livery St. | 1865 |
| Sword cutler | Thomas Sadle | St. Luke's Row | 1803–1810 |
| Steel toy (hardware) maker Sword cutler | Charles Sanders | Bradford St. | 1774–1779 |
|  | Sanders (Charles) & Bennett (Richard) | Bradford St. | 1780–1782 |
|  | Charles Sanders | Bradford St. | 1783–D1784 |
| Gun, pistol and rifle maker Sword maker Edged tool maker | William Sargant | | Bc1780–1802 |
|  | William Sargant | Lionel St. | 1803–1814 |
|  | Woolley (James) & Sargant (William) | 74 Edmund St. | 1815–1817 |
|  | • Successor to Woolley, Deakin & Dutton (see James Woolley). | | |
|  | • Had a Board of Ordnance contract for 1,000 rifleman's swords in 1815. | | |
|  | Woolley (James), Sargant (William) & Crane (George) | 74 Edmund St. | 1818–1820 |
|  | • Sword maker to the Board of Ordnance and the East India Company. | | |
|  | • Also made edged tools. | | |
|  | Woolley (James) & Sargant (William) | 74 Edmund St. | 1821–1825 |
|  | Woolley (James), Sargant (William) & Fairfax (Samuel) | 74 Edmund St. | 1826–1834 |
|  | Sargant (William) & Son | 74 Edmund St. | 1835–1837 |
|  | • Also made bayonets. | | |
|  | • Partner and son: William Lucas Sargant. | | |

| | W.L. (William Lucas) & H. (Henry) Sargant | 74 Edmund St. | 1838–1839 |
|---|---|---|---|
| | • In 1839, the consulate in London purchased 600 brass-hilted British Model 1821 cavalry sabers from W.L. & H. Sargant and sent them to the U.S. Ordnance Department for testing and examination. | | |
| | Sargant Brothers | 74 & 75 Edmund St. | 1840–1849 |
| | • Partners: William Lucas and Henry Sargant. | | |
| | Sargant Brothers | 74 Edmund St. & Steam Mills, Charlotte St. | 1850–1851 |
| | William Lucas Sargant | 74 Edmund St. | 1852–1867 |
| | William Lucas Sargant | 35 Whittall St. | 1868–1879 |
| Sword cutler | Seydel & Co. | 7 1/2 St. Mary's Row | 1873–1879 |
| Sword grip maker | Sarah Shaw | Park St. | 1815–1820 |
| Sword maker | John Sherrard | | 1710–1711 |
| | Sherrard (John) & Woolley (Thomas) | | 1712–1715 |
| | • In August 1715, made 300 small swords for a Ben Tanner at 5 shillings each. | | |
| | John Sherrard | | 1716 |
| Bladesmith | Francis Smith | | 1682–1690 |
| Sword cutler | Thomas Swatkins | | 1800 |
| Silversmith Hilt maker | Joseph Taylor | Aston | 1790–1799 |
| | • Made silver sword hilts, hilt parts, and sword furniture. | | |
| Sword and bayonet scabbard maker | John Taylor | Tower St. | 1810–1820 |
| | • Leather worker and scabbard maker. | | |
| Sword cutler | Terry (James) & Oakes (John) | 120 Suffolk St. | 1847–1849 |
| | John Oakes | 6 Exeter Row | 1850–1856 |
| Sword blade maker | John Thomas | Lichfield St. | 1760–1780 |
| | • Operated a forge. | | |
| Steel ring maker Steel hilt maker Steel toy (hardware) maker Sword cutler | William Thompson | Coleshill St. | 1803–1821 |
| | Samson Tomlinson (see William Bickley) | | |
| Sword blade maker Hilt maker | Elias Vallant | 55 Smallbrook St. | 1767–1782 |
| | Elias Vallant | 50 Smallbrook St. | 1783–D1788 |
| Bayonet maker | Joseph Wall | | 1852–1853 |
| | • Made sword bayonets. | | |

| | | | |
|---|---|---|---|
| Silversmith<br>Clock maker<br>Hilt maker<br>Sword cutler | S. (Samuel) Walker<br>• Sold small swords to Henry Upson of London (see Henry Upson). | | 1800–1820 |
| | James Watt<br>(see Matthew Boulton) | | |
| Silversmith<br>Hilt maker<br>Sword cutler | J.B. Watts & Co.<br>• Also sold bayonets. | 35 & 36 Livery St. | 1861–1864 |
| Gun and pistol<br>maker<br>Sword and<br>bayonet<br>maker | Robert Wheeler<br>Robert Wheeler & Son<br>• Son and partner: John Wheeler.<br>John Wheeler | Snow Hill<br>Snow Hill<br><br>49 Whittall | 1804–1808<br>1809–1846<br><br>1847–1850 |
| Hilt maker<br>Sword cutler | Thomas Wilcox<br>Thomas Wilcox | 78 Hill St.<br>71 Hill St. | 1777–1799<br>1800–1808 |
| Silversmith<br>Hilt maker<br>Sword cutler | John Wilkes | | 1762–1765 |
| Silversmith<br>Hilt maker<br>Sword cutler | Willmore (Willmoor) (Joseph) & Alston (Henry)<br>• Made silver sword hilts and shoe buckles.<br>Joseph Willmore | Colmore Row<br><br>Colmore Row | 1775–1802<br><br>1803–1843 |
| Goldsmith<br>Hilt maker<br>Sword cutler | Thomas Willmore (Willmoor) | | 1779–1800 |
| Sword maker<br>Silversmith<br>Hilt maker<br>Gun ramrod<br>maker<br>Bayonet and<br>knife maker | James Woolley<br>James Woolley<br>• Sword factory.<br>James Woolley<br>• Sword factory.<br>James Woolley & Co.<br>• Shop and warehouse.<br>• Sword maker to the East India Company.<br>• Made M1788 and M1796 light and heavy cavalry sabers.<br>• Made steel and silver dagger hilts and guards, scabbard mounts, sword hilts and guards, and gun furniture.<br>• Had 54 silver sword hilts assayed, 1790–1792.<br>• The Birmingham city directory of 1791 (Pyes) lists Woolley as a manufacturer of all kinds of swords, sword hilts, bayonets, ram rods, machetes, etc.<br>• Woolley had his own mill for grinding sword blades at Perry (near Birmingham).<br>Woolley (James) & Deakin (Thomas)<br>Woolley (James), Deakin (Thomas) & Co.<br>• Had 29 silver sword hilts assayed, 1805 and 1806.<br>• Partners: John and Joseph Dutton.<br>Woolley (James), Deakin<br>(Thomas), Dutton (John & Joseph)<br>& Johnston (Richard)<br>• Exported cavalry sabers to the United States. | <br>Newhall St.<br><br>5 Old Square<br><br>74 Edmund St.<br><br><br><br><br><br><br><br><br><br>74 Edmund St.<br>74 Edmund St.<br><br><br>74 Edmund St. | Bc1759–1779<br>1779–1785<br><br>1785–1790<br><br>1790–1798<br><br><br><br><br><br><br><br><br><br>1798–1805<br>1805–1806<br><br><br>1806–1811 |

| | | | |
|---|---|---|---|
| | **Woolley (James), Deakin (Thomas) & Dutton (John & Joseph)** | 74 Edmund St. | 1811–1815 |
| | • Also made edged tools, frying pans, and agricultural tools. | | |
| | • Succeeded by Woolley & Sargant (see Woolley & Sargant). | | |
| | (see Woolley & Deakin, London) | | |
| | (see John Dutton, London) | | |
| | (see Thomas Deakin) | | |
| | (see Richard Johnson, London) | | |
| | **Coales (Robert) & Woolley (James) & Co.** | | |
| | (see Robert Coales) | | |
| Steel maker<br>Steel wire drawer | **Woolley (James), Deakin (Thomas) & Pinkley (Henry)** | Deritend Mills | 1808–1811 |
| Leather worker | **John Woolley** | New John St. | 1800–1820 |
| | • Made leather powder flasks, shot belts, and scabbards. | | |
| Sword cutler | **Thomas Woolley** | | 1790–1800 |
| Silversmith<br>Hilt maker<br>Sword cutler | **William Woolley** | | 1805–1814 |
| | • Probably a relative of James Woolley. | | |
| | **Thomas Worley** | | |
| | (see John Sherrard) | | |
| Chape maker | **William Wright** | 15 Church St. | 1777–1779 |
| | **William Wright** | 39 Church St. | 1780–1788 |
| | • Probably made leather scabbard mounts for swords, including scabbard mouthpieces, carrying rings, and tips (chapes). | | |
| Hilt maker | **George Wyon** | | 1790–1800 |
| Hilt ornamentor | **Thomas Yates** | | 1890–1900 |

## BRISTOL

| | | | |
|---|---|---|---|
| Army and navy tailor<br>Sword cutler | **Edward M. Dyer** | 51 Park St. | 1886–1887 |
| | **E. (Edward) M. Dyer** | 44 Park St. | 1888–1923 |
| Sword maker | **William Greene** | | 1630–1640 |
| | • Repaired swords for the Board of Ordnance, 1631. | | |
| Sword cutler | **R. Thornley** | | 1890 |
| Sword maker | **William Willett** | | 1630–1642 |
| | • Repaired swords for the Board of Ordnance, 1634. | | |

## CAMBRIDGE

| | | | |
|---|---|---|---|
| Sword maker | J. (John) Henshaw | | 1779–1796 |
| Sword maker | Matthew Jesson | Parish of St. Leonard | 1612–1627 |

## CHATHAM

| | | | |
|---|---|---|---|
| Silversmith<br>Sword cutler | John Booth<br>• Made presentation sword for Admiral John Markhaw. | | 1800 |
| Army and navy<br>cap, helmet and<br>accoutrement<br>maker<br>Sword cutler | Hebbert & Co. | Cross St. | 1903–1912 |
| Army and navy<br>clothier<br>and outfitter<br>Sword cutler | F. (Frederick) & H. (Henry) Newcombe<br>F. (Frederick), H. (Henry)<br>& W. (William) Newcombe<br>Frederick, Henry John Newcombe<br>F. (Frederick) & H. (Henry) Newcombe<br>F. (Frederick) & H. (Henry) Newcombe | 101 High St.<br>101 High St.<br><br>101 High St.<br>133 High St.<br>133 High St. & 5 Railway St. | 1830–1857<br>1858–1866<br><br>1867–1889<br>1890–1897<br>1898–1965 |

## DEVONPORT
### (PLYMOUTH DOCK UNTIL 1824)

| | | | |
|---|---|---|---|
| Army and navy<br>tailor and<br>outfitter<br>Sword cutler | Richard Rowley Adams<br>Thomas Adams<br>William Adams<br>William Adams<br>William Adams<br>William Adams | 41 Fore St.<br>27 Bedford St.<br>16 East St. & 27 Bedford St.<br>16 East St. & 70 Bedford St.<br>16 East St. & 29 & 30 Bedford St.<br>16 East St. | 1824–1849<br>1850–1855<br>1856<br>1857–1869<br>1870–1812<br>1813–1875 |
| Army and navy<br>tailor<br>Sword cutler | John Adams & Son<br>• Successor to Batten & Adams (see Batten & Adams).<br>John Adams & Son<br>• Succeeded by J. (James) Gieve & Son Ltd. | 50 Fore St.<br><br>44 Fore St. | 1870–1872<br><br>1873–1899 |
| Army and navy<br>tailor<br>Sword cutler | James Batten<br>James Batten<br>John Batten<br>Batten (John) & Adams (John)<br>Batten (John) & Adams (John)<br>• Succeeded by John Adams & Son (see John Adams & Son). | Fore St.<br>47 Fore St.<br>47–49–50 Fore St.<br>47–49–50 Fore St.<br>44 Fore St. | 1822<br>1823–1829<br>1830–1849<br>1850–1865<br>1866–1869 |
| Army and navy<br>outfitter<br>Sword cutler | Cross & Morgan<br>Cross, Morgan & Co. | 47 Union St.<br>47 Union St. | 1889–1905<br>1906 |
| Army and navy<br>outfitter<br>Sword cutler | G. Cullum<br>G. Cullum & Co.<br>H.M.B. Cullum | 51 Queen St.<br>51 Queen St.<br>49 & 50 Queen St. | 1870–1888<br>1889–1896<br>1897–1905 |

|  |  |  |  |
|---|---|---|---|
|  | H.M.H. Cullum | 49 to 51 Queen St. | 1906–1913 |
|  | G. Cullum & Co. | 49 to 51 Queen St. | 1914–1929 |
|  | Cullum (G.) & Co. | 49 to 51 Queen St. | 1930–1931 |
|  | H.M. Cullum & Co. | 49 to 51 Queen St. | 1932–1934 |
|  | Cullum (H.M.) & Co. | 49 to 51 Queen St. | 1935–1936 |
| Army and navy tailor and outfitter Draper (Cloth maker) Mercer (Cloth dealer) Sword cutler | J. (James) Gieve & Sons Ltd. • Successor to John Adams & Son (see John Adams & Son). Gieve (James), Matthews (Henry G.) & Seagrove (Edwin Augustus) Gieves Ltd. Gieves Ltd. Gieves Ltd. Gieves Ltd. (see James Gieve, London) (see William Seagrove, Portsea) (see Seagrove & Co., Portsmouth) | 44 Fore St. 44 Fore St. 63 George St. 8 Alton Terrace 6 Alton Terrace & 11 North Hill 66 Mutley Plain | 1900–1906 1907–1911 1912–1947 1948–1949 1950–1956 1957–1966 |
| Navy outfitter Sword cutler | Samuel Roger Gould S. (Samuel) R. (Roger) Gould | 106 Fore St. 106 Fore St. | 1870–1909 1910 |
| Navy tailor Sword cutler | Orlando & James Joliffe Joliffe (Orlando & James) & Sons | 32 Tavistock St. 32 Tavistock St. | 1889–1896 1897–1935 |
| Navy outfitter Sword cutler | William Edward Legge | 43 Station Rd. | 1900–1919 |
| Army and navy tailor and outfitter Sword cutler | James Mackey James Mackey James Mackey James Mackey & Co. James Mackey & Co. James Mackey & Co. | Catherine St. 20 Catherine St. & 46 George St. 23 Catherine St. 23 Catherine St. 6 Market St. 76 Fore St. | 1850–1855 1856–1863 1864–1869 1870–1874 1875–1881 1882–1893 |
| Sword cutler | Scorey & Stevens |  | 1827–1830 |
| Sword cutler | H. Shannon Shannon (H.) & Sons | 21 Fore St. 21 Fore St. | 1850–1859 1860 |
| Sword retailer | Philips Stonehouse |  | 1850–1870 |
| Navy outfitter Sword cutler | Stumbles & Son Stumbles & Son Stumbles & Son | 13 Catherine St. 43 Fore St. 242 Peverell Park Rd. | 1873–1896 1897–1947 1948–1950 |
| Army and navy tailor Sword cutler | John Symons John Symons | 74 Fore St. 76 Fore St. | 1823–1849 1850 |

## GOSPORT

|  |  |  |  |
|---|---|---|---|
| Army and navy outfitter Sword cutler | James Frisby Frisby (James) & Son James Frisby & Son | 15 North Cross St. 15 North Cross St. 15 North Cross St. | 1823–1854 1855–1866 1867–1872 |

| | J. (James) E. (Edward) Frisby | 15 North Cross St. | 1873–1875 |
|---|---|---|---|
| Army and navy clothier and outfitter<br>Sword cutler | Frederick Highatt | 70 High St. | 1855–1905 |
| Army and navy outfitter<br>Sword cutler | Thomas Walton<br>Eliza Walton<br>E. (Eliza) Walton | 62 High St.<br>62 High St.<br>62 High St. | 1852–1862<br>1863–1904<br>1905–1914 |
| Army and navy tailor, clothier and outfitter<br>Sword cutler | Henry Woodrow<br>Henry & Alfred Woodrow<br>Woodrow (Alfred) & Row (Samuel)<br>Alfred Woodrow | 42 & 50 High St.<br>112 High St.<br>112 High St.<br>61 High St. | 1830–1851<br>1852–1854<br>1855–1856<br>1857–1865 |

## GUERNSEY

| | | | |
|---|---|---|---|
| Sword cutler | J. Rouqier | High St. | 1895 |

## KENT

| | | | |
|---|---|---|---|
| Sword maker | J. (John) Coupar | Maidstone | 1816–1822 |

## LANDPORT

| | | | |
|---|---|---|---|
| Navy outfitter<br>Sword cutler | Jacob Friedeberger<br>Harry A. Friedeberger<br>(see Portsea listing) | 32 London Rd.<br>32 London Rd. | 1914–1917<br>1918–1929 |
| Army and navy outfitter<br>Sword cutler | Alfred Hatton Hancox<br>Alfred Hatton Hancox<br>Hancox (Alfred Hatton) & Co | 105 Commercial Rd.<br>109 Commercial Rd.<br>109 Commercial Rd. | 1878–1887<br>1888–1899<br>1900–1921 |
| Army and navy outfitter<br>Sword cutler | B. Joseph & Co.<br>(see Portsea listing) | | 1863–1865 |

## LITCHFIELD

| | | | |
|---|---|---|---|
| Sword cutler | J. Joseph Barlow | | 1821 |

## LIVERPOOL

| | | | |
|---|---|---|---|
| Army and navy outfitter<br>Sword cutler | James Davis | Centre Royal Arms | 1856–1860 |
| Gold and silver | George Kenning | 2 Monument Place | 1872–1895 |

| | | | |
|---|---|---|---|
| lacemen | George Kenning & Son | 2 Monument Place | 1896–1897 |
| Embroiderer | George Kenning & Son | 23 Williamson St. | 1898–1934 |
| Sword cutler | George Kenning & Son | 50 Sir Thomas St. | 1835–1942 |
| | George Kenning & Son | 57 White Chapel St. | 1943–1948 |
| | George Kenning & Son | 16 & 18 Harknes St. | 1949–1958 |
| | (see London and Glascow, Scotland, listings) | | |
| Sword cutler | J. Richardson | | 1810–1820 |
| Army and navy | Silver & Co. | 11 St. George, Crescant | 1832–1833 |
| clothier and | Silver, Hayter & Co. | 10 St. George, Crescant | 1834–1836 |
| outfitter | Silver, Hayter & Co. | 5-10-11 St. George, Crescant | 1837–1838 |
| Sword cutler | Silver, Hayter, Wrenn & Co. | 10 & 11 St. George, Crescant | 1839–1856 |
| | John Wrenn & Co. | 10 & 11 St. George, Crescant | 1857–1893 |
| | John Wrenn & Co. | 11 St. George, Crescant | 1894–1895 |
| | (see London and Portsmouth listings) | | |
| Sword cutler | Robert Skinner | | 1840–1850 |
| Sword cutler | G.H. Sunderland | 28 John St. | 1856–1860 |

## MORICE TOWN

| | | | |
|---|---|---|---|
| Military tailor Sword cutler | John Copplestone | 2 Navy Row | 1850–1866 |
| Army and navy tailor and outfitter Sword cutler | Joseph Henry Trounce | Lower Portland Place | 1866–1869 |
| | Joseph Henry Trounce | 2 Albert Road | 1870–1873 |

## NEWCASTLE UPON TYNE

| | | | |
|---|---|---|---|
| Sword cutler | Thomas Carnforth | | 1680–1720 |

- In 1703, offered to buy 20 dozen German-made sword blades from Herman Mohll of Shotley Bridge who was returning from Solingen with more than 40 dozen blades. (see Chapter 4)

| | | | |
|---|---|---|---|
| Goldsmith Sword cutler merchant | William Rumsey the Elder | Sandhill | 1670–1700 |

- Sold military goods to local garrisons.
- In June 1689, sold 500 muskets, 240 pikes, halberds, drums, cartouche boxes, 6 1/2 barrels of gun powder, and 650 pounds of matchlock musket match (cord) to Colonel Beaumond's regiment at Carlisle.
- Partners: John Blakiston and George Morten.

## PLYMOUTH DOCK
## (RENAMED DEVONPORT IN 1824)

| | | | |
|---|---|---|---|
| Silversmith | Banks (Ralph) | 49 Fore St. | 1775–1797 |
| Engraver | Ralph Banks | 49 Fore St. | 1798–D1812 |

| Trade | Name | Address | Dates |
|---|---|---|---|
| Sword cutler | Rebecca Banks | 49 Fore St. | 1812 |
| | • Widow of Ralph Banks. | | |
| | George Banks | 49 Fore St. | 1813–1829 |
| | • Son of Ralph Banks. | | |
| | George Banks | 45 St. Aubyn St. | 1830 |
| | Cooper (John) & Banks (George) | 45 St. Aubyn St. | 1831–1860 |
| Sword maker | Abraham Colmer | | 1625–1635 |
| | • Repaired swords for the Board of Ordnance, 1630. | | |
| Surgical instrument maker | William Cox | 86 Fore St. | 1812–1821 |
| | William Charles Cox | 86 Fore St. | 1822–1829 |
| | William Charles Cox | 87 Fore St. | 1830–1839 |
| Sword cutler | William Charles Cox | 87 to 89 Fore St. | 1840–1849 |
| | William Charles Cox | 87 to 89 Fore St. & 35 Cheapside St. | 1850–1861 |
| | William Charles Cox | 87 to 89 & 93 Fore St. & 35 Cheapside St. | 1862–1865 |
| | William Charles Cox | 87 to 89 Fore St. & 35 Cheapside St. | 1866–1881 |
| | Cox (William Charles) & Coombes (John) | 87 to 89 Fore St. & 35 Cheapside St. | 1882–1888 |
| | John Coombes | 87 to 89 Fore St. | 1889–1918 |
| | Messrs Coombes | 87 to 89 Fore St. | 1919–1939 |
| Silversmith | William Dunsford | Fore St. | 1812–1822 |
| Sword cutler | William Dunsford | Fore St. & 56 Market St. | 1823–1829 |
| | William Dunsford | Fore St. | 1830 |
| Sword cutler | W. George | | 1802 |
| Sword maker | Nicholas Wills | | 1625–1630 |
| | • Repaired swords for the Board of Ordnance, 1628. | | |

## PORTSEA

| Trade | Name | Address | Dates |
|---|---|---|---|
| Navy outfitter | J. Baker & Co. | | 1938–1940 |
| Sword cutler | • Successor to Henry Friedeberger (see Henry Friedeberger). | | |
| Army and navy outfitter | Bilney & Ashdowne | 20 Lion Terrace | 1887 |
| | • Successor to Whiteman & Co. (see Whiteman & Co.). | | |
| Sword cutler | Bilney & Ashdowne | 100 St. George Square | 1888–1891 |
| | Bilney & Rowlands | 100 St. George Square | 1892–1894 |
| | Bilney & Co. | 100 St. George Square | 1895–1900 |
| | Bilney & Co. | 20 Ordnance Row | 1901–1911 |
| Army and navy tailor | James Cracknell | 49 Queen St. (Portsea) | 1854–1905 |
| | Erastus W. Cracknell | 49 Queen St. (Portsea) | 1906–1913 |
| Sword cutler | Erastus W. Cracknell | 23a Elm Grove (Southsea) | 1914–1929 |
| | Erastus W. Cracknell | 25 Elm Grove (Southsea) | 1930–1935 |
| | Erastus W. Cracknell | 23 Elm Grove (Southsea) | 1936–1938 |
| | E. (Erastus) W. Cracknell & Son | 83 Elm Grove & 10 High St. | 1939–1942 |
| | E. (Erastus) W. Cracknell & Son | 7 Highbury Building | 1943–1968 |
| Army and navy tailor and outfitter | Dickensen (George) & Matthews (Henry) | 66 Queen St. | 1845–1834 |
| | • Succeeded by Henry Matthews (see Henry Matthews). | | |

| Sword cutler | | | |
|---|---|---|---|
| Silversmith | James Dudley (Sr.) | Grand Parade, Portsmouth | 1790–1804 |
| Jeweler | James Dudley (Sr.) | 79 High St., Portsmouth | 1805–1819 |
| Sword cutler | Joseph Dudley | 78–79–84 High St., Portsmouth | 1820–1829 |
| | Joseph Dudley | 79 & 80 High St., Portsmouth | 1830–1871 |
| | Samuel & Georgina Dudley | 79 & 80 High St., Portsmouth | 1872–1888 |
| | James Dudley (Jr.) | 79 & 80 High St., Portsmouth | 1889–1897 |
| | James Dudley (Jr.) | 59 Osborne Rd. (Southsea) | 1898–1906 |
| | Dudley (James Jr.) & Cox (Henry) Ltd. | 59 Osborne Rd. (Southsea) | 1907–1931 |

• James Dudley died in 1914.

| | Dudley (James Jr.) & Cox (Henry) Ltd. | 83 Osborne Rd. | 1932–1938 |
|---|---|---|---|

• Dimmer Ltd. purchased Dudley & Cox in 1938.

| Hatter | Thomas Ellyett | 129 Queen St. | 1815–1841 |
|---|---|---|---|
| Gold and silver lacemen | Thomas Ellyett & Charles Dignace | 129 Queen St. | 1842–1849 |
| Navy outfitter | Frederick Ellyett | 129 Queen St. | 1850–1855 |
| Sword cutler | | | |

| Goldsmith | Ezekiel Emanuel | 3 Common Hard | 1823–1829 |
|---|---|---|---|
| Jeweler | Ezekiel & Emanuel Emanuel | 3 Common Hard | 1830–1879 |
| Sword cutler | | | |

• Partner: Moses Emanuel, 1830–1833.
• Appointed goldsmith and jeweler by appointment to Her Majesty the Queen, 1855–1879.
(see Portsmouth listing)

| Goldsmith | Moses Emanuel | 2 Common Hard | 1817–1829 |
|---|---|---|---|
| Jeweler | Ezekiel & Emanuel Emanuel | 3 Common Hard | 1830–1879 |
| Sword cutler | | | |

• Partner: Moses Emanuel, 1830–1833.

| Navy outfitter | J. (Jacob) Friedeberger | Havant St. | 1848–1858 |
|---|---|---|---|
| Sword cutler | J. (Jacob) Friedeberger | 68 Hanover St. | 1859–1862 |
| | J. (Jacob) Friedeberger | 81 Queen St. | 1863–1913 |
| | J. (Jacob) Friedeberger | 81 & 82 Queen St. | 1914–1917 |
| | Harry A. Friedeberger | 81 & 82 Queen St. | 1918–1937 |

• Succeeded by J. Baker & Co. (see J. Baker & Co.).
(see Lanport listing)

| Army and navy outfitter and tailor | Gillot & Hassell | 27 The Hard | 1907–1958 |
|---|---|---|---|
| Sword cutler | (see London listing) | | |

| Navy outfitter | Goldman Brothers | 106–107 Queen St. | 1901 |
|---|---|---|---|
| Sword cutler | Soloman Goldman | 106–107 Queen St. | 1902 |
| | Soloman Goldman | 93 Queen St. | 1903–1909 |
| | Soloman Goldman | 936 Queen St., 4 & 5 Half Moon St. | 1910–1917 |
| | Soloman Goldman | 1 Havant St. & 936 Queen St. & 4 & 5 Half Moon St. | 1918–1920 |
| | Soloman Goldman | 4 & 5 Half Moon St. | 1921 |
| | Goldman (Soloman) & Jeffries (Richard) | 4 & 5 Half Moon St. | 1922–1933 |
| | Richard Jeffries | 4 & 5 Half Moon St. | 1934–1935 |
| Sword cutler | Emanuel Hard | | 1850–1860 |
| Sword cutler | Judah Hart | | 1791–1798 |

| | | | |
|---|---|---|---|
| Navy outfitter | R.G. James & Co. | 13 Queen St. | 1897–1901 |
| Sword cutler | R.G. James & Co. | 13 & 14 Queen St. | 1902–1915 |
| | Richard Jeffries (see Goldman Brothers) | | |
| Army and navy outfitter Sword cutler | B. Joseph & Co. (see London listing) | 92 & 93 Queen St. | 1863–1865 |
| Army and navy tailor and outfitter Sword cutler | Larcom & Veysey Larcom & Veysey • Sold Wilkinson Sword Company swords. | 52 Queen St. 51 & 52 Queen St. | 1882–1886 1887–1915 |
| Army and navy tailor and outfitter Sword cutler | Henry Matthews • Successor to Dickensen & Matthews (see Dickensen & Matthews). H. (Henry) G. Matthews & Co. Matthews (Henry G.) & Son Matthews (Henry G.) & Co. • Successor to John Stone at Camden Alley address. | 66 Queen St. 66 Queen St. 66 Queen St. 6 & 9 Camden Alley | 1855–1871 1872–1880 1881–1886 1887–1903 |
| Goldsmith Silversmith Jeweler Sword cutler | John Moses Alexander Moses Alexander & Louis Moses | 25 The Hard 26 The Hand 26 The Hand | 1847–1856 1857–1858 1859–1888 |
| Silversmith Sword cutler | William Price | | 1791–1798 |
| Jeweler Sword cutler | George Price | 3 Camden Alley | 1823–1824 |
| Army and navy tailor and outfitter Sword cutler | William Seagrove William Seagrove William Seagrove William & Edward Seagrove William & Edwin Seagrove Edwin Augustus Seagrove Seagrove (Edwin Augustus) & Co. • Succeeded by Gieve, Matthew & Seagrove. (see James Gieve's Devonport, London, and Portsmouth listings) | 32 Harant St. 28 Common Hard 22 Common Hard 22 Common Hard 22 & 23 Common Hard 22 & 23 Common Hard 22 & 23 Common Hard | 1822–1829 1830–1838 1839–1849 1850–1851 1852–1866 1867–1891 1892–1906 |
| Watch maker Silversmith Sword cutler | Charles Edward Smithers Charles Edward Smithers Charles Edward Smithers | Queen St. 45 Queen St. 186 & 45 Queen St. | 1823–1829 1830–1854 1855–1863 |
| Army and navy tailor Sword cutler | Edward Totterdell | 62 Queen St. | 1841–1863 |
| Army tailor and outfitter Sword cutler | Mrs. Henry Whiteman William Henry Whiteman William Whiteman Whiteman (William) & Co. • Succeeded by Bilney & Ashdowne (see Bilney & Ashdowne). | Lion Terrace 68 Lion Terrace 68 Lion Terrace 20 Lion Terrace | 1872–1874 1875–1877 1878–1885 1886–1896 |

## PORTSMOUTH

| | | | |
|---|---|---|---|
| Silversmith<br>Sword cutler | **Robert Baker**<br>• Made presentation swords. | High St. | 1812 |
| Sword cutler | **W. Chappell & Co. Ltd.**<br>• Sold Wilkinson Sword Company swords. | 267 Commercial Rd | 1950 |
| Sword cutler | **W. Davidson** | 123 Union St. | 1900 |
| Watchmaker<br>Jeweler | **George Dimmer**<br>• Successor to Ezekial and Emanuel Emanuel. | 101 High St. | 1880–1889 |
| Silversmith<br>Sword cutler | **George Dimmer**<br>Dimmer (George) & Sons Ltd.<br>Dimmers Ltd.<br>Dimmers Ltd.<br>Dimmers Ltd. | 44 Palmerston Rd.<br>44 Palmerston Rd.<br>44 Palmerston Rd.<br>45 Osborne Rd.<br>66 Osborne Rd. | 1890–1913<br>1914–1929<br>1930–1940<br>1941–1955<br>1956–1968 |
| | **Dudley**<br>(see Portsea listing) | | |
| Sword maker | **Phillip Elmes**<br>• Repaired swords for the Board of Ordance, 1636–1658. | | 1630–1660 |
| Goldsmith<br>Silversmith<br>Sword cutler | **Ezekiel & Emanuel Emanuel**<br>• Succeeded by George Dimmer (see George Dimmer). | 101 High St. | 1830–1879 |
| Silversmith<br>Jeweler<br>Sword cutler<br>Sword belt<br>maker | **Thomas Henry Fiske**<br>• Successor to William Price Read (see William Price Read). | 59 High St. | 1824–1839 |
| Army and navy<br>outfitter<br>Sword cutler | **Fraser & Davis**<br>Fraser & Davis | 99 High St.<br>78 & 79 High St. | 1857–1862<br>1863–1890 |
| Army and navy<br>tailor and<br>outfitter<br>Sword cutler | **Joseph Galt**<br>Joseph Galt<br>Joseph Galt<br>Joseph James Galt<br>Galt (Joseph James), Gieve (James) & Co.<br>Galt (Joseph James) & Gieve (James)<br>• Succeeded by Gieve & Son. | 63 High St.<br>73 High St.<br>73 & 111 High St.<br>111 High St.<br>111 High St.<br>111 High St. | 1823–1838<br>1839–1849<br>1850–1854<br>1855–1862<br>1863–1880<br>1881–1886 |
| Silversmith<br>Sword cutler | **William Gibbons** | 2 St. Mary's St. | 1783–1830 |
| Army and navy<br>clothier<br>Sword cutler | **Joseph Gibson** | 67 High St. | 1823–1830 |
| Army and navy<br>tailor and | **Gieve (James) & Son**<br>• Successors to Galt & Gieve (see Joseph Galt). | 111 High St. | 1887–1899 |

| Trade | Name | Address | Dates |
|---|---|---|---|
| outfitter<br>Sword cutler | J. (James) Gieve & Sons Ltd.<br>Gieve (James), Matthews (Henry G.) & Seagrove (Edwin Fugustus) | 111 High St.<br>110 & 111 High St.<br>& 22 & 23 The Hard<br>& 44 St. Thomas St. | 1900–1906<br>1907–1910 |
| | Gieve (James), Matthews (Henry G.) & Seagrove (Edwin Fugustus) | 22 The Hard | 1911 |
| | Gieves Ltd.<br>• Also had shops in London and Davenport.<br>• Sold Wilkinson Sword Company swords.<br>(see London and Devonport listings) | 22 The Hard | 1912–1969 |
| Army and navy outfitter<br>Sword cutler | Guy & Eames<br>• Successor to Jane Selby. | 49 High St. | 1863–1875 |
| Mercer (clothes dealer)<br>Outfitter<br>Sword cutler | William Hammond | High St. | 1790–1808 |
| | Hebbert & Co.<br>(see London listing) | | |
| Sword cutler | S. Hough | 29 High St. | 1810–1838 |
| Sword cutler | Thomas Jarvoice | | 1791 |
| Outfitter | Henry G. Matthews<br>• Sold Wilkinson Sword Company swords.<br>(see Gieve & Son) | | 1890–1906 |
| Sword cutler | William Mitchell | 32 High St. | 1830–D1851 |
| Navy tailor and outfitter | Betsy Mitchell<br>• Widow of William Mitchell. | 32 High St. | 1851–1854 |
| Woolen draper (maker) | Henry James Mitchell<br>• Son of William Mitchell. | 32 High St. | 1855–1858 |
| Sword cutler | William Mountaine | | 1791–1800 |
| Silversmith<br>Sword cutler | Philip Nathan | | 1790–1798 |
| Sword maker | John Neave<br>• Repaired swords for the Board of Ordnance. | | 1600–D1635 |
| | Eleanor Neave<br>• Widow of John Neave.<br>• Repaired swords for the Board of Ordnance. | | 1635–1640 |
| Army and navy tailor and outfitter<br>Sword cutler | Edmund Neck<br>Edmund Neck | 16 Ordnance Row<br>8 & 16 Ordnance Row | 1823–1829<br>1830 |
| Silversmith<br>Jeweler<br>Sword cutler<br>Goldsmith<br>Silversmith | William Read<br>William Read & Nephew (John)<br>John Read<br>William Price Read<br>• Succeeded by Thomas Henry Fiske. | Charlotte Row<br>Charlotte Row<br>Charlotte Row<br>59 High St. | 1775–1799<br>1800–1804<br>1805–1830<br>1781–1823 |

| Role | Name | Address | Dates |
|---|---|---|---|
| Watch maker<br>Jeweler<br>Sword cutler | | | |
| Sword cutler | **Michael Richardson**<br>• Furbisher of arms at Portsmouth at 60 pounds a year, 1675–1685.<br>• Sword cutler to the Board of Ordnance, 1685–1690. | | 1670–1690 |
| Army and navy tailor and outfitter<br>Sword cutler | **Seagrove (Adwin Augustus) & Co.**<br>• Succeeded by Gieve, Matthews & Seagrove.<br>(see James Gieve under Portsmouth, Devonport, and London listings)<br>(see William Seagrove, Portsea) | 30 Pearl Building, Commercial Rd. | 1892–1906 |
| Army and navy tailor and outfitter<br>Sword cutler | **John Charles Selby**<br>**Jane Selby**<br>• Widow of John Charles Selby.<br>**Jane Selby**<br>**Jane Selby**<br>• Succeeded by Guy & Eames (see Guy & Eames). | 85 High St.<br>85 High St.<br><br>50 High St.<br>49 High St. | 1830–D1851<br>1851–1856<br><br>1857–1858<br>1859–1862 |
| Silversmith<br>Sword cutler | **William Shoveller** | | 1791 |
| Army and navy outfitter<br>Sword cutler | **S. (Stephen) W. (Winkwork) Silver & Co.**<br>**S. (Stephen) W. (Winkwork) Silver & Co.**<br>(see London listing) | 106 High St.<br>23 The Hard | 1901–1939<br>1940–1962 |
| Sword cutler | **George Spencer** | | 1902–1905 |
| Army and navy accoutrement maker<br>Sword cutler | **William Stephens** | 13 Warblington St. | 1830–1840 |
| Army and navy tailor and outfitter<br>Sword cutler | **Trayler & Co.** | 103 High St. | 1881–1905 |
| Navy outfitter | **E. Walton & Co. Ltd.**<br>(see Gosport listing) | 18 Ordnance Row & Victory Rd. | 1915–1920 |
| Silversmith<br>Hilt maker<br>Sword cutler | **William Watts** | | 1791–1800 |
| Armourer<br>Sword maker | **Thomas Whitehorn**<br>• Repaired swords for the navy. | | 1640–1654 |
| Silversmith<br>Hilt maker<br>Sword cutler | **Wolfe & Soloman** | | 1785–1800 |
| Silversmith<br>Hilt maker<br>Navy agent | **Gershom Woolf** | | 1791–1800 |

| | | | |
|---|---|---|---|
| Jeweler<br>Silversmith<br>Goldsmith<br>Sword cutler | Seiaske Zachariah | High St. | 1835–1865 |

## SHEFFIELD

| | | | |
|---|---|---|---|
| Army<br>accoutrement<br>maker<br>Sword cutler | Frederick Barnes & Co.<br>Frederick Barnes & Co.<br>Frederick Barnes & Co.<br>Frederick Barnes & Co.<br>Frederick Barnes & Co.<br>(see Birmingham and London listings) | 6 Union Lane<br>13 Howard St.<br>25 Carver St.<br>222 Solly St.<br>220 Solly St. | 1839<br>1840–1848<br>1849–1856<br>1857–1859<br>1860–1904 |
| Sword cutler | C. (Charles) Congreve | | 1849–1859 |
| Cutlery<br>manufacturer<br>and dealer<br>Sword cutler | Joseph T. Deakin<br>Deakin (Joseph T.), Ecroyd (Henry) & Co.<br>• Founder and president: Joseph T. Deakin.<br>• Ecroyd retired in 1872.<br>Joseph T. Deakin & Ernest G. Reuss<br>Deakin (Joseph T.) Sons & Co.<br>• Sons and partners: Frank and Walter Deakin.<br>• The factory on West Street called Tiger Works shipped large quantities of cutlery to South America.<br>• Sold butcher knives, pocket cutlery, daggers, spear point knives, razors, and short swords.<br>• Joseph Deakin died in 1898.<br>Deakin, Sons & Co.<br>• Factory on West Street (Tiger Works). | 76 Arundel St. | B1849–1867<br>1868–1872<br><br><br>1873–1895<br>1896–1900<br><br><br><br><br>1901–c1915 |
| Tailor<br>Sword cutler | John Dixon & Son | | 1900–1920 |
| Silversmith<br>Electroplater<br>Hilt maker<br>Sword cutler | John Frederick Fenton<br>Hukin (Henry) & Fenton (John Frederick)<br>John Frederick Fenton<br>Fenton Brothers<br>• Partners and brothers: John Frederick Fenton and Frank Fenton (B1837).<br>• Silversmiths and electroplaters.<br>Fenton Bros.<br>• Called South Moor Works.<br>• In c. 1873, opened a London office at 22 Bartletts Building.<br>• John Frederick Fenton died in 1883; Frank Fenton died in 1884.<br>• John Frederick Fenton's sons Samuel and Alfred John took over the company.<br>• Samuel Fenton died in 1893; Alfred John Fenton died c. 1895.<br>Fenton Bros. Ltd.<br>• William Stainforth was director and manager.<br>• Made silver-hilted knives and swords. | <br>Cadman Lane<br>Cadman Lane<br>Norfolk Lane<br><br><br>66 Porter St. & Earl St. | B1820–1849<br>1850–1855<br>1856–1859<br>1860–1872<br><br><br>1873–1895<br><br><br><br><br><br>1896–1938 |
| Knife and<br>sword maker | George Gray<br>• Made bowie knives for the U.S. market.<br>Clement Gray<br>• Made dirks, knives, and swords.<br>• Bought out by John & William Ragg in 1912 but kept Clement Gray name.<br>• His catalog advertised swords for British, Indian and colonial officers of the army and navy. | Rockingham Lane<br><br>12 Elden St. | 1818–1870<br><br>1871–1955 |

| | | | |
|---|---|---|---|
| Knife and sword maker | J. (Joseph) Hawksworth | Club Gardens | 1841–1890 |
| Sword cutler | H. Hill | Norfolk St. | 1912–1915 |
| Knife, machette and sword maker | Samuel Kitchin<br>S. (Samuel) & J. (John) Kitchin<br>• Factory called Soho Works.<br>• Made table and butcher knives, bowie and hunting knives, machettes and short swords, daggers, shoe knives, pocket knives, and razors. | Summerfield St. | c1855–1867<br>1868–1938 |
| | S. & J. Kitchin<br>S. & J. Kitchin<br>• The Kitchin's sold out in 1987.<br>• Made pattern 1897 royal engineers swords. | Broadfield & Saxon Rd., Heeley<br>Chesterfield | 1939–c1967<br>c1968–1987 |
| Tool maker<br>Knife, dirk and sword maker | John Lockwood<br>• Filer maker.<br>• Partner and son: William Lockwood.<br>William Lockwood Sr.<br>Lockwood Brothers<br>• William's sons William Jr., John, Joseph, and Charles Lockwood ran the company.<br>• They enlarged the tool line.<br>• They bought John Sorby & Son, a tool manufacturer at Spital Hill.<br>• Began cutlery production, including dirks, knives, and short swords.<br>• Sold knives and short swords to the South American market.<br>• George F. Lockwood, son of John Lockwood, took over the company, c. 1870. | Ecdesfield<br><br><br>74 Arundel St., Sheffield<br>74 Arundel St., Sheffield | 1767–1797<br><br><br>1798–c1840<br>c1840–c1931 |
| Engraver<br>Silversmith<br>Cutler<br>Sword and knife hilter<br>Knife maker | Joseph Mappin (Sr.)<br>Joseph Mappin (Jr.)<br>Arundel (John) & Mappin (Joseph Jr.)<br>• Made pen, spring, sporting, and table knives.<br>• Arundel retired in 1836.<br>• Son Frederick Thorpe Mappin (B1821) apprenticed with his father, 1835.<br>Joseph Mappin (Jr.) & Co.<br>• Joseph Mappin Jr. died in 1841.<br>• His son Frederick Thorpe Mappin took over the company.<br>• In 1845, opened a London office.<br>• In 1845, purchased cutler William Sampson & Son.<br>• Eventually his three brothers, Edward, Joseph Charles, and John Mappin, joined the company.<br>Joseph Mappin & Co.<br>Mappin Brothers<br>• Opened a factory called Queen's Cutlery Works in 1851. It was bordered by Bakers Hill, Flat Street, and Little Pond Street.<br>• Eventually employed 500 workers.<br>• They expanded into silver dinnerware, plated ware, dishes and tableware, carving knives, and silver-hilted knives and swords.<br>• Frederick Thorpe Mappin left in 1859 and joined steel maker Jurton & Sons.<br>• John Newton Mappin left in 1859 and opened his own silver electroplating and cutlery business (see John Newton Mappin).<br>• Edward and Joseph Charles Mappin continued the business.<br>• In 1899, they were purchased by a London firm called Goldsmith & Silversmith Co. but kept the Mappin Bro. name.<br>• In 1903, the company was purchased by Mappin & Webb but continued the name Mappin Bros. until 1910. | Fargate<br>Mulberry St.<br>Mulberry St.<br><br><br><br>16 Mulberry St.<br><br><br><br><br><br>32 Norfolk St.<br>32 Norfolk St. | c1770–1809<br>1810–c1820<br>c1821–1836<br><br><br><br>1837–1845<br><br><br><br><br><br>1846–1850<br>1851–1910 |

| | | | |
|---|---|---|---|
| Silversmith<br>Cutler<br>Knife maker<br>Sword and knife<br>hilter | **Mappin (John Newton) & Co.**<br>• Factory called "Royal Cutlery Works" was on Norfolk St. opposite St. Pauls church and down the street from Mappin Bros.<br>• Opened a London office in 1861 (see)<br>• Made silver dinnerware and plated ware, pocket knives, and silver-hilted knives and swords | Norfolk St. | 1859–1863 |
| | **Mappin (John Newton) & Webb (Henry)**<br>• Factory: Royal Cutlery Works. | Norfolk St. | 1864–1967 |
| | **Mappin & Webb** | 78 Bull St. | 1968–2003 |
| Sword maker | **Thomas Moseley** | 51 Broad St. & 92 Cricket Rd. | 1845–1915 |
| Sword cutler | **S.C. Mower** | 78 Ball St. | 1904–1910 |
| Sword maker | **H.H. Nicholson** | Meadow Works | 1835–1850 |
| Knife and<br>sword maker | **John Rogers**<br>• Granted a "six point star" mark, 1692.<br>• Eventually became Joseph Rogers & Sons, knife maker. | Holy Croft | 1690–1730 |
| Silversmith<br>Electroplater<br>Knife and<br>sword maker | **John Round**<br>**John Round & Son**<br>**John Round & Son Ltd.** | Tudor St.<br>Tudor St.<br>Tudor St. | 1847–1864<br>1865–1873<br>1874–1932 |
| Knife, bayonet<br>and sword<br>maker | **Sanderson Bros.**<br>**Sanderson Bros. & Neubold** | Darnell Works, Darnell Rd.<br>Newhall Road | 1876–1899<br>1900–1998 |
| Knife and<br>sword maker | **Saynor Cutlery**<br>**Samuel & John Saynor**<br>• Made knives, swords, shoe buckles, skates, scissors, and razors.<br>**Samuel Saynor (Sr.)**<br>**Samuel Saynor (Jr.)**<br>**Saynor & Cooke**<br>**Saynor, Cooke & Ridal**<br>• Purchased by Needham, Veall & Tyzack in 1948. | Bank St.<br>Bank St.<br><br>Scargill Croft<br>Edward St.<br>Edward St.<br>Edward St. (Paxton Works) | 1738–c1780<br>c1780–c1815<br><br>c1815–c1825<br>c1825–c1850<br>c1850–1876<br>1876–1948 |
| Knife and<br>sword maker | **John Taylor**<br>• Granted "eye witness" mark, 1838.<br>• Made pen, pocket, and sportsman's knives and short swords.<br>• Purchased by Thomas Brown Nedham, 1855. | St. Philips Rd. | 1820–1855 |
| Knife and<br>sword maker | **Thomas Turton** | Sheaf Works | 1845–1885 |
| Sword cutler | **Wade & Butcher** | | 1861–1865 |
| Sword cutler | **William Walker** | | 1653 |
| Sword cutler | **Watson & Gillott** | | 1910–1915 |
| Sword cutler | **Alfred Williams** | | 1915 |

## SUFFOLK

| | | | |
|---|---|---|---|
| Sword cutler | **C. Annakin** | 17 Blake St., York | 1850–1860 |

| | | | |
|---|---|---|---|
| Sword cutler | Worthan Manon | | 1845–1855 |
| Sword cutler | Salmon & Tillett | | 1830 |

## WINCHESTER

| | | | |
|---|---|---|---|
| Tailor | William Flight | St. Thomas St. | 1750–1789 |
| Military tailor | John Flight | St. Thomas St. | 1790–1824 |
| Sword cutler | William Flight | 123 High St. | 1825–1835 |
| | William Pike Flight | 123 High St. | 1836–1838 |
| | William Pike Flight | 40 High St. | 1839–1849 |
| | William Pike Flight | 72 High St. | 1850–D1854 |
| | Mrs. Elizabeth Flight | 72 High St. | 1854–1866 |

  • Widow of William Pike Flight.

| | | | |
|---|---|---|---|
| | Frederick William Flight | 72 High St. | 1867–1874 |
| | Frederick William Flight | 90 High St. | 1875–1919 |
| | Flights Ltd. | 90 High St. | 1920 |

## WOODSTOCK, OXFORDSHIRE
## (8 MILES NORTH OF OXFORD)

| | | | |
|---|---|---|---|
| Steel hilt maker | George Eldbridge | | 1740–1780 |

  • Had a factory making high-quality polished steel watch chains, watch keys, buckles, tie buckles, and sword hilts.
  • Other factory owners in Woodstock making sword products included Henry Grantham, John Medcalfe, and Edward Staunton.

## WOOLWICH

| | | | |
|---|---|---|---|
| Sword cutler | Shirley Brooks Ltd. | | 1921–1936 |
| Sword cutler | J. Daniels & Co. | Artillery Place | 1830–1835 |
| Sword cutler | Daniel Mills | Greens End | 1830–1835 |
| Sword cutler | W.W. White | Francis St. | 1850–1900 |

  • Sold M1821 light cavalry sabers.

## YARMOUTH

| | | | |
|---|---|---|---|
| Sword cutler | R. Ward | | 1807–1810 |

# Chapter 10

# English Sword Makers, Cutlers, and Dealers Who Exported Swords to the United States

## LONDON

| | | | |
|---|---|---|---|
| Military outfitter | **Isaac, Campbell & Co.** • Exported British M1853 (variant) cavalry sabers and M1822 (variant) foot officers swords to the Confederacy during the American Civil War. | 71 Jermyn St. | 1861–1864 |
| Sword cutler | **David Davies** • Exported eaglehead pommel swords to the United States. | 10 St. James St. | 1796–1801 |
| Silversmith Sword cutler | **John Feesey** • Made the "state" silver-hilted small sword of U.S. president George Washington. | Pall Mall | 1759–1796 |
| Sword and blade maker | **J. (John) Harrison** • Sold sword blades to George W. Simons & Co., Philadelphia, PA. | | 1840–1876 |
| Goldsmith Silversmith Jeweler Sword cutler | **John Salter** • Large assembler and exporter of swords, including eaglehead pommel infantry and artillery officers swords, to United States dealers Lemuel Wells, A.W. Spies, and Richards, Upson & Co. of New York before the War of 1812. | 34–35 Strand | 1801–1824 |
| Sword maker Sword belt maker | **Francis Thurkle** • Large maker and exporter of swords, including eaglehead pommel infantry and artillery officer swords, to the United States. | 15 Great New Street Square | 1791–D1801 |
| Sword, gun and button exporter | **Henry Upson** • Possibly a partner in the company of Samuel Williams, a London-based international trader. Upson was a relative of George Upson, who was part of the New York military outfitter Richards, Upson & Co. He exported swords to Richards, Upson & Co. before and even early into the War of 1812, though there was a U.S. embargo on all English goods. The swords were purchased from Osborn & Gunby, London and Birmingham. | | 1790–1820 |

# BIRMINGHAM

| | | | |
|---|---|---|---|
| Gun maker<br>Silversmith<br>Steel toy (hardware) maker<br>Sword maker | **Thomas Bate**<br>• Exported swords, including eaglehead pommel officers swords, to the United States. | Lancaster St. Summer Lane | 1814–1820 |
| Gun maker<br>Sword maker | **R. (Richard) D. Bolton & Co.**<br>• Exported swords, including eaglehead pommel infantry and artillery officers swords, to the United States. | | 1790–1815 |
| Hilt maker<br>Sword hardware maker<br>Machinery maker<br>Industrialist<br>Merchant | **Matthew Boulton**<br>• A good friend of William Henry Jr., a large gun and sword maker in Pennsylvania. Henry named his largest gun factory at Bushkill after Boulton (Boulton Works, opened in 1812).<br>• Boulton may have sold Henry gun machinery for Henry's Boulton Works.<br>• It is possible he sold swords to the United States before he died in 1809. | | B1728–D1809 |
| Sword cutler | **John Cooper**<br>• Exported swords to the United States. | Bartholomew St. | 1782–1802 |
| Saw, file and tool maker<br>Steel toy (hardware) maker<br>Gun maker<br>Sword blade maker<br>Sword maker | Thomas Gill (Sr.)<br>Thomas Gill (Sr.)<br>Thomas Gill (Sr.)<br><br>Thomas (Jr.),<br>James & John Gill<br>Thomas (Jr.) & James Gill (Jr.)<br>• James Jr. died in 1808.<br>Thomas (Jr.) Gill<br><br>• Many swords, including eaglehead pommel officers swords, that were exported to the United States had the Gill or "G" mark on the blades. | 95 Dale End<br>92 Dale End<br>Jennons Row, near<br>St. Bartholomew Chapel<br>Jennors Row, near<br>St. Bartholomew Chapel<br>Jennors Row, near<br>St. Bartholomew Chapel<br>Jennors Row, near<br>St. Bartholomew Chapel | 1774–1776<br>1777–1782<br>1783–D1801<br><br>1801–1802<br><br>1803–1808<br><br>1809–D1826 |
| Sword maker | John Gill<br>John Gill<br><br>John Gill<br>Elizabeth Gill<br>• Many swords, including eaglehead pommel officers swords, that were exported to the United States had the Gill or "G" mark on the blades. | 20 Masshouse Lane<br>20 Masshouse Lane &<br>St. Bartholomew Square<br>20 Masshouse Lane<br>20 Messhouse Lane | 1803–1814<br>1815<br><br>1816–D1826<br>1827–1837 |
| Steel button maker<br>Steel buckle maker<br>Steel toy (hardware) maker<br>Steel hilt maker<br>Sword maker | Joseph Harvey<br>Joseph Harvey<br>• Exported eaglehead pommel officers swords to the United States. | 16 Upper Priory<br>Park St. | 1800–1814<br>1815–1820 |
| Gun maker<br>Gun lock maker<br>Sword cutler | Soloman Jackson<br>Soloman Jackson<br>• Exported swords, including eaglehead pommel infantry and artillery officers swords, to the United States. | Lionel St.<br>Summer Lane | 1803–1814<br>1815 |

| | | | |
|---|---|---|---|
| Gun maker<br>Sword cutler | **Thomas Ketland Sr.** | | 1767–D1806 |

- Exported muskets and musket locks to his sons Thomas Ketland Jr. and John Ketland, who had set up temporary residence in Philadelphia, Pennsylvania, from 1789–1800.
- May have exported swords to them also.

| | | | |
|---|---|---|---|
| | **Ketland & Co.** | | 1804–1871 |

- Between 1804–1812, exported many eaglehead pommel infantry and artillery officers swords to dealers in New York, NY, such as George Upson and Richards, Upson & Co.
- Also exported a complete line of pistols, muskets, and fowling pieces to such U.S. dealers as Richards, Upson & Co.

| | | | |
|---|---|---|---|
| Sword maker | **Robert Mole (Sr.) & Son (Robert Mole Jr.)** | 171-172 Broad St. Islington | 1856–1874 |

- During the American Civil War (1861–1865), Mole sold brass-hilted M1853 cavalry sabers to the Confederacy and iron-hilted M1821 cavalry sabers to Tiffany & Co., New York, NY. Tiffany then sold them (approximately 6,000) to the U.S. government.

| | | | |
|---|---|---|---|
| Gun maker<br>Sword maker<br>Accoutrement<br>maker | **Osborn (Henry) & Gunby (John)**<br>**Henry Osborn** | Bordesly St. Mills<br>Bordesly St. Mills | 1808–1820<br>1821–1838 |

- Both companies exported swords, including eaglehead pommel infantry and artillery officers swords, to the United States.

| | | | |
|---|---|---|---|
| Steel toy<br>(hardware)<br>makers<br>Sword cutler | **Reddell (John) & Bate (Thomas)**<br>**J. (Joseph) H. Reddell & Co.** | Carey's Court, Dale End<br>Carey's Court, Dale End | 1803–1813<br>1814–1821 |

- Both companies exported swords, including eaglehead pommel officers swords, to the United States.

| | | | |
|---|---|---|---|
| Pistol and rifle<br>maker<br>Sword maker | **W.L. (William Lucas) & H. (Henry) Sargant** | 74 Edmund St. | 1838–1839 |

- In 1839, the U.S. consulate in London purchased 600 brass-hilted Model 1821 cavalry sabers from them and sent them to the U.S. Ordanance Department for testing and examination.

| | | | |
|---|---|---|---|
| Sword maker | **Woolley (James), Deakin (Thomas),**<br>**Dutton (John & Joseph) & Johnston (Richard)** | 74 Edmund St. | 1807–1811 |

- Exported cavalry sabers and sword blades to the United States before the War of 1812.

# Chapter 11

# English Armourers

In England during the Middle Ages, metalsmiths who made knightly armour, horse armour, and hand weapons such as swords, axes, bills, maces, halberds, lances, and early hand and shoulder guns were called armourers. The Armourers Company of London (established in 1347) and the Greenwich (Almain) Royal Armouries (established in 1511) had sword makers among their members. Some of these armourer/swordsmiths later became members of the Hounslow Heath sword and blade making center and the Cutlers Company of London.

## THE GREENWICH (ALMAIN) ROYAL ARMOURIES

Established in 1511 by King Henry VIII (1509–1547), who imported foreign armourers from Brussels, Belgium; Milan, Italy; Landshut, Bavaria; Nuremburg, Bavaria; Cologne, Westphalia; Augsburg, Swabia; and Solingen, Prussia.

The German armourers were called "Allmayne" or "Almain" armourers. The shops were first set up in London at Southwark, then quickly moved to Greenwich. Foreign armourers continued to immigrate to Greenwich during the reign of Henry VIII.

Some swordsmiths from Solingen, Prussia, immigrated to Greenwich in 1629. Most went to Hounslow.

The Greenwich Royal Armouries were closed in 1644.

### Master Armourers at Greenwich

|  | Master | Active |
|---|---|---|
| John Blewbery | 1511–1514 | 1511–1514 |
| Martin Van Royne (Old Martyn) | 1515–1518 | 1515–1544 |
| Erasmas Kirkener (Kyrkenor) (Member of the Armourers Company of London) | 1519–1561 | 1513–D1567 |
| Jakob Topf | 1562–1566 | 1530–1587 |

|  | Master | Active |
|---|---|---|
| John Kirke (Kelke) (Apprenticed to John Lindsay) | 1567–1575 | 1552–1635 |
| Jacob Halder (Master Jacobi) | 1576–1606 | 1553–1606 |
| William Pickering (Pupil of Jakab Topf) (Member of the Armourers Company of London) | 1607–1617 | 1567–D1618 |
| Thomas Stevens | 1618–1627 | 1618–1631 |
| Nicholas Sherman (Member of the Armourers Company of London) | 1628–1644 | 1628–1644 |

### Armourers at Greenwich
(many were swordsmiths)

| | |
|---|---|
| William Austin | |
| John Beatty | |
| Alies Hynde Boreman | 1599–1609, D1645 |
| Baltesar Bullato (from Milan) | 1532 |
| Peter Brock (Solingen) | 1603 |
| William Clerc | 1530 |
| Hans Clerc | 1530 |
| Hans Clynkendager (Clinkedag) | 1542–1544 |
| John Clynkendager | 1525–1529 |
| Thomas Cowper | 1559 |
| William Carter (London) | 1528 |
| Richard Carter (London) | 1528–1540 |
| Peter Crouche (Crochet, Crowche, Croehe) (London) | 1537–1547 |
| Francis Crouche (Crochet, Crowche, Croehe) (London) | 1528–1529 |
| William Crouche (Crochet, Crowche, Croehe) (London) | 1629–1636 |
| John Crouche (Crochet, Crowche, Croehe) (London) | 1515–1520 |
| Thomas Dael | 1515 |
| William Darwin | 1613 |
| John Dawson | 1609 |
| Thomas Dolling | 1608 |
| Mathew Dericke (Dedikes, Dethyke, Dirike) | 1544–1574 |
| Robert Dericke | 1524 |
| John Diconson | 1528 |
| Peter Ferers (from Brussells) | 1512–1518 |
| James Fuller | 1559 |
| Filippo De Grampis (from Milan) | 1514 |
| John Garrett | 1559–1601 |
| Leonarde Guynell | — |
| Martyn Herste | 1554–1574 |

| | |
|---|---|
| Richard Hoane | 1600 |
| Johannes Hoppe the Younger (from Solingen) | 1629–1633 |
| (Moved to Hounslow in 1633) | |
| Jasper Kemp | 1544 |
| Roger Keymer | 1571 |
| Kornelys | 1515 |
| Glorann Angelo de littis (from Milan) | 1514 |
| John Lindsey | 1540–1552 |
| Hans Mery | 1514 |
| Hans Mightner | 1559–1574 |
| John Pickering (son of William) | B1587–D1635 |
| Tobias Pickering (son of William) | B1594–1625 |
| Nighel Pope (Pipe) | 1559 |
| John Polston | 1552 |
| M. Pixe | ——— |
| John Richmond | ——— |
| (Tarys) Carries Spirarde (Spiradi) | 1554–1574 |
| John Stevens (Stephens) | 1571–1618 |
| Thomas Stevens (Stephens) | 1618–1631 |
| John Stile | 1524 |
| Anton Snyster | 1514 |
| Peter Van Ureland (engraver) | 1515 |
| Paul Van Ureland (engraver) | 1514–1519 |
| Copyn Jacob de Walt (from Brussels) | 1512–1526 |
| Thomas Wollwarde | 1530–1541 |
| Johann Wolf (from Landshut) | 1538–1542 |

## THE ARMOURERS COMPANY OF LONDON

First established in 1347 as the Company of Helmers. King Edward III (1347–1377) established the Armourers Company of London c. 1360 and authorized the company mark.

### List of Known Armourers

| | |
|---|---|
| John Ashton (Aston) | 1629–1631 |
| Edward Aynesley | 1629–1660 |
| Dethe Blaunche (earliest member) | 1369 |
| Thomas Baker | 1547 |
| Gyles Bechell | 1588 |
| George Brownfeide | ——— |
| Christopher Carlton | 1581 |
| Thomas Cootles | 1581 |
| Thomas Cope | 1660 |
| Thomas Cox | 1660 |
| William Coxe | 1629–1631 |
| James Chadwick | 1656 |
| John Crompton | 1544 |
| Peter Crouche (Crochet, Crowche) (worked at Greenwich also) | 1537–1547 |
| Francis Crouche (Crochet, Crowche) (worked at Greenwich also) | 1528–1529 |
| John Crouche (worked at Greenwich also) | 1515–1520 |

| | |
|---|---|
| William Crouche (worked at Greenwich also) | 1629–1636 |
| Hans Rudolph Deutch (Damascener) | 1569–D1571 |
| Hugh Doxsy | 1581 |
| Rowland Foster | 1629–1631 |
| John Franklin | 1629–1631 |
| Thomas Geminus | 1524–1563 |
| Thomas Gynne | 1564 |
| Richard Hamkyn | 1547 |
| Richard Hartford | 1590 |
| Johann Hennes | 1565–1571 |
| Richard Hoden | 1686 |
| Thomas Hurst | 1588 |
| Richard Hutton | 1581 |
| Henry Keene | 1660 |
| Erasmas Kirkener (Master at Greenwich also) | 1513–D1567 |
| Richard Klinge (sword maker) | 1660 |
| Joris Vander Kotelyne | 1550–1562 |
| Thomas Lincohn | 1694–1696 |
| Peter Lovat | ——— |
| Richard Loxson (Master) | 1581 |
| Nicholas Marshall | 1629–1631 |
| Roger Morgan | 1520 |
| Anthony Moult (Clerk) | 1575 |
| Thomas Parker | 1581 |
| John Pasfield (Master) | 1583–1595 |
| John Phillipes (sword maker) | 1578 |
| William Pickering (Master) (worked at Greenwich also) | 1567–D1618 |
| John Pickering (son of William) (worked at Greenwich also) | B1587–D1695 |
| Tobias Pickering (son of William) (worked at Greenwich also) | B1594–1625 |
| John Richards (Master) | 1565–1573 |
| William Rigg | 1604 |
| John Sewall | 1590–1604 |
| John Skinner | 1581 |
| Thomas Sparke | 1575–1581 |
| John Stevens (Stephens) (worked at Greenwich also) | 1571–1618 |
| Thomas Stevens (Stephens) (Master) (worked at Greenwich also) | 1618–1631 |
| Roger Tyndal (Master) | 1567 |
| Richard Woode | 1590 |
| Edward Yonger (Royal Damascener, sword furbisher, sword slipper, sword cutler) (worked at Greenwich also) | 1640–D1689 |

## OTHER ARMOURERS IN ENGLAND

| | |
|---|---|
| Hugh de Auggey | 1322 |
| Robert Amadas (etcher) | 1513 |
| Hans Albert | 1515 |
| John Basyn | 1524–1544 |

| | |
|---|---|
| Richard Bridde (helm maker) | 1347 |
| Alex Bawdesonne | 1547 |
| Rauffe Brande | 1520 |
| James Cooper | 1627–1629 |
| John Cooper | 1627–1629 |
| Thomas Conoun (helm maker) | 1347 |
| Copeland | 1529 |
| Richard Cutler | 1520 |
| Hans Cutler | 1520 |
| Thomas Carpenter | ——— |
| Edward Daniele | 1547 |
| John Daniele | 1547 |
| Nicholas de Farringdon | 1322 |
| Roger Faulkenor | 1625–1631 |
| William Gurre | 1511–1538 |
| Johan Hill (Hyll) | 1434 |
| Geofrey Hornel (Horne) | 1516–1518 |
| Richard Hotton | 1592 |
| Hans Hunter | 1547 |
| David Le Hope (armourer, bladesmith) | 1321 |
| John Lasy | 1533 |
| Will de la Mare | 1672 |
| Jemyn Oliver | 1514–1544 |
| Richard Pellande | 1520 |
| John Phillipes | 1578 |
| Francis Pellysonne | 1524–1544 |
| Giles Pitwell | 1516–1544 |
| John de Pounde | 1520 |
| Francis Poyes | 1525–1544 |
| John Purday | 1562 |
| Francis Speldrup | 1532 |
| Robert De Shirwode (Sherwood) (helm maker) | 1347 |
| Gerard de Toumay | 1337–1344 |
| Thomas White | 1416 |
| John Vandelf | late 15th century |

# Rulers of Scotland

| Dates of Reign | Ruler | Born | Died |
|---|---|---|---|
| 843–858 | Kenneth I MacAlpine | ? | c858 |
| 858–862 | Donald I | ? | 862 |
| 862–877 | Constantine I | ? | 877 |
| 877–878 | Aodh | ? | 878 |
| 878–889 | Cyric | ? | 889 |
| 878–889 | Eocha | ? | 889 |
| 889–900 | Donald II | ? | 900 |
| 900–943 | Constantine II | ? | 952 |
| 943–954 | Malcolm I | ? | 954 |
| 954–962 | Indulf | ? | 966 |
| 962–966 | Duff (Dubh) | ? | 967 |
| 966–971 | Colin (Cuilean) | ? | 971 |
| 971–995 | Kenneth II | ? | 995 |
| 995–997 | Constantine III | ? | 997 |
| 997–1005 | Kenneth III | ? | 1005 |
| 1005–1034 | Malcolm II | c954 | 1034 |
| 1034–1040 | Duncan I | c1001 | 1040 |
| 1040–1057 | Macbeth | ? | 1058 |
| 1057–1058 | Lulach | ? | 1059 |
| 1058–1093 | Malcolm III Canmore | c1031 | 1093 |
| 1093–1093 | Donald Bane | c1033 | 1097 |
| 1093–1094 | Duncan II | c1060 | 1094 |
| 1094–1097 | Donald Bane | c1033 | c1100 |
| 1097–1107 | Edgar | c1074 | 1107 |
| 1107–1124 | Alexander I | c1078 | 1124 |
| 1124–1153 | David I | c1084 | 1153 |

| Dates of Reign | Ruler | Born | Died |
|---|---|---|---|
| 1153–1165 | Malcolm IV, the Maiden | 1142 | 1165 |
| 1165–1214 | William, the Lion | 1143 | 1214 |
| 1214–1249 | Alexander II | 1198 | 1249 |
| 1249–1286 | Alexander III | 1241 | 1286 |
| 1286–1290 | Margaret, Maid of Norway | 1283 | 1290 |
| | **First Interregnum** | | |
| 1292–1296 | John de Baliol | c1250 | 1315 |
| | **Second Interregnum** | | |
| 1306–1329 | Robert I, the Bruce | 1274 | 1329 |
| 1329–1371 | David II | 1324 | 1371 |
| 1371–1390 | Robert II | 1316 | 1390 |
| 1390–1406 | Robert III | c1337 | 1406 |
| 1406–1437 | James I | 1394 | 1437 |
| 1437–1460 | James II | 1430 | 1460 |
| 1460–1488 | James III | 1451 | 1488 |
| 1488–1513 | James IV | 1473 | 1513 |
| 1513–1542 | James V | 1512 | 1542 |
| 1542–1567 | Mary, Queen of Scots | 1542 | 1587 |
| 1567–1625 | James VI (James I of England) | 1566 | 1625 |

# Chapter 13: Scottish Words Pertaining to Swords and Sword Guards

The following is a list of Scottish words pertaining to swords and sword guards as shown in guild records and other documents.

*Inglis, londoun gairdis*–English or London guards
*Slittit gairdis*–Slitted guards
*Schellit gairdis*–Shell guards
*Ribbit gairdis*–Ribbed guards
*Huntin gairdis*–Hunting guards
*Quheild gairdis*–Wheel guards
*Lolane gairdis*–Lowland guards
*Clos, clois gairdis*–Claw guards
*Hieland, hylane, hyland gairdis*–Highland (basket) guards
*Cort gairdis*–Court guards
*Slycht, slyt ribbit gairdis*–Slightly ribbed guards
*Braid slittit gairdis*–Broad slitted guards
*Skaillis gairdis*–Saucer guards
*Rydling swerdie*–Riding sword
*Braid swerdie*–Broad sword
*Harnes swerdie*– ? sword
*Rapperis, rapir swerdie*–Rapier sword
*Dampil swerdie*– ? sword
*Shearing swerdie*– ? sword
*Swerd sleper*–Sword slipper

# Chapter 14: Scottish Royal Armourers and Cutlers

### ROYAL ARMOURERS
(Sword Slippers)

| Name | Location | Dates Served |
|---|---|---|
| John Moncur the Elder | Dundee | 1455–1460 |
| William Moncur | Edinburgh | 1461–1471 |
| John Moncur the Younger | Edinburgh | 1472–1474 |
| John Tait | Edinburgh | 1475–1493 |
| Alan Cochrane | Edinburgh | 1494–1525 |
| William Smethberd (Smeberd) | Edinburgh | 1526–1538 |
| Thomas Softlaw the Elder | Edinburgh | 1539–1547 |
| John Simpson the Elder | Glasgow | 1715–1717 |
| Thomas Gemmill | Glasgow | 1718–1737 |
| John Simpson the Younger | Glascow | 1738–1749 |

### ROYAL CUTLERS

| Name | Location | Dates Served |
|---|---|---|
| Robert Selkirk | Edinburgh | 1497–1512 |
| William Rae the Elder | Edinburgh | 1512–1565 |
| William Rae the Younger | Edinburgh | 1565–1578 |
| James Young | Edinburgh | 1578–1580 |

# Chapter 15

## Scottish Sword Makers, Dealers, Cutlers, and Craftsmen Who Mounted Swords

During the late Middle Ages, the term "armourer" applied to metal workers who made armour and arms (mostly edged weapons). By the mid-sixteenth century, after the decline of knights in full armour, the terms armourer and sword maker began to be used interchangeably in Scotland. In Scotland, most sword makers were actually "sword slippers," or sword assemblers.

The term "slipper" comes from the German word *Scheifer*. A schiefer was a grinder, polisher, and sharpener of sword blades—in other words, a blade finisher. Therefore, a sword slipper made hilts (hand guards), finished blades, and fit the blades to the hilts. Very few sword blades were made in Scotland; most were imported from Solingen, Prussia. Others came from Spain, Italy, and France.

In Scotland, "Lorimers" made small metal accessories such as bits, buckles, spurs, and lance and pike heads. Edinburgh was the exception—its Lorimers made the hilts, and sword slippers had to buy hilts from them. Cutlers were knife and dirk makers, although some made swords too. Buckle makers were leather workers who made shields (targes) and knife and sword sheaths and scabbards. Bowars were wood workers who made bows, arrows, lance and pike shafts, clubs, and gunstocks.

In Scotland, all workers in metal (i.e., hammerers)—such as sword slippers, blacksmiths, cutlers, goldsmiths, silversmiths, and pewterers—belonged to the Hammermans Guild (i.e., union). One had to be a member in order to work as a free (freeman) metal worker allowed to practice his craft. When a freeman became known as an excellent artisan who had reached full potential in his craft, he could be voted as a master of the Hammermans Guild for a one-year term. Many served several terms.

The basket-hilted sword became the national style in Scotland. Basket hilts were put on infantry and cavalry broadswords and hangers (short swords) as well as many presentation swords. The cities of Stirling and Edinburgh were well-known for their production of basket-hilted swords. Three armourer/sword slippers from Stirling—John Allen the Elder, John Allen the Younger, and James Grant—became known as the best and most imaginative basket hilt makers. Three armourer/sword slippers from Glasgow—John Simpson the Elder, John Simpson the Younger, and Thomas Gemmill—were also well known.

The following is a listing of Scottish armourers, armourer/sword makers, sword slippers, cutlers, blade makers, sword guard (hilt) makers, sheath makers, silversmiths and goldsmiths who mounted swords, silversmiths who made sword guards (hilts), retailers, outfitters and contractors who sold swords, and laceman and tailors who sold swords.

# ABBOTSHELL

| | | |
|---|---|---|
| Armourer | James Watt | 1676–1706 |

# OLD ABERDEEN

| | | |
|---|---|---|
| Armourer | John Allen<br>• Also made daggers and dirks. | 1606–1613 |
| Armourer | John Anderson | 1702–1712 |
| Armourer | Alexander Archibald | 1691–1701 |
| Armourer | Robert Baxter | 1604–1614 |
| Armourer | William Blackhall | 1677–1690 |
| Armourer | George Bruce<br>• Made basket-hilted broadswords. | 1684–D1772 |
| Armourer<br>Cutler | James Clerk (Clarke)<br>• Made broadswords, short swords, small swords, knives, dirks, and scabbards. | 1669–1715 |
| Armourer<br>Cutler | John Clerk (Clarke) | 1669–1707 |
| Armourer | Thomas Clerk (Clarke) | 1695–1725 |
| Armourer | James Cruikshank | 1616–1626 |
| Silversmith<br>Goldsmith<br>Sword slipper | Francis Cruikshank<br>• Son of Robert Cruikshank)<br>• Apprenticed to goldsmith George Cooper, 1731–c. 1738.<br>• Mounted silver basket-hilted broadswords. | 1731–1750 |
| Armourer | George Cruikshank | 1621–1651 |
| Silversmith<br>Goldsmith | Robert Cruikshank<br>• Made silver basket-hilted broadswords. | 1697–1732 |
| Armourer | George Divinnes | 1628–1638 |
| Armourer | John Drum | 1602–1612 |
| Armourer | George Drummond | 1741–1751 |
| Armourer | John Gray | 1694–1704 |
| Armourer | Mathew Guild | 1569–1621 |
| Armourer | Thomas Hendrie | 1693–1702 |
| Armourer | Andro (Andrew) Horne | 1630–1660 |

| | | |
|---|---|---|
| Armourer | **William Laing** | 1707–1717 |
| Armourer | **Patrick Leithe** | 1581–1591 |
| Blacksmith<br>Silversmith<br>Guard maker | **William Lindsay**<br>• Made silver basket guards for broadswords. | 1660–1742 |
| Armourer<br>Sword maker | **Johnathan Lowe**<br>Made broadswords and small swords. | 1732–1742 |
| Armourer | **Robert Lowe** | 1692–1735 |
| Sword slipper | **Mauritius** | 1388–1408 |
| Armourer | **Alexander Morray** | 1591–1601 |
| Armourer | **Andrew Mylne** | 1610–1620 |
| Armourer | **Alexander Paterson (the Elder)** | 1618–D1634 |
| Armourer<br>Sword maker | **Alexander Paterson (the Younger)**<br>• Made rapiers and two-handed swords. | 1674–D1715 |
| Armourer<br>Sword maker | **Alexander Paterson III**<br>• Made broadswords, some with silver hilts. | 1708–D1723 |
| Armourer | **Hendrick Patersoun (Paterson)** | 1633–1653 |
| Armourer | **Robert Patersoun (Paterson)** | 1603–1613 |
| Silversmith<br>Guard maker | **George Robertson**<br>• Made silver basket guards for broadswords. | 1708–1727 |
| Armourer<br>Sword maker | **William Robertson**<br>• Made small swords and broadswords. | 1715–1734 |
| Silversmith | **Robert Ross** | 1741–1751 |
| Silversmith<br>Guard maker | **George Walker**<br>• Made silver basket guards for broadswords. | 1681–1722 |
| Silversmith<br>Sword slipper | **William Scott the Elder**<br>• Moved to Banff in 1688; died in 1701.<br>• Joined the Hammermans Guild of Elgin, 1701.<br>• Mounted silver basket-hilted broadswords. | 1666–1688 |
| Silversmith<br>Sword slipper | **William Scott the Younger**<br>• Son of William Scott the Elder.<br>• Moved to Banff in 1699; died in 1748.<br>• Had shop in Elgin, 1701.<br>• Mounted silver basket-hilted broadswords. | 1691–1699 |

## AYRSHIRE COUNTY

| | | |
|---|---|---|
| Silversmith<br>Cutler | **David Biggart (Bigert) of Kilmaurs**<br>• Moved from Kilmaurs to Ayershire, c. 1690.<br>• Made silver-hilted hangers and dirks.<br>(see Kilmaurs and Irvine listings) | c1690–1710 |
| Sword slipper | **John Wallace** | 1573–1793 |

## BANFF

| | | |
|---|---|---|
| Silversmith<br>Sword slipper | **William Scott the Elder**<br>• Mounted silver basket-hilted broadswords.<br>(see Aberdeen and Elgin) | 1688–D1701 |
| Silversmith<br>Sword slipper | **William Scott the Younger**<br>• Mounted silver basket-hilted broadswords.<br>(see Aberdeen and Elgin listings) | 1699–D1748 |

## BRECHIN

| | | |
|---|---|---|
| Sword slipper | **George Davidson** | 1759–1769 |

## CANONGATE
## (LATER PART OF EDINBURGH, C. 1784)

| | | |
|---|---|---|
| Silversmith<br>Guard maker | **James Aytoun the Younger**<br>• Made silver basket guards for broadswords. | 1708–1717 |
| Sword slipper | **Thomas Currie** | 1621–1628 |
| Silversmith<br>Guard maker | **Peter Cuthbertson**<br>• Apprentice and journeyman to Colin Mitchell, 1734–1753 (Mitchell died in 1753).<br>• Made silver basket guards for broadswords. | 1753–1770 |
| Sword slipper | **George Fouller (Foullair, Foulle)** | 1584–1626 |
| Sword slipper | **John Fouller (Foullair, Foulle)** | 1617–1622 |
| Sword slipper | **Andrew Grube** | 1568–1588 |
| Armourer<br>Sword maker | **Andro (Andrew) Hay**<br>• Master Armourer, 1622. | 1616–1631 |
| Sword slipper | **Thomas Huchesoune (Hucheson) the Elder** | 1588–1598 |
| Sword slipper | **Thomas Huchesoune (Hucheson) the Younger** | 1613–1656 |
| Silversmith<br>Guard maker | **Peter Inglis**<br>• Made silver basket guards for broadswords. | 1716–1727 |
| Sword slipper | **John Johnstoune (Johnston)** | 1574–1601 |

| | | |
|---|---|---|
| Armourer<br>Sword maker | **Patrick Lawrisoun the Elder**<br>• Master Armourer, 1614. | 1604–D1631 |
| Armourer<br>Sword maker | **Patrick Lawrisoun the Younger**<br>• Son of Patrick Lawrisoun the Elder.<br>• Made broadswords with basket guards. | 1631–1652 |
| Armourer | **John Lorimer** | 1630–D1652 |
| Armourer | **John McGraw** | 1727–1731 |
| Silversmith<br>Guard maker | **Colin Mitchell**<br>• Apprenticed to Peter Inglis, 1717–1727.<br>• Made silver basket guards for broadswords.<br>• Made some for Walter Allen and John Allen the Elder of Stirling.<br>• Also made chapes and mountings. | 1727–D1753 |
| Silversmith<br>Guard maker | **David Mitchell**<br>• Relative of Colin Mitchell.<br>• Made silver basket guards for broadswords. | 1698–1708 |
| Silversmith<br>Guard maker | **Peter Mitchell**<br>• Relative of Colin Mitchell.<br>• Made silver basket guards for broadswords. | 1728–1738 |
| Silversmith<br>Guard maker | **Thomas Mitchell**<br>• Relative of Colin Mitchell.<br>• Made silver basket guards for broadswords. | 1724–1734 |
| Cutler<br>Sword slipper | **Patrick Mitchell**<br>• Made swords and daggers (dirks). | 1613–1637 |
| Armourer | **Thomas Rennik** | 1642–1647 |
| Armourer | **Henry Scharpe** | 1642–1647 |
| Armourer | **James Scott** | 1607–D1637 |
| Sword slipper | **William Smyth (Smith)**<br>• Made broadswords with basket guards. | 1714–1732 |
| Sword slipper | **James Stanelie** | 1642–1647 |
| Sword slipper | **Walter Stoddart (Stoduord) the Elder** | 1624–1643 |
| Sword slipper | **Walter Stoddart (Stoduord) the Younger** | 1642–1652 |
| Sword slipper | **William Stoddart (Stoduord)** | 1635–1652 |
| Sword slipper | **William Vaus (Waus)** | 1580–D1589 |
| Sword slipper | **John Wair** | 1615–1620 |
| Sword slipper | **James Wardrop**<br>• Son of Patrick Wardrop. | 1636–1646 |

| | | |
|---|---|---|
| Sword slipper | **John Wardrop**<br>• Son of William Wardrop. | 1677–1687 |
| Sword slipper | **Lewes Wardrop**<br>• Son of Patrick Wardrop. | 1633–1652 |
| Sword slipper | **Patrick Wardrop** | 1610–1642 |
| Sword slipper | **William Wardrop**<br>• Son of Patrick Wardrop. | 1637–1677 |
| Sword slipper | **Duncan Wilson**<br>• Son of William Wilson. | 1637–1652 |
| Armourer | **Robert Wilson** | 1629–1639 |
| Sword slipper | **Thomas Wilson**<br>• Son of William. | 1660–1690 |
| Sword slipper | **William Wilson**<br>• Made rapiers. | 1613–1653 |

## DALKEITH

| | | |
|---|---|---|
| Armourer | **William Waldie** | 1670–D1694 |

## DOUNE

| | | |
|---|---|---|
| Sword slipper | **John Allan the Elder**<br>(see Glascow and Stirling listings) | 1710–1714 |
| Armourer | **John Macleod** | 1711–1750 |

## DUMFRIES

| | | |
|---|---|---|
| Sword slipper | **Johne Greir** | 1598–1618 |

## DUNDEE

| | | |
|---|---|---|
| Armourer | **George Anderson** | 1635–1646 |
| Sword slipper | **Andrew Bonar the Elder** | 1599–1622 |
| Sword slipper | **Andrew Bonar the Younger** | 1622–1632 |
| Sheath maker | **David Carnwath**<br>• Son of James Carnwath. | 1624–D1648 |
| Sheath maker | **James Carnwath** | 1611–1624 |
| Armourer<br>Sword maker | **Allan Dioneis (Dynnes)** | 1590–1595 |

| Trade | Name | Dates |
|---|---|---|
| Armourer<br>Sword maker | **George Dioneis**<br>• Son of John Dioneis. | 1602–1612 |
| Armourer<br>Sword maker | **John Dioneis**<br>• Master Armourer, 1587. | 1577–D1618 |
| Armourer<br>Sword maker | **Patrick Dioneis**<br>• Son of John Dioneis. | 1618–1628 |
| Silversmith<br>Sword slipper<br>Cutler | **Charles Dickson the Elder**<br>• Mounted silver-hilted hangers and dirks. | 1721–1730 |
| Silversmith<br>Sword slipper<br>Cutler | **Charles Dickson the Younger**<br>• Mounted silver-hilted hangers and dirks. | 1730–1735 |
| Sword slipper | **John Farguhar** | 1661–1671 |
| Sheath maker | **James Fleming** | 1643–1653 |
| Sword slipper | **Henry Hobard (Hobbert)**<br>• Son of John Hobard. | 1695–1705 |
| Sword slipper | **John Hobard (Hobbert)** | 1650–1670 |
| Armourer<br>Sword maker | **David Hunter**<br>• Master Armourer, 1587. | 1577–D1598 |
| Armourer<br>Sword maker | **James Hunter**<br>• Son of David Hunter. | 1598–1608 |
| Armourer<br>Sword maker | **William Hunter**<br>• Son of David Hunter. | 1592–1602 |
| Silversmith<br>Sword slipper<br>Cutler | **Alexander Johnstone**<br>• Apprenticed to Charles Dickson the Younger, 1734–1739.<br>• Mounted silver-hilted hangers and dirks. | 1739–1755 |
| Armourer | **James Knight (Knicht)** | 1660–1680 |
| Sword slipper | **Hew Lindsay** | 1545–1565 |
| Sword slipper | **David McKenzie**<br>• Also a gunsmith. | 1707–D1728 |
| Sword slipper | **James McKenzie**<br>• Son of David McKenzie.<br>• Also a gunsmith. | 1728–1738 |
| Armourer | **John Moncur the Elder**<br>• Royal Armourer, 1455–1460. | 1445–D1460 |
| Sword slipper | **Gilbert Mortimer** | 1600–1605 |

| | | |
|---|---|---|
| Sword guard maker | **Richard Pait**<br>• Master Armourer, 1587. | 1577–1602 |
| Sword guard maker | **Walter Pait**<br>• Master Armourer, 1582. | 1577–1597 |
| Armourer | **Robert Peterson** | 1617–1623 |
| Armourer<br>Sword maker | **Patrick Petillok the Elder**<br>• Master Armourer, 1587. | 1577–D1599 |
| Armourer<br>Sword maker | **Patrick Petillok the Younger** | 1599–D1610 |
| Armourer | **Patrick Robertson** | 1670–1680 |
| Sword slipper | **Alexander Smith the Elder** | 1636–D1650 |
| Sword slipper | **Alexander Smith the Younger** | 1684–1701 |
| Sword slipper | **Andrew Smith the Elder** | 1587–1611 |
| Sword slipper | **Andrew Smith the Younger** | 1609–1650 |
| Sword slipper | **David Smith**<br>• Son of Andrew the Elder. | 1611–1621 |
| Sword slipper | **James Smith** | 1661–1671 |
| Sword slipper | **John Smith**<br>• Son of Alexander the Younger. | 1701–1711 |
| Sword slipper | **Thomas Smith the Elder**<br>• Son of Andrew the Elder. | 1611–1621 |
| Sword slipper | **Thomas Smith the Younger**<br>• Son of Andrew the Younger. | 1650–1660 |
| Armourer<br>Sword maker | **Richard Wilkie (Wilky)**<br>• Son of William Wilkie the Elder. | 1588–1599 |
| Armourer<br>Sword maker | **William Wilkie (Wilky) the Elder**<br>• Master Armourer, 1587. | 1577–1594 |
| Armourer<br>Sword maker | **William Wilkie (Wilky) the Younger**<br>• Son of William Wilkie the Elder. | 1594–1604 |

## DUNFERMLINE

| | | |
|---|---|---|
| Armourer | **Patrick Allen** | 1643–1688 |
| Armourer | **Robert Allen**<br>• Son of Patrick Allen. | 1670–1680 |

| | | |
|---|---|---|
| Armourer<br>Sword maker | **Gilbert Kerr** | **1624–1644** |
| Armourer<br>Sword maker | **James Kerr**<br>• Son of Gilbert Kerr. | **1634–1640** |
| Armourer<br>Sword maker | **John Kerr the Elder**<br>• When he died, he had 20 swords and 60 sword blades in stock. | **1560–D1588** |
| Armourer<br>Sword maker | **John Kerr the Younger**<br>• Son of Gilbert Kerr. | **1633–1670** |
| Armourer<br>Sword maker | **Patrick Kerr** | **1560–1580** |
| Armourer | **John Stevenson** | **1588–1663** |
| Armourer | **David Torrie** | **1628–1640** |

## EDINBURGH

| | | |
|---|---|---|
| Sword slipper | **John Abell** | **1633–1638** |
| Armourer | **Robert Abercrombie** | **1598–1603** |
| Armourer | **Jakkis Alexander** | **1528–1548** |
| Armourer<br>Sword maker | **John Allan** | **1649–1673** |
| Armourer | **James Anderson** | **1590–1600** |
| Armourer | **John Anderson** | **1606–1616** |
| Military outfitter | **William Anderson & Sons**<br>• Located at 14 George Street. | **1910–1956** |
| Armourer | **Alexander Archibald** | **1701–1711** |
| Silversmith<br>Guard maker | **James Aytoun the Elder**<br>• Made silver basket guards for broadswords.<br>(see Cannongate listing) | **1636–1650** |
| Silversmith<br>Goldsmith<br>Brass founder<br>Guard maker | **William Aytoun**<br>• Made brass and silver basket guards for broadswords. | **1718–1750** |
| Sword slipper | **Charles Bally** | **1545–1575** |
| Armourer<br>Sword maker | **George Barbour**<br>• Master Armourer, 1573. | **1560–D1578** |
| Sword retailer | **James Bassindin** | **1510–1541** |

| | | |
|---|---|---|
| Sword slipper | **Robert Baxter**<br>• Made basket-hilted broadswords. | **1613–1618** |
| Sword slipper | **William Baxter**<br>• Made basket-hilted broadswords. | **1579–D1640** |
| Silversmith<br>Sword slipper | **Harry Beathune**<br>• Apprenticed to James Penman, 1694–1704.<br>• Mounted silver half and full basket-hilted broadswords. | **1704–1740** |
| Armourer | **Henry Bell** | **1652–1662** |
| Armourer | **John Bennet** | **1678–1688** |
| Sword retailer | **Billin & Mann** | **1827–1835** |
| Armourer | **James Blair** | **1624–1670** |
| Armourer | **Hew Blair**<br>• Son of James Blair. | **1670–1680** |
| Armourer<br>Sword maker | **Robert Borthwick** | **1589–1600** |
| Armourer<br>Sword maker | **David Boy**<br>• Master Armourer, 1698, 1701, 1702, 1703, 1704, and 1706. | **1685–1710** |
| Armourer<br>Sword maker | **Charles Brackenrig**<br>• Master Armourer, 1663 and 1674. | **1646–1680** |
| Armourer | **James Brown the Elder** | **1585–1601** |
| Lorimer<br>Sword guard maker | **James Brown the Younger** | **1644–1674** |
| Sword slipper | **John Brown** | **1644–1654** |
| Armourer | **John Brunton** | **1689–1699** |
| Sword retailer | **Brown & Tregilgas** | **1840–1870** |
| Lorimer<br>Sword guard maker | **John Callender**<br>• In 1596, with partner Robert Lyell, contracted to make sword guards for armourer/sword slippers of Edinburgh at 2 shillings each. | **1593–1615** |
| Armourer | **James Campbell** | **1695–1700** |
| Armourer | **William Campbell** | **1677–1687** |
| Sheath maker | **Walter Carmichael** | **1578–D1582** |
| Armourer | **John Carrington** | **1655–1675** |
| Lorimer<br>Sword guard maker | **John Carword**<br>• Made basket guards for broadswords. | **1580–1600** |

| | | |
|---|---|---|
| Sword blade maker | • Made Damascus sword blades. | |
| Sword and sword blade retailer | **David Christie**<br>• Sold sword blades to John Wright the Elder. | 1580–1590 |
| Sword retailer | **J. Christie & Sons** | 1870–1890 |
| Sword slipper | **Mungo Cleland**<br>• Made basket-hilted broadswords. | 1661–D1686 |
| Armourer | **William Clunies** | 1680–1690 |
| Armourer<br>Sword maker | **Alan Cochrane**<br>• Royal Armourer, 1494–1525.<br>• Made armour, swords, and steel saddles. | 1481–D1529 |
| Sword slipper | **Andrew Cornall (Cornwall)** | 1548–D1572 |
| Sword slipper | **John Cornall (Cornwall)**<br>• Brother of Andrew Cornall. | 1560–1570 |
| Armourer<br>Sword maker | **John Cowan**<br>• Son of Robert Cowan.<br>• Made basket-hilted broadswords.<br>• Master Armourer, 1713–1716. | 1713–1726 |
| Armourer<br>Sword maker | **Robert Cowan**<br>• Made basket-hilted broadswords. | 1671–1676 |
| Armourer | **Alexander Cranston (Cranstoune)** | 1625–1635 |
| Armourer | **Robert Cranston (Cranstoune)**<br>• Master Armourer, 1629, 1631, and 1632. | 1600–D1647 |
| Armourer<br>Sword maker | **George Crawford (Crauford)**<br>• Apprenticed to John Hislop, 1635–1642.<br>• Master Armourer, 1660, 1662, 1664, 1666, 1669, and 1670. | 1635–D1681 |
| Armourer<br>Sword maker | **John Crawford**<br>• Master Armourer, 1579. | 1552–D1581 |
| Silversmith<br>Goldsmith<br>Jeweler<br>Guard maker | **A. (Alexander) Cunningham (Cunynghame)**<br>**Mrs. Alexander Cunningham**<br>• Located at South Bridge.<br>• Widow of Alexander Cunningham.<br>• Made silver basket guards for broadswords. | 1799–D1812<br>1812–1813 |
| Lorimer<br>Sword guard maker | **Thomas Cunningham** | 1569–D1580 |
| Sheath maker | **John Curror** | 1617–1627 |
| Armourer | **George Cuthbert** | 1696–1706 |
| Armourer | **Hector Davidson** | 1574–1598 |

| | | |
|---|---|---|
| Sword slipper | **Martin Davidson** | 1546–1571 |
| Armourer | **John Davie** | 1667–1677 |
| Armourer | **John Dickson** | 1606–1616 |
| Armourer | **Gilbert Dickson** | 1525–1540 |
| Armourer | **Andrew Doge (Dook)** | 1633–D1682 |
| Armourer | **James Doge (Dook)** | 1635–1640 |
| Armourer<br>Sword maker | **Archibold Douglas (Douglass)**<br>• Master Armourer, 1678, 1679, 1680, 1681, 1683, 1684, 1688, and 1696. | 1662–1693 |
| Armourer<br>Sword maker | **John Douglas the Elder**<br>• Apprenticed to John Hislop.<br>• Made basket-hilted broadswords and rapiers. | 1643–1646 |
| Armourer<br>Sword maker | **John Douglas the Younger**<br>• Son of Archibold Douglas.<br>• Master Armourer, 1709, 1714, 1717, and 1723.<br>• Made basket-hilted broadswords. | 1706–D1725 |
| Armourer<br>Sword maker | **John Douglas III**<br>• Son of John Douglas the Younger.<br>• Made basket-hilted broadswords.<br>• Master Armourer, 1725, 1726, 1729, and 1739–1741. | 1725–1768 |
| Armourer<br>Sword maker | **Alexander Duff**<br>• Made basket-hilted broadswords.<br>• Master Armourer, 1717, 1721–1724, 1727, and 1729. | 1713 – 1739 |
| Armourer<br>Sword maker | **John Duff**<br>• Master Armourer, 1701–1705 and 1709–1712. | 1691–1722 |
| Armourer | **Thomas Fairly**<br>• Master Armourer, 1682. | 1672–1696 |
| Armourer | **John Falyon** | 1820–1850 |
| Armourer | **George Finlayson** | 1668–1678 |
| Armourer | **John Fotheringham** | 1606–1616 |
| Silversmith<br>Cutler | **R.W. Forsyth**<br>• Mounted silver-mounted dirks.<br>(see Glascow listing) | 1897–1929 |
| Sword slipper | **Robert Foular** | 1555–1578 |
| Lorimer<br>Sword guard maker | **Alexander Freland (Freeland)** | 1584–1594 |
| Armourer | **Joseph Freland** | 1590–1600 |

| | | |
|---|---|---|
| Armourer | **Robert Freland** | **1502–1535** |
| Clothier<br>Sword cutler | **Gibson, Thomson & Craig** | **1803–1820** |
| Sword slipper | **James Gifford (Giffart)** | **1667–D1702** |
| Armourer | **Hew Gilmour (Gilmure)** | **1589–1616** |
| Sword<br>and sword<br>blade retailer | **Peter Grant**<br>• Sold sword blades to John Wright the Elder. | **1580–1590** |
| Armourer | **George Greenlaw (Ginlaw)** | **1683–1693** |
| Armourer | **John Greir** | **1593–1598** |
| Sword retail | **Grieve, Oliver & Co.** | **1840–1860** |
| Armourer<br>Sword maker | **Alexander Grube**<br>• When Grube died in 1585, he had 115 sword guards, 37 sword blades, and 13 swords in stock. | **1550–D1585** |
| Armourer | **James Guthrie** | **1614–1646** |
| Armourer | **Patrick Haistie** | **1640–1650** |
| Armourer | **John Haliburton** | **1676–1683** |
| Armourer | **John Hall** | **1670–1680** |
| Armourer<br>Sword maker | **John Hamilton**<br>• Master Armourer, 1706, 1707, and 1708.<br>• A large sword and saber maker.<br>• When Hamilton died in 1712, he had 316 sword blades, 128 sword handles, 18 sword mountings, 141 sword guards, 48 swords, and 43 sabers in stock.<br>• Made basket-hilted broadswords. | **1685–D1712** |
| Armourer | **Robert Hamilton** | **1601–1611** |
| Sword slipper | **Andrew Henderson (Henryson)** | **1551–1572** |
| Armourer | **George Henderson (Henryson)** | **1592–1602** |
| Sword slipper | **John Henderson (Henryson) the Elder** | **1595–1616** |
| Sword slipper | **John Henderson (Henryson) the Younger** | **1606–D1631** |
| Armourer<br>Sword maker | **David Hislop (Hyslop)** | **1595–1605** |
| Armourer<br>Sword maker | **George Hislop the Elder**<br>• When Hislop died in 1605, he had 23 swords, 107 blades, and 100 saucer guards in stock. | **1565–D1605** |

| | | |
|---|---|---|
| Armourer<br>Sword maker | **George Hislop the Younger** | 1595–D1633 |
| Armourer<br>Sword maker | **John Hislop**<br>• Son of George Hislop the Younger.<br>• When Hislop died in 1646, he had 9 broadswords and 15 sword guards in stock. | 1633–D1646 |
| Armourer<br>Sword maker | **Mungo Hislop**<br>• Son of Thomas Hislop. | 1600 |
| Armourer<br>Sword maker | **Nicol Hislop**<br>• Son of George Hislop the Elder. | 1569–1574 |
| Armourer<br>Sword maker | **Philip Hislop**<br>• Son of Thomas Hislop | 1622–1632 |
| Armourer<br>Sword maker | **Thomas Hislop**<br>• Son of George Hislop the Elder.<br>• When Hislop died in 1600, he had 10 riding swords, 2 hand-and-a-half knightly swords, 13 rapiers, 1 broadsword, and 7 sword blades in stock. | 1587–D1600 |
| Armourer<br>Sword maker | **Walter Hislop (Hyslop)** | 1589–1599 |
| Armourer | **John Hodge** | 1647–1657 |
| Armourer | **Frances Hunter** | 1541–1551 |
| Army clothier<br>Army contractor<br>Sword cutler | **George Hunter**<br>• Located at 12 Parliament Street. | 1797–1802 |
| | **George Hunter**<br>• Located at 16 South Bridge.<br>• Clothier to the king. | 1803–1808 |
| | **Hunter (George) & Boyd (James)**<br>• Located at 16 South Bridge. | 1809–1810 |
| | **George Hunter and Co.**<br>• Located at 16 South Bridge. | 1811–1814 |
| | **George Hunter and Co.**<br>• Located at 23 Princes Street. | 1815–1822 |
| Cutler<br>Sword slipper | **James Hunter**<br>• Made midshipmen's dirks and cold stream guard officers swords. | 1780–1810 |
| Armourer<br>Sword maker | **James Hunter**<br>• Master Armourer, 1570.<br>• When he died, he had 62 swords and 125 blades in stock. | 1560–D1580 |
| Sword slipper | **Nicoll Hunter** | 1596–1602 |
| Armourer | **John Hutcheson** | 1646–1656 |
| Armourer | **Robert Jardine** | 1589–1610 |
| Armourer | **John Johnston** | 1721–1731 |

| | | |
|---|---|---|
| Armourer<br>Sword maker | **Alexander Kello**<br>• Son of John Kello.<br>• Master Armourer, 1641. | 1636–D1646 |
| Armourer<br>Sword maker | **Andrew Kello** | 1671–1681 |
| Armourer<br>Sword maker | **James Kello** | 1640–1670 |
| Armourer<br>Sword maker | **John Kello**<br>• Master Armourer, 1628, 1629, 1631, 1632, and 1636. | 1599–D1639 |
| Armourer | **John Kerr** | 1680–1690 |
| Silversmith<br>Cutler | **Rand H.B. Kirkwood**<br>• Mounted silver-hilted swords and dirks. | 1880–1890 |
| Armourer | **Antony Lamb** | 1673–1683 |
| Armourer | **Laurence Leitch** | 1674–1684 |
| Sword slipper | **Thomas Letyll** | 1561–1571 |
| Armourer<br>Sword maker | **Alexander Lindsay**<br>• Son of John Lindsay.<br>• Master Armourer, 1643. | 1642–D1648 |
| Armourer<br>Sword maker | **Andrew Lindsay** | 1590–D1640 |
| Armourer<br>Sword maker | **John Lindsay**<br>• Son of Andrew Lindsay. | 1636–D1646 |
| Armourer<br>Sword maker | **William Lindsay** | 1596–1616 |
| Sword slipper | **William Listoun** | 1592–1598 |
| Sword slipper | **John Lowis**<br>• Bought sword guards from George Thomson the Elder. | 1575–1600 |
| Lorimer<br>Sword guard maker | **Robert Lyell (Lyall)**<br>• Made shell and ribbed guards.<br>• In 1596, with John Callender, contracted with armourer/sword slippers of Edinburgh to make sword guards at 2 schillings each. | 1583–1606 |
| Sword slipper | **F.A. Lyon**<br>• Located on Leith Street. | 1796–1850 |
| Sword slipper | **Robert Macaulay the Elder** | 1609–1648 |
| Sword slipper | **Robert Macaulay the Younger**<br>• Son of Robert Macaulay the Elder. | 1638–1643 |

| | | |
|---|---|---|
| Cutler<br>Silversmith<br>Sword slipper | **John MacLeod**<br>• Located at 17 College Street (1813–1819) and 2 College Street (1819–1837).<br>• Made dirks for the Gordon Highlanders.<br>• Assembled and refurbished silver basket-hilted broadswords. | 1813–1837 |
| Armourer<br>Sword maker | **John Maine (Meane) the Elder**<br>**John Maine (Meane) the Younger**<br>• Made basket-hilted broadswords.<br>• Master Armourer, 1725–1727, 1739, and 1740. | 1685–1717<br>1717–1751 |
| Silversmith<br>Guard maker | **David Marshall** | 1774 |
| Silversmith<br>Guard maker | **Charles Marshall** | 1786 |
| Silversmith<br>Guard maker | **James Marshall (Sr.)** | 1788–1810 |
| Silversmith<br>Guard maker | **Francis Marshall & Sons**<br>• Location: Northbridge.<br>• Sons and partners: James (Jr.) and Walter Marshall<br>• Made very elaborate silver basket guards for swords.<br>**Marshall (Francis) & Sons (James Jr. & Walter)**<br>• Location: Northbridge.<br>**J. (James Jr.) & W. (Walter) Marshall**<br>• Location: Northbridge.<br>**Walter Marshall & Sons**<br>• Location: Northbridge. | 1804–1823<br><br><br><br>1824–1856<br><br>1857–1870<br><br>1871–1880 |
| Armourer | **John Marr** | 1629–1639 |
| Armourer | **Henry Marr** | 1743–1753 |
| Armourer | **William Martin (Martine)** | 1701–1716 |
| Armourer<br>Sword maker | **James Masch (Mash) the Elder**<br>• Made basket-hilted broadswords.<br>• Master Armourer, 1691–1694.<br>• Made 12 bayonets and 12 sabers on April 21, 1694, and delivered them to Edinburgh Castle. | 1681–D1704 |
| Armourer<br>Sword maker | **James Masch (Mash) the Younger**<br>• Son of James Masch the Elder.<br>• Made basket-hilted broadswords.<br>• Master Armourer, 1715–1720. | 1713–1731 |
| Sword slipper | **Alexander Matheson** | 1558–1568 |
| Armourer<br>Sword maker | **John McCall (McKall)**<br>• Made basket-hilted broadswords.<br>• Master Armourer, 1662, 1672, 1673, 1679, 1680, 1681, 1685, and 1686. | 1698–1710 |
| Armourer<br>Sword maker | **David McCall (McKall)**<br>• Made basket-hilted broadswords. | 1698–1710 |

|             |                                                      |           |
|-------------|------------------------------------------------------|-----------|
|             | • Master Armourer, 1699–1700.                        |           |
| Armourer    | **James McCulloch**                                  | 1601–1611 |
| Armourer    | **Mongo McCulloch**                                  | 1615–1625 |
| Silversmith | **James McEver**                                     | 1790–1800 |
| Sword slipper | • Mounted silver "lion hilt" swords.               |           |
| Silversmith | **Charles McKay**                                    | 1774–1787 |
| Guard maker | **William McKay**                                    | 1788–1799 |
|             | **James McKay (Sr.)**                                | 1800–1803 |
|             | **James McKay (Sr.) & James McKay (Jr.)**            | 1804–1816 |
| Silversmith | **James McKay (Jr.)**                                | 1817–1867 |
|             | • Made silver basket guards for swords.              |           |
| Armourer    | **John Medilmest**                                   | 1547–1561 |
| Armourer    | **James Middletoune (Midleteine)**                   | 1650–D1691 |
|             | • Master Armourer, 1664, 1667, 1668, 1671, 1674, 1675, 1676, 1678, and 1679. |  |
| Silversmith | **Colin MacKenzie (McKenzie)**                       | 1695–1721 |
|             | • Apprenticed to James Penman, 1688–1695.            |           |
|             | • Mounted silver half basket-hilted broadswords.     |           |
|             | • Moved to Inverness, 1721.                          |           |
| Sword slipper | **James Miller**                                   | 1592–1615 |
|             | • Made hunting swords.                               |           |
| Armourer    | **Andrew Moncur (Muncur)**                           | 1484–1539 |
|             | • Master Armourer, 1494.                             |           |
| Armourer    | **John Moncur (Muncur) the Younger**                 | 1455–D1474 |
|             | • Royal Armourer, 1472–1474.                         |           |
| Armourer    | **William Moncur (Muncur)**                          | 1450–D1471 |
|             | • Royal Armourer, 1461–1471.                         |           |
| Goldsmith   | **John Mortimer**                                    | 1844–1856 |
| Silversmith | • Mounted silver basket-hilted broadswords.          |           |
| Sword slipper | (see London listing)                               |           |
| Sword slipper | **Hew Mortoun**                                    | 1585–1589 |
| Blade maker | **H. (Hendrik) Moyes**                               | 1800–1810 |
| Armourer    | **Lewis Muir (Muire)**                               | 1664–1674 |
| Armourer    | **David Muir (Muire)**                               | 1643–1680 |
|             | • Master Armourer, 1660–1663.                        |           |
| Sword slipper | **Patrick Murdock**                                | 1561–D1575 |
| Armourer    | **John Neill**                                       | 1494–1504 |

| | | |
|---|---|---|
| Lorimer<br>Sword guard maker | **James Newlands** | **1601–1611** |
| Silversmith<br>Sword slipper | **Lawrence & Ebenezer Oliphant**<br>**James Oliphant**<br>• Mounted silver basket-hilted broadswords and sabers. | **1737–1759**<br>**1760–1780** |
| Sheath maker | **John Paterson**<br>• Master Armourer, 1576. | **1538–1580** |
| Armourer | **John Paterson** | **1740–1764** |
| Armourer | **John Peddie** | **1679–1716** |
| Silversmith<br>Sword slipper | **James Penman**<br>• Mounted basket-hilted broadswords. | **1673–1707** |
| Sheath maker | **James Penstoun** | **1646–1651** |
| Sheath maker | **William Pook** | **1560–D1589** |
| Cutler | **William Rae (Ra) the Elder**<br>• Royal Cutler, 1512–1565. | **1494–D1565** |
| Cutler | **William Rae (Ra) the Younger**<br>• Royal Cutler, 1565–1578. | **1550–D1578** |
| Armourer | **Andrew Rainie** | **1684–1694** |
| Armourer | **Walter Ramsay** | **1700–1710** |
| Sword slipper | **Edward Rannald (Rannaldsone)** | **1555–1569** |
| Sword slipper | **George Rannald (Rannaldsone)** | **1571–1607** |
| Sword slipper | **John Rannald (Rannaldsone)**<br>• Son of George Rannald. | **1607–1620** |
| Sword slipper | **William Rannald (Rannaldsone)** | **1675–1706** |
| Armourer<br>Sword maker | **Alexander Robertson (Robieson)**<br>• Made basket-hilted broadswords.<br>• Master Armourer, 1687, 1688, 1695, 1696, 1697, and 1699. | **1674–1705** |
| Armourer | **James Robertson (Robieson)**<br>• Son of John. | **1638–1643** |
| Armourer | **John Robertson (Robieson)** | **1642–1652** |
| Armourer | **William Robertson (Robieson)**<br>• Son of John Robertson. | **1639–1649** |
| Silversmith<br>Brass founder<br>Goldsmith | **John Rollo**<br>• Apprenticed to Harry Beathune, 1724–1730.<br>• Mounted silver- and brass-hilted broadswords. | **Bc1710–1737** |

| | | |
|---|---|---|
| Sword slipper | • Mounted silver-hilted hangers and cutlasses.<br>• Mounted a silver basket-hilted broadsword for the Duke of Perth. | |
| Sword maker | **William Sandilands** | **1624–1644** |
| Armourer | **Thomas Schort** | **1524–D1543** |
| Sword slipper | **John Scott**<br>• Made basket-hilted broadswords. | **1661–1681** |
| Armourer | **David Scroggie** | **1588–1598** |
| Cutler<br>Sword maker | **Robert Selkirk**<br>• Royal Cutler, 1497–1512.<br>• Made and repaired swords and daggers (dirks). | **1477–D1512** |
| Sword slipper | **Andrew Simpson (Symsone)** | **1600–1615** |
| Sword slipper | **William Simpson (Symsone)** | **1593–1613** |
| Armourer | **John Smethberd (Smeberd)** | **1538–1548** |
| Armourer<br>Sword maker | **William Smethberd (Smeberd)**<br>• Made swords and steel saddles.<br>• Royal Armourer, 1526–1538. | **1505–1551** |
| Armourer | **Alexander Smith** | **1688–1698** |
| Armourer<br>Sword maker | **Andrew Softlaw (Soffla)**<br>• When Softlaw died in 1583, he had 20 swords and 20 sword guards in stock. | **1577–D1583** |
| Armourer<br>Sword maker | **Archibald Soflaw**<br>• Master Armourer, 1639. | **1627–D1661** |
| Armourer<br>Sword maker | **George Softlaw**<br>• Son of James Softlaw. | **1588–1598** |
| Armourer<br>Sword maker | **Hew Softlaw**<br>• Master Armourer, 1666. | **1656–D1682** |
| Armourer<br>Sword maker | **James Softlaw the Elder**<br>• When he died, he had 10 sword blades in stock. | **1530–D1564** |
| Armourer<br>Sword maker | **James Softlaw the Younger**<br>• Master Armourer, 1576. | **1564–D1585** |
| Armourer<br>Sword maker | **Thomas Softlaw the Elder**<br>• Royal Armourer, 1539–1547. | **1528–1547** |
| Armourer<br>Sword maker | **Thomas Softlaw the Younger**<br>• Master Armourer, 1576. | **1564–D1585** |
| Armourer<br>Sword maker | **Thomas Softlaw the Elder**<br>• Royal Armourer, 1538–1541. | **1528–1547** |

| | | |
|---|---|---|
| Armourer<br>Sword maker | **Thomas Softlaw the Younger**<br>• Master Armourer, 1640. | 1626–1666 |
| Armourer<br>Sword maker | **William Softlaw**<br>• Master Armourer, 1630. | 1610–D1632 |
| Sword retailer | **J. (James) Stewart**<br>• Located at 88 George Street. | 1840–1860 |
| Armourer | **William Stewart** | 1662–1672 |
| Sheath maker<br>Cutler<br>Knife maker | **Charles Stradgeon (Sturgeon)** | 1584–1626 |
| Sword slipper | **William Sword the Elder** | 1698–D1721 |
| Sword slipper | **William Sword the Younger**<br>• Son of William Sword the Elder. | 1721–1726 |
| Armourer | **Archibald Tait (Taitt)** | 1648–1658 |
| Armourer | **George Tait (Taitt)** | 1615–1620 |
| Silversmith<br>Goldsmith<br>Brass founder<br>Guard maker | **James Tait (Taitt)**<br>• Made brass and silver sword guards. | 1704–1733 |
| Armourer<br>Sword maker | **John Tait (Taitt)**<br>• Royal Armourer, 1475–1493. | 1465–1499 |
| Armourer<br>Sword maker | **Robert Tait (Taitt)**<br>• Master Armourer, 1695–1698.<br>• Made basket-hilted broadswords. | 1676–1711 |
| Silversmith<br>Sword slipper | **William Taylor**<br>**William & John Taylor**<br>**John Taylor**<br>• Mounted silver-hilted small swords. | 1753–1769<br>1770<br>1771–1780 |
| Cutler<br>Sword slipper | **Alexander Thomson the Elder**<br>• Apprenticed to royal cutler James Young, 1546–1559.<br>• Bought sword guards from George Thomson.<br>**Alexander Thomson the Younger**<br>• Son of Alexander Thomson the Elder.<br>• Master Cutler, 1628 and 1633. | 1560–D1601<br><br><br>1601–1635 |
| Armourer<br>Sword maker | **Alexander Thomson III**<br>• Son of John Thomson.<br>• Made basket-hilted broadswords. | 1685–1695 |
| Lorimer<br>Sword guard maker | **George Thomson the Elder**<br>• Master Lorimer, 1568–1573.<br>• Sold sword guards to sword slipper John Lowis<br>  and cutler Alexander Thomson the Elder. | 1563–D1592 |

| | | |
|---|---|---|
| Lorimer<br>Sword guard maker | **George Thomson the Younger**<br>• Son of George Thomson the Elder. | **1612–D1646** |
| Lorimer<br>Sword guard maker | **Henry Thomson**<br>• Son of George Thomson the Elder. | **1589–1599** |
| Armourer<br>Sword maker | **James Thomson the Elder** | **1611–1621** |
| Armourer<br>Sword maker | **James Thomson the Younger**<br>• Made basket-hilted broadswords.<br>• Master Armourer, 1662, 1665, 1669, and 1670. | **1638–D1673** |
| Armourer<br>Sword maker | **James Thomson III**<br>• Son of William Thomson.<br>• Master Armourer, 1673, 1674, and 1677. | **1660–D1679** |
| Armourer<br>Sword maker | **John Thompson**<br>• Master Armourer, 1674–1676.<br>• Made basket-hilted broadswords, 1662–D1682. | **1662–D1682** |
| Lorimer<br>Sword guard maker | **Thomas Thomson**<br>• Son of George Thomson the Elder. | **1588–1598** |
| Armourer<br>Sword maker | **William Thomson**<br>• Master Armourer, 1640. | **1616–D1660** |
| Armourer | **Archibald Todrig** | **1642–1652** |
| Damascener | **Hew Vaus (Vauss, Vas, Vaws)**<br>• Probably a German.<br>• Made scabbards, guards, and Damascus blades.<br>• Master Damascener, 1526–1586. | **1566–D1626** |
| Armourer | **Patrick Vaus (Vauss, Vas, Vaws)** | **1673–1683** |
| Armourer | **William Vaus (Vauss, Vas, Vaws)** | **1570–D1625** |
| Armourer<br>Sword maker | **Thomas Voy**<br>• Master Armourer, 1677–1682.<br>• Made basket-hilted broadswords. | **1659–D1690** |
| Armourer<br>Sword maker | **John Wadie (Waldie)**<br>• Made basket-hilted broadswords.<br>• Master Armourer, 1705, 1707, and 1708. | **1689–1718** |
| Armourer | **James Wark** | **1589–1599** |
| Armourer | **Thomas Wark** | **1590–1600** |
| Sword slipper | **David Watson (Watsone)** | **1643–1653** |
| Sword maker | **George Watson (Watsone)** | **1543–D1571** |
| Lorimer | **James Welands (Wilands) the Elder** | **1568–1598** |

| | | |
|---|---|---|
| Sword guard maker | • Royal Lorimer, 1588–1598.<br>• Master Lorimer, 1588. | |
| Lorimer<br>Sword guard maker | **James Welands (Wilands) the Younger** | 1601–1631 |
| Armourer<br>Sword slipper | **William Whalley (Wheylie)** | 1676–1696 |
| Goldsmith<br>Sword slipper | **John White & Co.**<br>• Mounted swords. | 1780–1810 |
| Armourer<br>Sword maker | **Andrew Whyte (White)** | 1608–1618 |
| Sheath maker | **James Whyte (White)** | 1590–1609 |
| Armourer<br>Sword maker | **James Whyte (White)** | 1622–1634 |
| Armourer<br>Sword maker | **Thomas Whyte (White) the Elder**<br>• Master Armourer, 1628–1633. | 1589–D1641 |
| Armourer<br>Sword maker | **Thomas Whyte (White) the Younger** | 1641–1650 |
| Sword slipper | **Robert Williamson** | 1640–1650 |
| Armourer<br>Sword maker | **Archibald Wilson** | 1659–1673 |
| Armourer<br>Sword maker | **John Wilson the Elder**<br>• Master Armourer, 1660, 1662, 1665, 1666, 1668, and 1672. | 1646–1688 |
| Armourer<br>Sword maker | **John Wilson the Younger** | 1640–D1661 |
| Lorimer<br>Sword guard maker | **James Wilson** | 1573–1583 |
| Armourer<br>Sword maker | **John Wilson III**<br>• Made basket-hilted broadswords. | 1661–D1702 |
| Armourer<br>Sword maker | **Thomas Wilson** | 1661–1671 |
| Cutler | **Wilson & Sharp**<br>• Located at 139 Princes Street.<br>• Made brass-mounted dirks. | 1875–1900 |
| Sheath maker | **John Wingate (Windyatt)**<br>• Master buckle maker, 1568, 1570, and 1573. | 1555–1583 |
| Sheath maker | **Thomas Wingate (Windyatt)** | 1581–1608 |

| | | |
|---|---|---|
| Sheath maker | **William Wingate (Windyatt)** | **1525–1550** |
| Armourer | **William Worrock (Warrock)** | **1683–1693** |
| Sword shipper | **John Wright the Elder (Wryte, Wricht)**<br>• Bought blades from retailers Peter Grant and David Christie<br>(see Peter Grant and David Christie).<br>• When Wright died in 1592, he had 100 sword blades and 10 swords in stock. | **1581–D1592** |
| Sword slipper | **John Wright the Younger**<br>• Son of John Wright the Elder. | **1617–1646** |
| Lorimer<br>Sword guard maker | **Alexander Young** | **1621–1631** |
| Sheath maker | **David Young** | **1574–1584** |
| Armourer | **Donald Young** | **1701–1711** |
| Cutler | **James Young**<br>• Royal Cutler, 1578–1580.<br>• Master Cutler, 1546, 1550, 1555, and 1560.<br>• Sword and dagger (dirk) maker. | **1540–D1586** |
| Sheath maker | **James Yule** | **1653–1663** |

## ELGIN

| | | |
|---|---|---|
| Armourer | **John Anderson** | **1715–1720** |
| Sword guard maker | **Anthony Cowie**<br>• In 1580, sold seven pair of sword guards to James Young. | **1520–1613** |
| Armourer | **John Grigor** | **1657–1667** |
| Silversmith<br>Sword slipper | **William Scott the Elder**<br>• Mounted silver basket-hilted broadswords.<br>(see Banff and Aberdeen listings) | **D1701** |
| Silversmith<br>Sword slipper | **William Scott the Younger**<br>• Mounted silver basket-hilted broadswords. | **1701–D1748** |
| Sword slipper | **James Young**<br>• In 1580, bought seven pair of sword guards from Anthony Cowie. | **1570–1603** |

## GLASCOW

| | | |
|---|---|---|
| Sword slipper | **James Adam**<br>• Journeyman to John Simpson the Younger, 1738–1748.<br>• Made basket-hilted broadswords. | **1738–1743** |
| Sword slipper | **James Allan**<br>• Journeyman to John Simpson the Elder, 1706–1710.<br>• Made basket-hilted broadswords. | **1700–1711** |

| Trade | Name | Dates |
|---|---|---|
| Sword slipper | **John Allan the Elder**<br>• Journeyman to John Simpson the Elder, 1702–1707.<br>• Moved to Doune, 1710 (see Doune listings).<br>• Moved to Stirling, 1714 (see Stirling listings).<br>• Made distinctive basket-hilted broadswords. | **1707–1710** |
| Sword retailer | **Robert Allison & Son** | **1860–1870** |
| Armourer | **James Anderson** | **1649–1659** |
| Outfitter<br>Sword retailer | **William Anderson & Sons**<br>(see Edinburgh listing) | **1910–1956** |
| Armourer | **William Baird** | **1628–1638** |
| Armourer | **James Barclay** | **1647–D1664** |
| Armourer | **John Baxler** | **1642–1652** |
| Sword slipper | **William Brown**<br>• Journeyman to John Simpson the Elder, c. 1739–1744.<br>• Made basket-hilted broadswords. | **1739–1749** |
| Sword slipper | **Duncan Buchanon**<br>• Made three-branch-hilt cavalry swords. | **1687–1706** |
| Sword slipper | **William Bulloch**<br>• Made basket-hilted broadswords. | **1702–1712** |
| Sword slipper | **William Burrell (Birrell)**<br>• Made basket-hilted broadswords.<br>• When he died in 1650, he had 7 sword blades, 140 sword guards, and 20 swords in stock. | **1618–D1650** |
| Sword slipper | **Alexander Cameron**<br>• Journeyman to John Simpson the Younger, c. 1743–1749.<br>• Made basket-hilted broadswords. | **1749–1786** |
| Silversmith<br>Sword slipper | **William Clerk**<br>• Mounted silver-hilted small swords and hangers.<br>• Bought blades from Toledo, Spain, and Solingen, Prussia. | **1693–1710** |
| Sword guard maker | **Robert Craig**<br>• Made brass basket guards for broadswords.<br>• Also made musket locks. | **1711–1740** |
| Sword slipper | **John Cumyng (Cumming)** | **1675–1685** |
| Sword slipper | **Donald Danzell** | **1676–1686** |
| Sword slipper | **John Davidson the Elder** | **1578–1588** |
| Sword slipper | **John Davidson the Younger**<br>• In June 1699, he and John Simpson the Elder bought sword guards from John Love.<br>• Supplied 250 horseman's sabers with basket hilts to the company of Scotland trading in Africa. | **1667–D1703** |

| | | |
|---|---|---|
| Sword slipper | **James Dewar**<br>• Journeyman to John Simpson the Younger, c. 1740–1749.<br>• Made basket-hilted broadswords. | **c1740–1749** |
| Sword slipper | **Andrew Dickie**<br>• Journeyman to John Simpson the Younger, c. 1742–1749.<br>• Made basket-hilted broadswords. | **1742–1749** |
| Sword slipper | **John Douglas** | **1640–1660** |
| Sword slipper | **Nicolas Edwards**<br>• Made basket-hilted broadswords. | **1693–1723** |
| Silversmith<br>Cutler | **R.W. Forsyth**<br>• Mounted silver mounted dirks.<br>(see Edinburgh listing) | **1897–1929** |
| Sword slipper<br>Cutler | **Duncan Ferguson**<br>• Knife and sword maker. | **1700–D1728** |
| Sword slipper<br>Cutler | **Fergus Ferguson**<br>• Knife and sword maker. | **1717–1727** |
| Blade maker | **James Gardiner** | **1690–1700** |
| Armourer<br>Sword maker | **Thomas Gemmill the Elder** | **1620–D1665** |
| Armourer<br>Cutler<br>Sword slipper | **Thomas Gemmill the Younger**<br>• Royal Armourer, 1717–1737.<br>• Succeeded John Simpson the Elder.<br>• Made dirks and bayonets.<br>• Continued to make the famous Glasgow-type basket-hilted broadswords. | **1709–D1737** |
| Sword slipper | **Patrick Glen**<br>• Made basket-hilted broadswords. | **1718–D1729** |
| Sword guard maker | **Alexander Govan** | **1650–1685** |
| Sword guard maker | **John Govan** | **1659–D1666** |
| Silversmith<br>Goldsmith<br>Sword slipper | **Adam Graham**<br>• Located at East Kilbridge, Lymekilns, Lanarkshire.<br>**Adam Graham**<br>• Located on King Street.<br>• Mounted silver-hilted hangers and small swords. | **B1742–1763**<br><br>**1763–D1824** |
| Cutler<br>Sword slipper | **Alexander Graham**<br>• Made swords and dirks. | **1776–1780** |
| Silversmith<br>Sword slipper | **Archibald Graham**<br>• Mounted silver-hilted swords. | **1761–1800** |
| Cutler<br>Sword slipper | **James Graham**<br>• Made dirks and officers swords. | **1779–1816** |

| | | |
|---|---|---|
| Silversmith<br>Sword slipper | **Robert Gray**<br>• Mounted silver-hilted swords. | 1776–1821 |
| Armourer | **John Hamilton** | 1689–1707 |
| Army cap,<br>accoutrement,<br>uniform maker<br>Sword retailer | **Hebbert & Co.** | 1850–1912 |
| Sword retailer | **Peter Henderson** | 1850–1875 |
| Sword slipper | **John Hodge** | 1660–1685 |
| Gold and silver laceman<br>Sword retailer | **George Kenning**<br>• Located at 145 Argle Street.<br>**George Kenning**<br>• Located at 9 West Howard Street.<br>**George Kenning & Son**<br>• Located at 9 Howard Street.<br>**Kenning (George) & Spencer (James) Ltd.**<br>• Located at 9 Howard Street.<br>**Kenning (George) & Spencer (James) Ltd.**<br>• Located at 52 St. Enoch Square. | 1874<br>1875–1906<br>1907–1954<br>1955–1958<br>1959–1961 |
| Sword slipper | **William Kerr** | 1671–1681 |
| Sword guard maker | **George Layng** | 1568–1579 |
| Highland tailor<br>Sword retailer | **John Leckie & Co.**<br>• Located at 116 Union Street.<br>**John Leckie & Co.**<br>• Located at 116–118–120 Union Street.<br>**Leckie, Graham & Co.**<br>• Located at 116–118–120 Union Street.<br>**Leckie, Graham & Co.**<br>• Located at 116 Union Street.<br>**Leckie, Graham & Co.**<br>• Located at 89 Renfield St. & & 56 Bath Street.<br>**Leckie, Graham & Co.**<br>• Located at 89 Renfield Street.<br>**Leckie, Graham & Co.**<br>• Located at 95 Renfield Street.<br>**Leckie, Graham & Co.**<br>• Located at 95 Renfield Street.<br>**Leckie, Graham & Co.**<br>• Located at 96 St. Vincent Street.<br>**Irving Bros. & Leckie, Graham**<br>• Located at 80 St. Vincent Street. | 1872–1874<br>1875–1878<br>1879–1881<br>1882–1902<br>1903–1909<br>1910–1912<br>1913–1922<br>1923–1932<br>1933–1935<br>1936–1957 |
| Goldsmith<br>Silversmith<br>Sword slipper | **George Luke (Louk)**<br>**James Luke the Elder**<br>**James Luke the Younger**<br>**John Luke the Elder**<br>**John Luke the Younger**<br>• Mounted silver-hilted swords and hangers. | c1680–1693<br>1693–1729<br>1729–1750<br>1680–1698<br>1694–1710 |

| | | |
|---|---|---|
| Sword guard maker | **John Love**<br>• In June 1699, sold sword guards to John Davidson the Elder and Younger. | 1673–D1689 |
| Silversmith<br>Sword slipper | **William Love**<br>• Mounted silver basket-hilted broadswords. | 1777–1815 |
| Sword slipper | **Alexander Magget**<br>• Journeyman to John Simpson the Younger, c. 1739–1749.<br>• Made basket-hilted broadswords. | c1749–1760 |
| Sword slipper | **Neil McLean**<br>• Location: Salt Market. | 1779–1819 |
| Sword slipper | **John Melvill**<br>• Journeyman to John Simpson the Younger, c. 1742–1749.<br>• Made basket-hilted broadswords. | 1749–1755 |
| Sword slipper | **Hugh Muir** | 1680–1690 |
| Sword slipper | **David Munro** | 1782–1792 |
| Sword slipper | **Adam Nicoll the Elder** | 1582–D1622 |
| Sword slipper | **Adam Nicoll the Younger**<br>• Son of Adam Nicoll the Elder. | 1622–1654 |
| Sword guard maker | **Alan Nicoll** | 1620–D1641 |
| Sword slipper | **Alexander Nicoll** | 1600–D1633 |
| Sword slipper | **Patrick Nicoll** | 1583–1593 |
| Sword guard maker<br>Brass founder | **George Noble**<br>• Made brass sword guards. | 1724–1740 |
| Sword slipper | **James Park (Parke)**<br>• When Park died in 1649, he had 56 sword blades and 102 sword guards in stock. | 1621–D1649 |
| Sword slipper | **James Pallock** | 1662–1672 |
| Sword slipper | **George Paton** | 1691–1701 |
| Sword guard maker | **Andrew Pont** | 1672–1680 |
| Armourer | **John Rankine** | 1694–1704 |
| Sword slipper | **Andrew Reid** | 1644–1654 |
| Sword slipper | **John Reid the Elder** | 1621–1655 |
| Sword slipper | **John Reid the Younger** | 1634–1644 |
| Armourer | **John Reid**<br>• Son of William Reid. | 1650–1660 |

| | | |
|---|---|---|
| Armourer | **William Reid** | 1601–1671 |
| Sword slipper | **William Robesone**<br>• Journeyman to John Simpson the Younger, c. 1716–1726.<br>• Made basket-hilted broadswords. | 1726–1730 |
| Sword guard maker | **James Robeson**<br>• Journeyman to John Scott the Elder, c. 1747–1757. | 1757–1767 |
| Sword guard maker | **Gavin Scott**<br>• Son of John Scott the Elder. | 1650–1670 |
| Sword guard maker | **James Scott**<br>• Son of John Scott the Elder. | 1649–1659 |
| Sword guard maker | **John Scott the Elder** | 1641–D1649 |
| Sword slipper | **John Scott the Younger**<br>• In August 1699, with John Simpson the Elder and Archibald Simpson, mounted horseman's saber blades for the company of Scotland trading in Africa. | 1672–D1702 |
| Sword slipper | **John Scott III**<br>• Son of John Scott the Younger.<br>• Made basket-hilted broadswords. | 1702–D1726 |
| Sword slipper | **John Scott IV**<br>• Son of John Scott III. | 1726–1737 |
| Sword slipper | **John Scott V**<br>• Son of John Scott IV. | 1745–D1779 |
| Sword slipper | **Thomas Scott** | 1649–D1664 |
| Sword slipper | **Archibald Simpson the Elder**<br>• Made broadswords and horseman's sabers with basket guards.<br>• In August 1699, with John Scott the Younger and John Simpson the Elder, mounted horseman's saber blades for the company of Scotland trading in Africa. | 1695–D1728 |
| Sword slipper | **Archibald Simpson the Younger**<br>• Son of Archibald Simpson the Elder.<br>• Bought basket hilts from John Simpson the Younger.<br>• Made basket-hilted broadswords. | 1725–D1739 |
| Sword slipper | **James Simpson**<br>• Son of Archibald Simpson the Elder. | 1739–1760 |
| Sword slipper | **John Simpson the Elder**<br>• Was the originator of the famous Glasgow-type basket guard.<br>• Also bought basket hilts from his son, John Simpson the Younger.<br>• Royal Armourer, 1715–1717.<br>• In June 1699, with John Davidson, supplied 250 horseman's sabers with basket guards to the company of Scotland trading in Africa.<br>• In August 1699, with Archibald Simpson the Elder and John Scott the Younger, mounted horseman's saber blades for the company of Scotland trading in Africa.<br>• His son John Simpson the Younger joined the business in 1711.<br>• Succeeded by Thomas Gemmill, 1717. | 1638–D1717 |

| | | |
|---|---|---|
| Sword slipper | **John Simpson the Younger** | **1711–D1749** |
| | • Son of John Simpson the Elder. | |
| | • Made broadswords with the famous Glascow-type basket guard originated by his father. | |
| | • Made naval cutlasses and small swords. | |
| | • Sold basket guards to John Simpson the Elder and Archibald Simpson the Younger. | |
| | • Bought iron from William Somervell. | |
| | • Replaced Thomas Gemmill as Royal Armourer, 1738–1749. | |
| | • Mounted some silver basket-hilted broadswords. | |
| Sword slipper | **John Simpson III** | **1725–1748** |
| | • Son of Archibald Simpson the Elder. | |
| | • Made basket-hilted broadswords. | |
| Sword slipper | **David Smith** | **1661–1671** |
| Sword slipper | **Nicolas Snodgras (Snodres)** | **1560–D1580** |
| Sword slipper | **Robert Snodgras (Snodres)** | **1580–1590** |
| | • Son of Nicolas. | |
| Gold and silver laceman<br>Sword retailer | **Edward Stillwell & Son** | **1875–1904** |
| Sword slipper | **Robert Syme (Sime)** | **1671–D1674** |
| Goldsmith<br>Silversmith | **Samuel Telfer** | **1749–1760** |
| | • Mounted silver-hilted swords. | |
| Sword slipper | **John Thomson** | **1670–1682** |
| Sword retailer | **Thomson & Son** | **1830** |
| Sword slipper | **Robert Walter** | **1671–1681** |
| Sword slipper | **Archibald Wilson (Wilsone)** | **1577–1587** |
| Sword slipper | **John Wilson (Wilsone)** | **1636–1646** |
| Sword slipper | **Robert Wilson (Wilsone)** | **1620–D1621** |
| Sword slipper | **Thomas Wilson (Wilsone)** | **1594–1605** |

## INVERNESS

| | | |
|---|---|---|
| Silversmith<br>Guard maker | **John Baillie** | **1727–1780** |
| | • Made silver basket sword guards. | |
| Sword slipper | **Patrick Beltan** | **1574–1594** |
| Armourer | **Robert Davidson** | **1596–1621** |
| Armourer | **Zacharie Dunbar** | **1596–1629** |
| Gun maker | **Patrick Grant** | **1709–1720** |

| | | |
|---|---|---|
| Sword slipper | | |
| Sword slipper | **Patrick Henderson**<br>• Made basket-hilted broadswords. | 1562–1582 |
| Goldsmith<br>Silversmith<br>Sword slipper | **Robert Innes**<br>• Mounted silver-hilted swords. | 1708–1730 |
| Armourer | **John Jameson** | 1601–1624 |
| Armourer | **Hugh Kemp** | 1711–1721 |
| Sword slipper<br>Gun maker | **Robert Low**<br>• The Disarmament Act of 1716 forced him to surrender 70 swords, 40 scabbards, 1 two-handed sword, and 3 guns. | 1696–1731 |
| Silversmith<br>Sword slipper | **Alexander MacLeod**<br>• Mounted silver-hilted swords. | 1830–1840 |
| Sword retailer | **Robert MacDougall** | 1840–1855 |
| Gun maker<br>Sword slipper<br>Cutler | **Charles McCulloch**     1715–D1746<br>• The Disarmament Act of 1716 forced him to surrender 7 swords, 15 guns, and 2 dirks. | |
| Sword slipper | **William McDonald**     1712–1750<br>• The Disarmament Act of 1716 forced him to surrender 31 swords, 1 pionard, and 63 sword blades. | |
| Silversmith<br>Sword slipper | **Colin McKenzie**<br>• Moved from Edinburgh in 1721.<br>• Made silver half and full basket-hilted broadswords.<br>(see Edinburgh listing) | 1721–1735 |
| Sword slipper | **James Petrie** | 1653–1673 |
| Armourer | **William Smith** | 1705–1715 |
| Armourer | **David Wood (Woid)** | 1590–D1605 |
| Armourer | **Richard Wood (Woid)** | 1606–1616 |
| Armourer | **Roger Wood (Woid)** | 1605–1624 |

## IRVINE

| | | |
|---|---|---|
| Cutler<br>Silversmith<br>Sword slipper | **Thomas Biggart (Bigert)**<br>• Made silver-hilted hangers and dirks.<br>• Probably a relative of David Biggart in nearby Kilmaurs. | 1720–1755 |

## KILMAURS

| | | |
|---|---|---|
| Cutler<br>Silversmith<br>Sword slipper | **David Biggart (Bigert)**<br>• Made silver-hilted hangers and dirks.<br>• Moved to Ayrshire County, c. 1690.<br>• Probably a relative of Thomas Biggart in nearby Irvine. | **1670–1690** |
| Cutler<br>Knife maker | **Robert Smith** | **1660–D1681** |

## PERTH

| | | |
|---|---|---|
| Cutler<br>Sword slipper | **George Adam** | **1768–1787** |
| Sword slipper | **Robert Atkyn (Atkynis)** | **1527–1556** |
| Armourer | **John Barbour** | **1647–1657** |
| Sword slipper | **James Beltan (Belton)**<br>• Bought sword guards from William Young. | **1518–1558** |
| Cutler<br>Sword slipper | **James Bennet** | **1733–1756** |
| Sword guard maker | **George Bow** | **1532–1575** |
| Armourer | **Robert Brachty (Brachtie)**<br>• Master Armourer, 1546, 1548–1550, 1554, and 1560. | **1446–1510** |
| Sword slipper | **William Bryson**<br>• Bought sword guards from William Young. | **1537–1558** |
| Armourer | **Andrew Cowie (Cowye)**<br>• Son of John Cowie. | **1628–D1682** |
| Armourer | **George Cowie (Cowye)** | **1641–1661** |
| Armourer | **John Cowie (Cowye)** | **1620–1649** |
| Sword guard maker | **James Davidson** | **1548–D1596** |
| Armourer | **Alexander Ferguson** | **1643–1653** |
| Cutler<br>Sword slipper | **Alexander Ferguson** | **1796–1806** |
| Cutler<br>Sword slipper | **John Ferguson** | **1808–1818** |
| Cutler<br>Sword slipper | **James Glass** | **1802–1812** |
| Sword guard maker | **William Giffen (Geffon)** | **1542–1561** |

| Trade | Name | Dates |
|---|---|---|
| Sword slipper | John Halbert (Hobert) | 1687–1697 |
| Armourer<br>Sword maker | Patrick Hair | 1568–1588 |
| Armourer<br>Sword maker | Robert Hair<br>• Master Armourer, 1546, 1548, 1549, and 1550. | 1535–1560 |
| Armourer<br>Sword maker | George Harlaw (Harlow)<br>• Master Armourer, 1560. | 1555–1570 |
| Armourer | Patrick Hay | 1674–1684 |
| Sword slipper | James Hunter | 1538–1548 |
| Armourer | John Hewison (Hewtsoune) | 1660–1670 |
| Armourer | David Kelour the Elder | 1518–1545 |
| Armourer | David Kelour the Younger | 1522–1556 |
| Sword slipper | Robert Lowdian | 1512–1532 |
| Sword slipper | Patrick Mar<br>• Made a cutlass (short sword) as his apprenticeship essay to become a freeman of the Hammermans Guild. | 1753–1763 |
| Cutler<br>Sword slipper | Walter Marshall | 1763–D1806 |
| Cutler<br>Sword slipper | William Marshall<br>• Son of Walter Marshall. | 1806–1822 |
| Sword slipper | Archibald Maxwell | 1544–1564 |
| Sword slipper | Patrick McEwen | 1688–1698 |
| Armourer | Henry Mershell<br>• Master Armourer, 1546, 1548–1550, and 1554. | 1546–1564 |
| Armourer | Henry Messer | 1549–1559 |
| Cutler | D. Mintyre Sons<br>• Made dirks. | 1790–1820 |
| Armourer | Alexander Murray (Murie) | 1575–1595 |
| Armourer | James Petrie | 1645–1655 |
| Sword guard maker | James Robertson | 1589–1820 |
| Sword guard maker | Patrick Tait<br>• Moved to Stirling, 1616. | 1601–1616 |
| Armourer<br>Sword maker | Archibald Wright (Wryght)<br>• Master Armourer, 1546, 1548, 1549, 1554, 1558, and 1568. | 1546–D1605 |

| | | |
|---|---|---|
| Armourer<br>Sword maker | **David Wright (Wryght)** | 1612–D1616 |
| Armourer<br>Sword maker | **John Wright (Wryght) the Elder** | 1570–1590 |
| Armourer<br>Sword maker | **John Wright (Wryght) the Younger**<br>• Son of Archibold Wright. | 1605–1615 |
| Sword guard maker | **Archibald Young**<br>• Son of William Young. | 1560–1611 |
| Sword guard maker | **William Young**<br>• Master Lorimer, 1546, 1548, 1549, 1554, 1560, and 1568.<br>• Sold sword guards to James Belton and William Bryson. | 1536–1584 |

## ST. ANDREWS

| | | |
|---|---|---|
| Armourer | **William Anderson** | 1615–D1656 |
| Cutler<br>Sword slipper | **Patrick Brown**<br>• Son of Thomas Brown. | 1569–1607 |
| Cutler<br>Sword slipper | **Thomas Brown** | 1545–D1569 |
| Armourer | **George Deas (Dais)** | 1592–1607 |
| Armourer | **David Dupling** | 1591–1611 |
| Cutler<br>Sword slipper | **John Edie** | 1611–1671 |
| Sword slipper | **Thomas Dougall** | 1588–1600 |
| Sword slipper | **Thomas Flukar** | 1552–1562 |
| Armourer | **Gilbert Fraser (Frysell)** | 1602–1619 |
| Sword slipper | **William Mungall** | 1592–1597 |
| Sword slipper | **David Sword**<br>• Son of James Sword the Elder. | 1586–1607 |
| Sword slipper | **James Sword the Elder** | 1548–D1608 |
| Sword slipper | **James Sword the Younger**<br>• Son of James Sword the Elder. | 1608–1618 |
| Sword slipper | **James Sword III**<br>• Son of Walter Sword. | 1633–1670 |
| Sword slipper | **John Sword** | 1550–D1607 |

| | | |
|---|---|---|
| Sword slipper | **Walter Sword**<br>• Son of John Sword. | 1607–D1633 |
| Sword slipper | **William Sword the Elder**<br>• Son of Walter Sword. | 1640–1650 |
| Sword slipper | **William Sword the Younger**<br>• Son of James Sword III. | 1660–D1694 |
| Sword slipper<br>Buckle maker | **William Sword III**<br>• Made swords, sword sheaths, and bayonets. | 1694–1710 |
| Armourer | **David Yonger** | 1542–D1560 |
| Armourer | **Walter Yonger**<br>• Son of David Yonger. | 1560–1601 |

## STIRLING

| | | |
|---|---|---|
| Sword slipper | **Andrew Allan**<br>• Journeymen to Walter Allan, c. 1746–1756.<br>• Made basket-hilted broadswords. | 1756–1770 |
| Sword slipper | **John Allan the Elder**<br>• Moved from Glasgow to Doune (8 miles from Stirling), 1710.<br>• Moved from Doune to Stirling, 1714.<br>• In Glasgow, he was journeyman to John Simpson the Elder, famous basket guard and broadsword maker, 1702–1707.<br>• Allan became famous for his Stirling-style basket guards and broadswords.<br>• Mounted some silver basket-hilted broadswords.<br>• Colin Mitchell of Cannongate made silver chapes, mountings, and basket guards for Allan. | 1714–1741 |
| Sword slipper | **John Allan the Younger**<br>• Son of John Allan the Elder.<br>• Made basket-hilted broadswords. | 1741–1755 |
| Sword slipper | **James Allan**<br>• Son of Andrew Allen.<br>• Made basket-hilted broadswords. | 1740–1760 |
| Sword slipper<br>Brass founder | **Walter Allan**<br>• Son of John Allan the Elder.<br>• In 1732, his essay piece submitted to gain entry to the Hammermans Guild was a hanger sword hilt.<br>• Made iron and brass basket-hilted broadswords.<br>• Became the most famous maker of basket-hilted broadswords in Scotland.<br>• He not only made a large number of basket-hilted broadswords but was known for the very ornate and complex nature of his hilts.<br>• Colin Mitchell of Cannongate made silver chapes, mountings, and basket hilts for Allan.<br>• Mounted some silver basket-hilted broadswords. | 1732–1765 |
| Armourer | **Thomas Christopher** | 1598–1618 |
| Sword slipper | **James Grant** | 1759–D1788 |

| | | |
|---|---|---|
| | • Journeyman to Walter Allan, 1749–1759. | |
| | • Made basket-hilted broadswords. | |
| Armourer | **Thomas Hutcheson (Hurchison)** | **1607–1618** |
| Sword slipper | **John Lawrie** | **1757–1765** |
| | • Journeyman to Walter Allan, c. 1747–1757. | |
| | • Made basket-hilted broadswords. | |
| Sword slipper | **Patrick McMartin (McMertine)** | **1757–1770** |
| | • Journeyman to Walter Allan, c. 1747–1757. | |
| | • Made basket-hilted broadswords. | |
| Sword slipper | **John McUrich** | **1758–1772** |
| | • Journeyman to Walter Allan, c. 1748–1758. | |
| | • Made basket-hilted broadswords. | |
| Sword slipper | **Alexander Renny** | **1563–1611** |
| Sword slipper | **James Stirratt (Starret)** | **1655–1665** |
| | • Brother of William Stirratt. | |
| Sword slipper | **William Stirratt (Starret)** | **1635–1655** |
| | • Brother of James Stirratt. | |
| Sword slipper | **William Stodart** | **1644–1660** |
| Sword guard maker | **Patrick Tait** | **1616–1626** |
| | • Came from Perth in 1616. | |
| Sword slipper | **William Thomson** | **1763–1773** |
| | • Apprenticed to Walter Allan, c. 1753–1763. | |
| Armourer | **James Williamson** | **1607–1617** |
| Sword slipper | **James Wilson** | **1597–1607** |

## TAIN

| | | |
|---|---|---|
| Goldsmith | **Hugh Ross** | **Bc1703–1723** |
| Silversmith | • Apprenticed to Robert Innes of Inverness, 1717–1723. | |
| Brass founder | **Hugh Ross** | **1723–D1782** |
| Guard maker | • Made silver and brass basket hilts. | |

# English Sword Photo Section

## Matthew Boulton Sword Hilts

All of these small sword hilt designs were taken directly from Matthew Boulton's Soho factory pattern book, c. 1762. All of the hilts were made of cut steel, highly burnished, and beautifully faceted. Sword cutlers (assemblers) could order hilts using Boulton's pattern number. (Courtesy of the Birmingham. City archives)

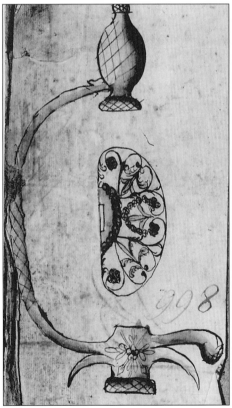
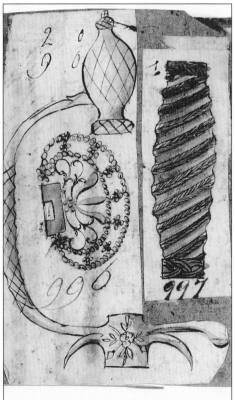

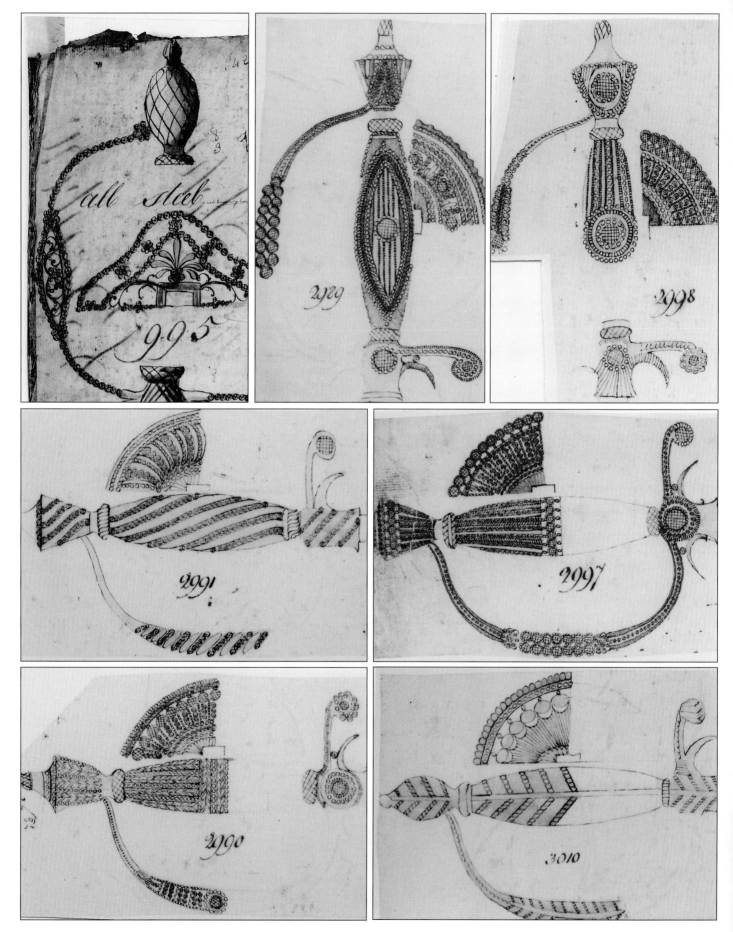

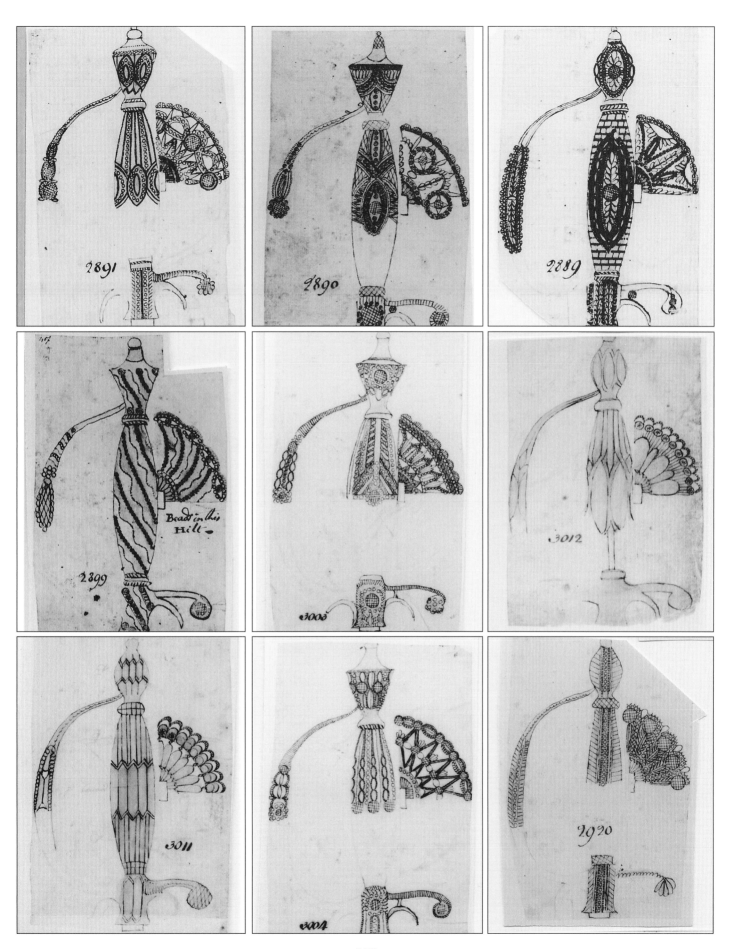

English Sword Photo Section

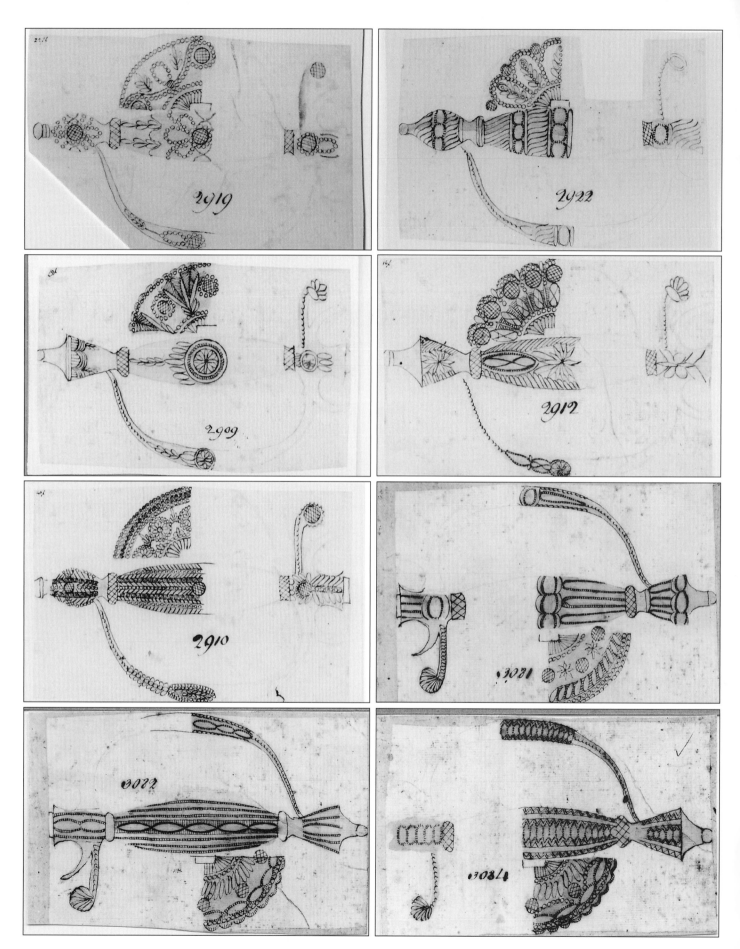

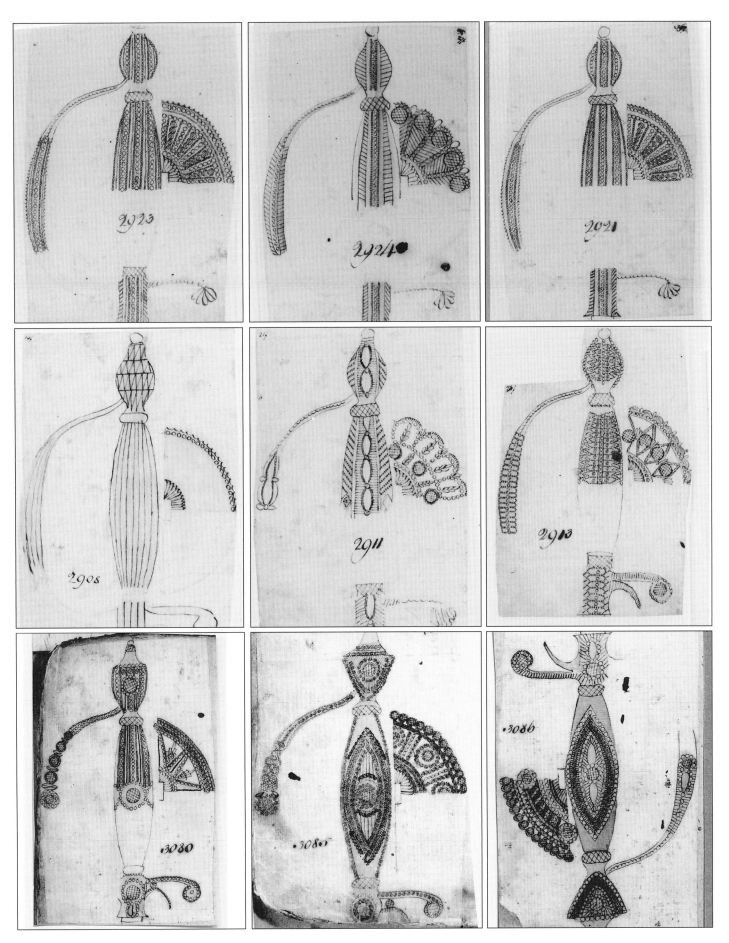
English Sword Photo Section

# English Hounslow Swords

The Hounslow sword blade mark. (Courtesy of the Hounslow Library)

Hounslow backsword with a very early basket hilt, c. 1630–1640. The blade is marked ME FACIT HOVN. (Gunnersbury Park Museum collection)

Hounslow rapier, c. 1630–1640. The blade is marked ME FECIT HVNSLO. (Courtesy of John Tofts White)

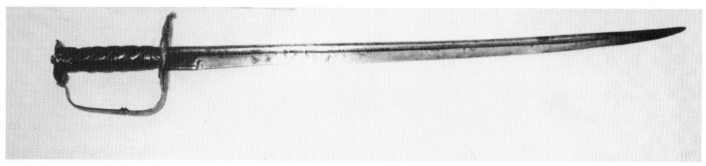

Hounslow hanger, c. 1640–1645. The blade is marked JOHAN KINNDT FECIT HOVNSLOE. (Neil Robert Grigg collection)

Hounslow hanger, c. 1630–1640. The blade is marked HOVNSLOE.
(Gunnersbury Park Museum collection)

Hounslow rapier, c. 1630–1640. The blade is marked HVNSLOE.
(Gunnersbury Park Museum collection)

Hilt of Hounslow hanger by Johan Kinndt (bottom of p. 274).
(Neil Robert Grigg collection)

Hounslow hanger, c. 1630–1640, with iron hilt, fish-skin grip covering, and 26 3/4 x 1 1/8 inch blade marked ME FECIT HOUNSLOE.
(Courtesy of John Tofts White)

Reverse of Hounslow hanger hilt on p. 275.
(Courtesy of John Tofts White)

Hounslow backsword, c. 1630–1640. The blade is marked HOUNSLOE.
(Gunnersbury Park Museum collection)

Hounslow backsword, c. 1630–1640. The blade is marked
ME FACIT HOVN. (Gunnersbury Park Museum collection)

Hounslow backsword for mounted troops with 40 inch blade marked
ME FECIT HOUNSLOE, c. 1630–1640. (Collection of Neil Robert Grigg)

Hounslow backsword, c. 1630–1640, with fish-skin grip covering with copper wire wrappings and 33 x 1 1/8 inch blade marked ME FECIT HOUNSLOE. (Courtesy of John Tofts White)

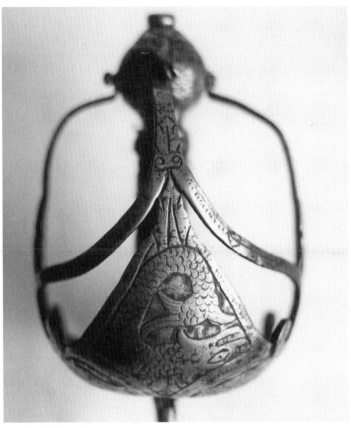

End of the hilt on the Hounslow backsword. (Courtesy of John Tofts White)

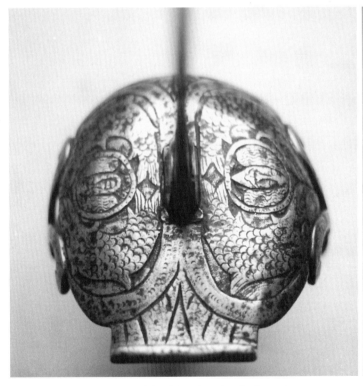

Bottom of the hilt of the Hounslow backsword. (Courtesy of John Tofts White)

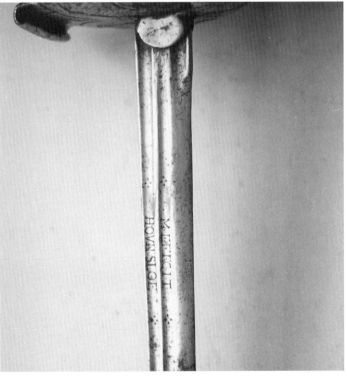

Blade markings on the Hounslow backsword. (Courtesy of John Tofts White)

# English Civil War Hangers

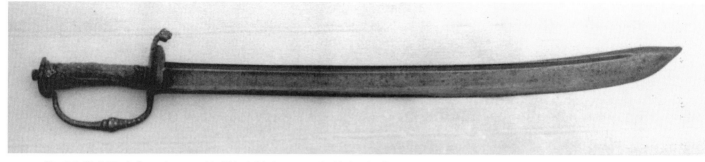

English Civil War infantry hanger with 25 inch blade engraved with heads of Lord Protector Oliver Cromwell and Gen. Sir Thomas Fairfax, 1645 "New Model" Army commander. Blade maker: Jan (Johann) Ohlig (Olich), Solingen, Germany. (James Forman collection)

English Civil War infantry hanger, c. 1645, with 27 3/8 inch blade. (George C. Neumann collection)

English Civil War infantry hanger, c. 1645, with 23 inch blade and chiseled steel hilt. Overall length is 28 inches. (Jan H. Zajac collection)

Hilt of English Civil War infantry hanger. (James Forman collection)

# English Mortuary Swords

English mortuary hilt broadsword, c. 1650, with 31 1/4 x 1 3/8 inch blade with three fullers and chiseled and pierced hilt with the face of executed King Charles I. (Collection of Sam Saladino)

Mortuary hilt broadsword, c. 1650, with 33 inch blade. (James Forman collection)

Hilt of the c. 1650 mortuary hilt broadsword. (James Forman collection)

Hilt of English mortuary hilt broadsword, c. 1650. (Collection of Sam Saladino).

English mortuary hilt backsword, c. 1650.
(James Forman collection)

Lower hilt of mortuary hilt backsword, c. 1650.
(James Forman collection)

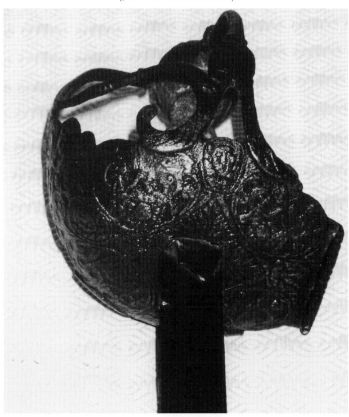

English mortuary hilt backsword, c. 1650. The Solingen, Germany, blade is marked with a running wolf. (Collection of Neil Robert Grigg)

Hilt of mortuary hilt backsword.
(Collection of Neil Robert Grigg)

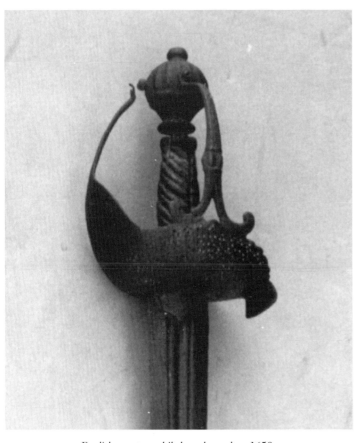

English mortuary hilt broadsword, c. 1650.
(Courtesy of LionGate Arms & Armour)

Hilt of mortuary hilt broadsword.
(Courtesy of LionGate Arms & Armour)

Reverse hilt of mortuary hilt broadsword.
(Courtesy of LionGate Arms & Armour)

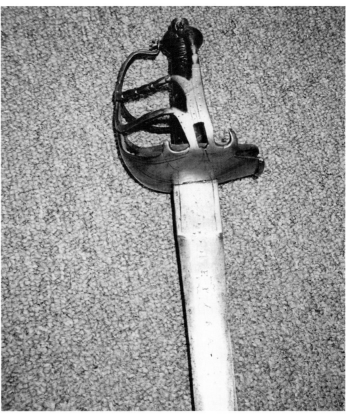

English mortuary hilt broadsword, c. 1650. The 33 inch blade is marked ANDREA FARARA with a running wolf mark. (G.H. Cook collection)

# English Cavalry Troopers Swords

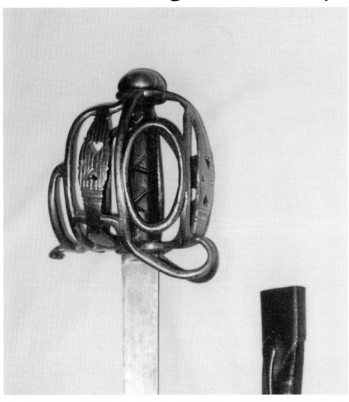

English pattern 1750 basket-hilted cavalry backsword with 38 x 1 1/4 inch straight blade and horseman's oval ring on the basket. (Author's collection)

Front of pattern 1750 cavalry sword hilt. (Author's collection)

Reverse of pattern 1750 cavalry sword hilt. (Author's collection)

English basket-hilted cavalry backsword, c. 1745, with 37 inch blade marked with SH in a fox with a horseman's oval ring on the basket. Maker: Samuel Harvey Sr., Birmingham. (G.H. Cook collection)

English dragoon backsword, c. 1755, with 32 5/8 x 1 5/8 inch straight blade. Features a unique cut steel plate hilt with disc guard and branches. (Author's collection)

Reverse hilt of c. 1755 English dragoon backsword. (Author's collection)

Close-up of hilt of c. 1755 English dragoon backsword. (Author's collection)

Top of hilt of c. 1755 English dragoon backsword. (Author's collection)

English three-quarter basket cavalry broadsword, c. 1760, with 34 inch blade. (Courtesy of Patrick Tougher)

English three-quarter basket cavalry broadsword, c. 1760–1765, with very wide 32 1/2 inch blade.
Maker: Samuel Harvey Sr., Birmingham. (Courtesy of Patrick Tougher)

English Dragoon backsword, c. 1755, with 36 1/4 inch straight blade.
Features a unique cutout steel plate hilt with disc guard and branches.
Maker: Samuel Harvey Sr., Birmingham. (G.H. Cook collection)

English three-quarter-basket cavalry sword hilt, c. 1750.
(Courtesy of LionGate Arms & Armour)

English basket-hilted dragoon backsword, c. 1760-1765, with oversized basket. Maker: Samuel Harvey Sr., Birmingham.
(Courtesy of Patrick Tougher)

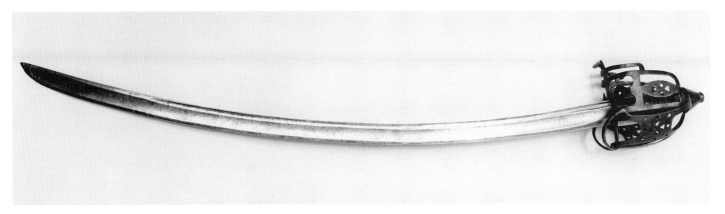

English basket-hilted dragoon saber for the Highland regiments, c. 1765-1775.
(Courtesy of Patrick Tougher)

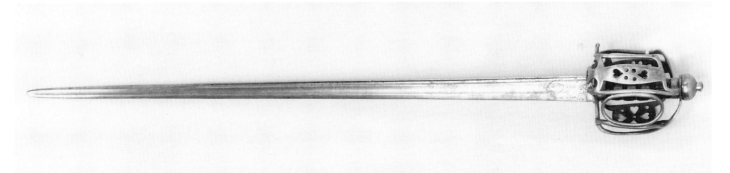

English basket-hilted Dragoon backsword, c. 1770, with 35 inch blade and oval horseman's ring on the basket.
(Courtesy of Patrick Tougher)

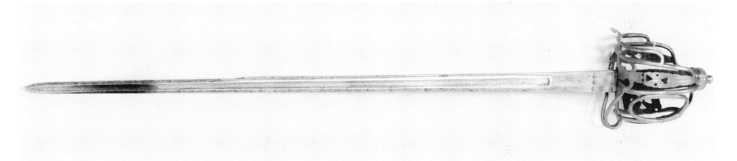

English basket-hilted cavalry backsword, c. 1780, with 36 inch blade and oval horseman's ring on the basket.
(Courtesy of Patrick Tougher)

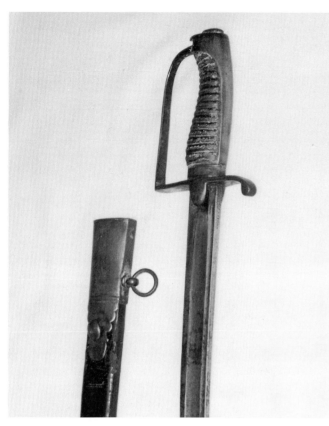

English brass-hilt cavalry saber, c. 1780, with 35 x 1 3/8 inch blade with a single fuller running to the tip. Maker: Woolley (James) & Co., Birmingham. (Author's collection)

Hilt of the c. 1780 brass-hilt cavalry saber. (Author's collection)

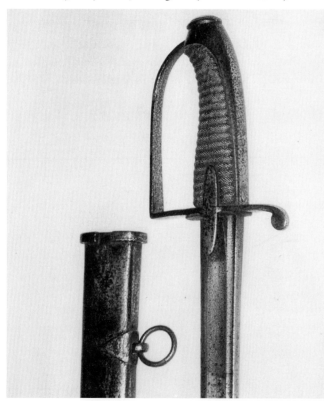

English pattern 1788 light cavalry saber (variation) with 35 1/4 x 1 3/8 inch blade with a single fuller running to the tip. Maker: Woolley (James) & Co., Birmingham. (Author's collection)

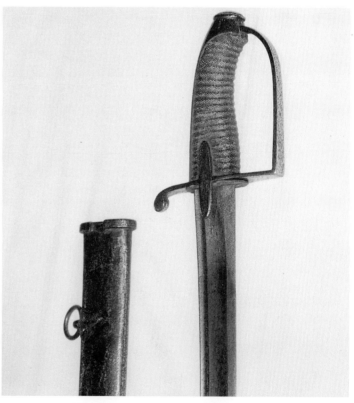

Hilt of the English pattern 1788 cavalry saber. (Author's collection)

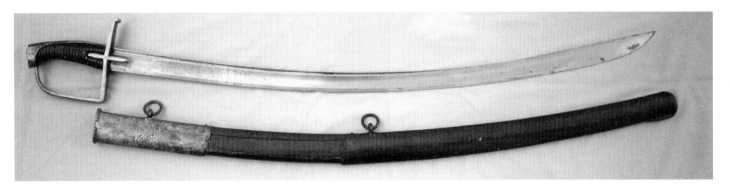

English pattern 1788 cavalry saber with Hussar scabbard with leather inserts. Retailer: J.J. Runkel, London. (Collection of Ulrich Dubi)

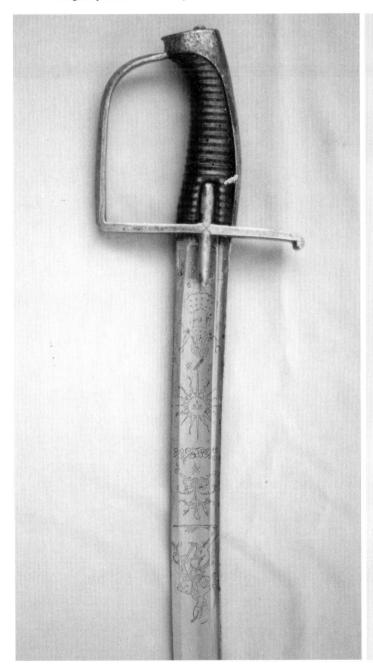

Hilt of the English pattern 1788 cavalry saber.
(Collection of Ulrich Dubi)

English pattern 1788 heavy cavalry sword (variation) with two-branch half-basket hilt and 36 x 1 1/4 inch straight blade with a single fuller running to the tip. Maker: Woolley (James) & Co., Birmingham. (Author's collection)

Hilt of English pattern 1788 heavy cavalry sword (variation) on p. 287. (Author's collection)

Reverse hilt of English pattern 1788 heavy cavalry sword (variation) on p. 287. (Author's collection)

English pattern 1796 light cavalry saber with 32 1/2 x 1 1/2 inch wide blade, which becomes wider at the hatchet point. (Author's collection)

Hilt of the pattern 1796 light cavalry saber. (Author's collection)

English pattern 1796 light cavalry saber. Maker: Durs Egg, London. (Jan H. Zajac collection)

Durs Egg marking on English pattern 1796 light cavalry saber scabbard. (Jan H. Zajac collection)

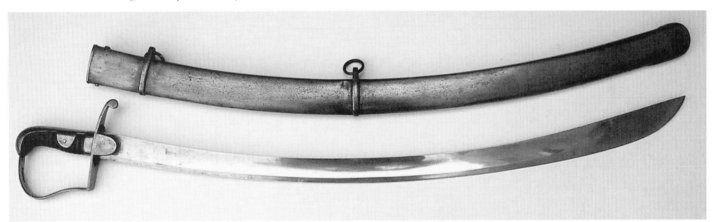

English pattern 1796 light cavalry saber with 33 x 1 1/2 inch blade engraved THO'S GILL'S WARRENTED 1799. Maker: Thomas Gill, Birmingham. (Richard J. Dellar collection)

Blade marking on the Gill pattern 1796 light cavalry saber. (Richard J. Dellar collection)

Hilt of the Gill pattern 1796 light cavalry saber on p. 289. (Richard J. Dellar collection)

Hilt of pattern 1796 heavy cavalry sword. (Richard J. Dellar collection)

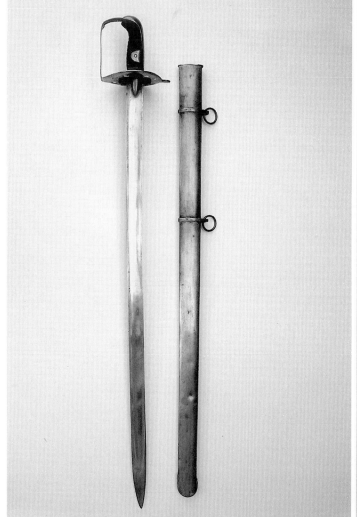

English pattern 1796 heavy cavalry sword with 34 1/2 x 1 1/2 inch straight blade with spear point variation. Maker: Thomas Craven, Birmingham. (Richard J. Dellar collection)

English pattern 1796 heavy cavalry saber with 34 3/4 x 1 1/2 inch straight blade with regulation hatchet point. (Author's collection)

Hilt of pattern 1796 heavy cavalry sword on p. 290.
(Author's collection)

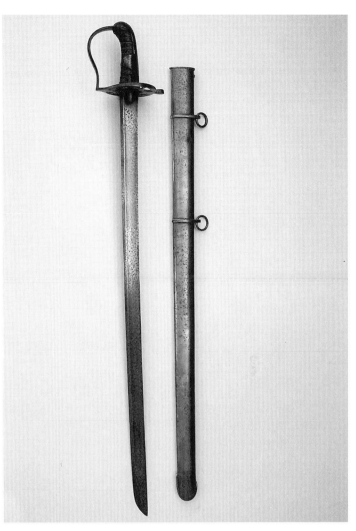

English pattern 1796 heavy cavalry sword with 34 3/4 x 1 1/2 inch blade with hatchet point, hilt disc guard with unusual cut outs (variation), and double langets. The blade is engraved on back J.J. RUNKEL SOHLINGEN. The scabbard is stamped OSBORN & GUNBY, BIRMINGHAM.
(Richard J. Dellar collection)

Hilt of pattern 1796 heavy cavalry sword (variation).
(Richard J. Dellar collection)

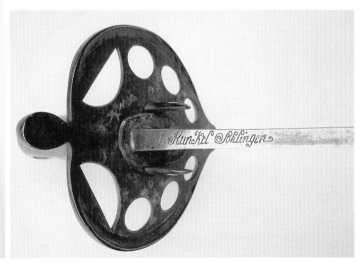

Runkel marking on the pattern 1796 heavy cavalry sword (variation).
(Richard J. Dellar collection)

Hilt disc guard of pattern 1796 heavy cavalry sword (variation) on p. 291. (Richard J. Dellar collection)

English pattern 1821 light cavalry saber with 34 3/4 x 1 1/4 inch blade with spear point. (Author's collection)

Hilt of pattern 1821 light cavalry saber. (Author's collection)

English pattern 1821 heavy cavalry sword with 35 1/2 x 1 1/4 inch straight blade with spear point. (Author's collection)

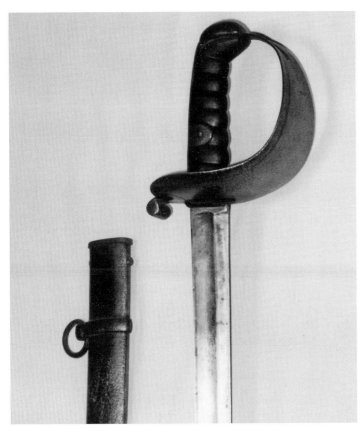

Hilt of pattern 1821 heavy cavalry sword on p. 292.
(Author's collection)

Top of hilt of pattern 1821 heavy cavalry sword on p. 292.
(Author's collection)

English pattern 1853 cavalry saber with 35 1/2 x 1 3/8 inch blade with spear point. Maker: Robert Mole & Son, Birmingham. (Author's collection)

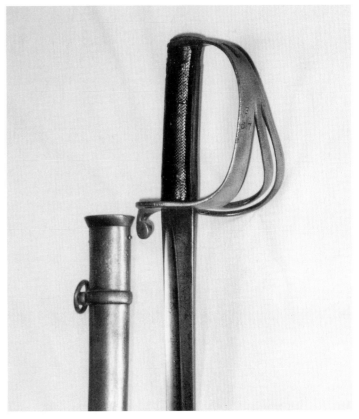

Hilt of pattern 1853 cavalry saber.
(Author's collection)

Top of hilt of pattern 1853 cavalry saber on p. 293.
(Author's collection)

English pattern 1864 cavalry saber with 35 1/2 x 1 3/8 inch blade with spear point. A Maltese cross is cut out in the sheet metal hilt.
Maker: Robert Mole & Son, Birmingham. (Author's collection)

Reverse hilt of pattern 1864 cavalry saber.
(Author's collection)

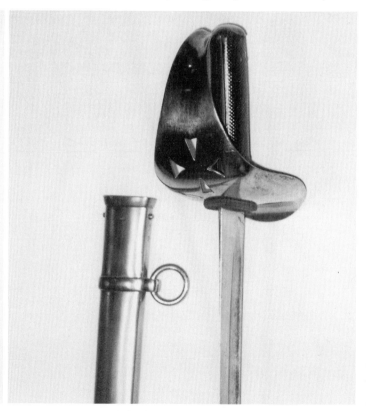

Rear hilt of pattern 1864 cavalry saber.
(Author's collection)

Top hilt of pattern 1864 cavalry saber on p. 294. (Author's collection)

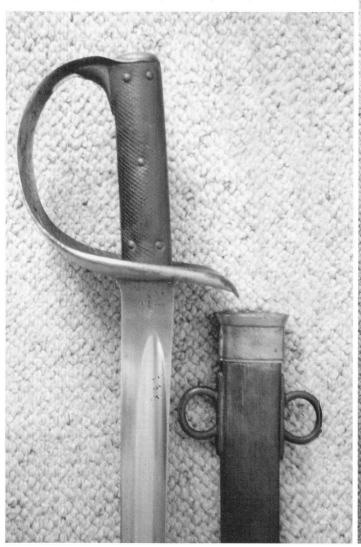

Hilt of pattern 1885 cavalry saber.
(Collection of Ulrich Dubi)

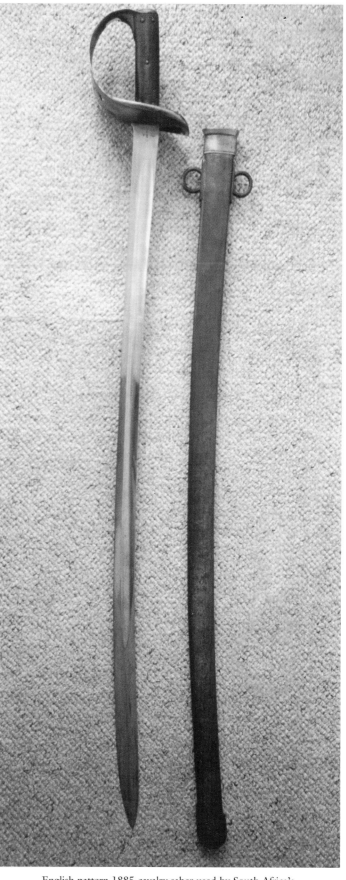

English pattern 1885 cavalry saber used by South Africa's Cape Mounted Rifles. (Collection of Ulrich Dubi)

English pattern 1890 cavalry saber with 34 1/2 x 1 1/4 inch blade with spear point. Maker: Enfield Armory. (Author's collection)

Reverse hilt of pattern 1890 cavalry saber. (Author's collection)

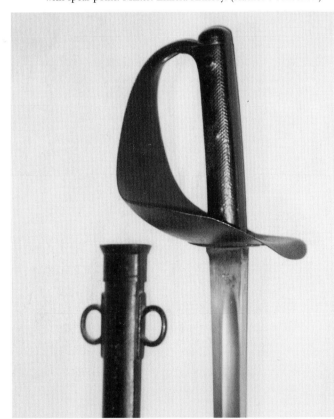

English pattern 1899 cavalry saber with 33 1/2 x 1 1/4 inch blade with spear point. Maker: Enfield Armory. (Author's collection)

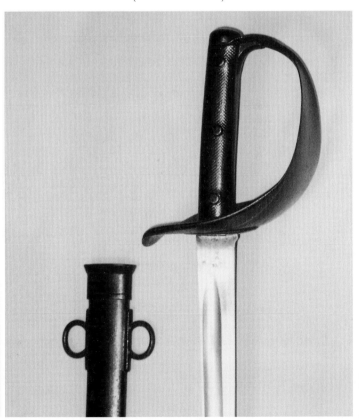

Reverse hilt of pattern 1899 cavalry saber. (Author's collection)

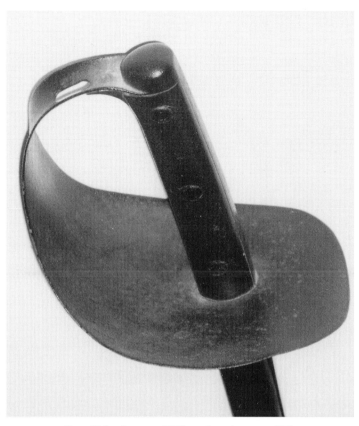

Top of hilt of pattern 1899 cavalry saber on p. 296. (Author's collection)

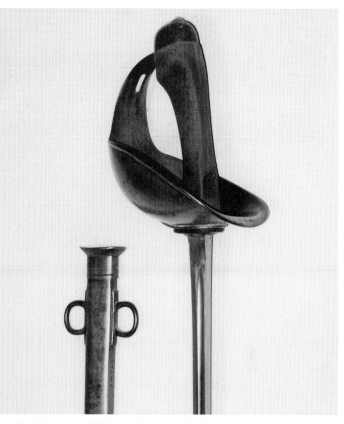

English pattern 1908 cavalry sword with 35 x 1 inch straight blade. Maker: Wilkinson Sword Co. (Author's collection)

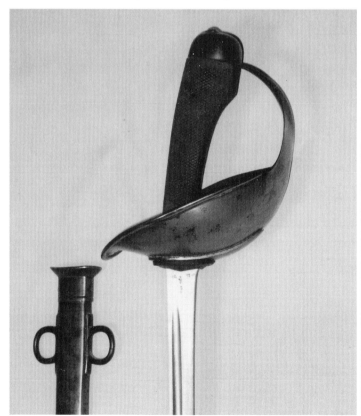

Reverse hilt of pattern 1908 cavalry sword. (Author's collection)

Top of hilt of pattern 1908 cavalry sword. (Author's collection)

# English Cavalry Officers Swords

English cavalry officer's broadsword, c. 1650, with very early full-basket hilt. Blade is marked JACOB AUS MEIGEN of Solingen, Prussia. (Collection of Mervyn Emms)

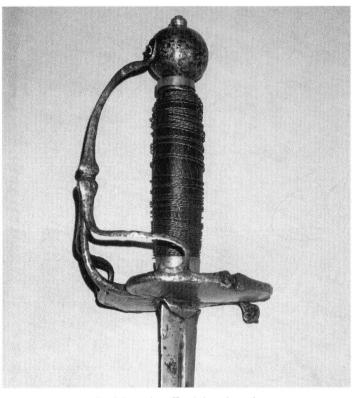

English cavalry officer's broadsword, c. 1680, with iron walloon hilt. (G.H. Cook collection)

English cavalry officer's broadsword, c. 1680, with iron walloon hilt and pierced double shell guards. Blade is marked IHN MINNI with a running wolf. (Jan H. Zajac collection)

Hilt of c. 1680 walloon broadsword. (Jan H. Zajac collection)

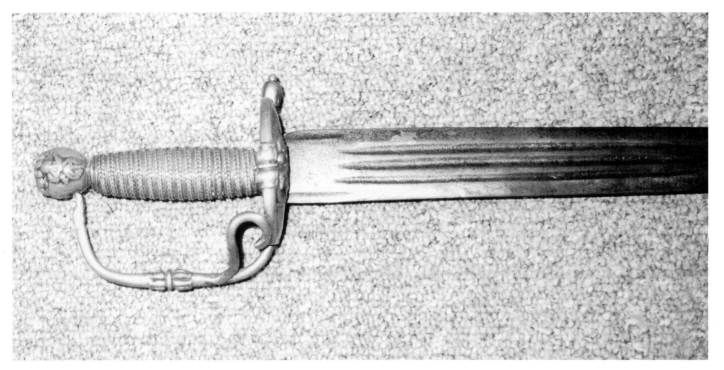

English cavalry officer's broadsword, c. 1700, with brass walloon hilt and 31 3/4 inch blade marked ANDREA FARARA. (G.H. Cook collection)

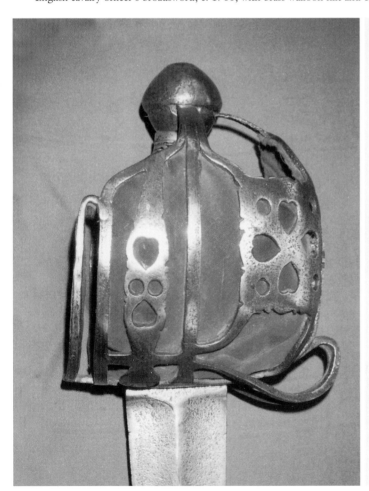

English basket-hilted dragoon officer's broadsword with c. 1750 brass basket hilt and c. 1700 early 37 3/4 x 2 inch blade with a median ridge. (G.H. Cook collection)

English dragoon officer's backsword, c. 1780. (C.M. Lewer-Allen collection)

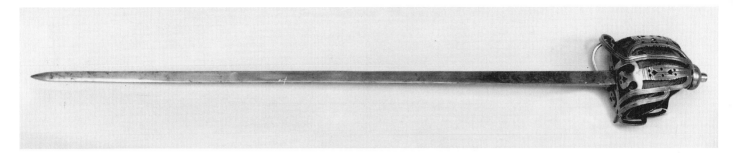

English basket-hilted dragoon officer's broadsword, c. 1755–1760, with 34 inch blade and basket with oval horseman's ring. (Courtesy of Patrick Tougher)

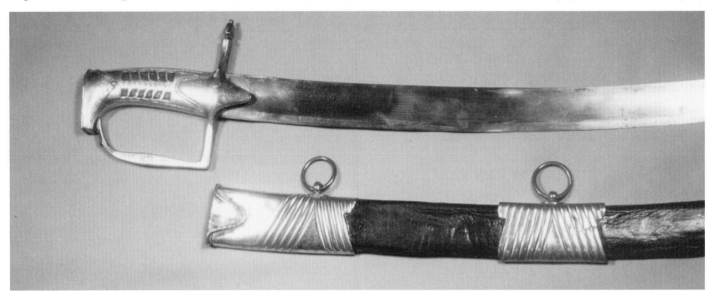

English pattern 1792 light dragoon officer's saber, 10th Light Dragoons. (Sheperd Paine collection)

English dragoon officer's half-basket backsword, c. 1770. The blade is marked ANDREA FARARA. (G.H. Cook collection)

English light dragoon officer's saber, c. 1786, 65th Light Dragoons. (C.M. Lewer-Allen collection)

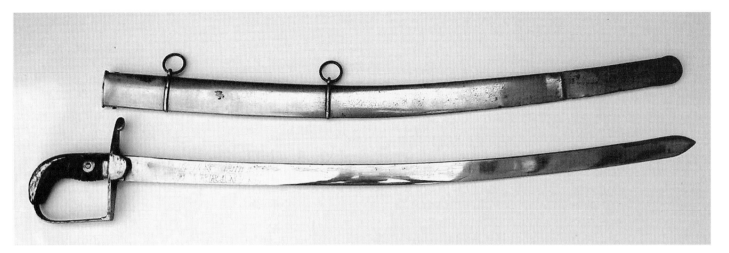

English pattern 1796 (variant) light cavalry officer's saber (variant blade with spear point) with fish-skin grip with copper wire and 32 1/4 x 1 3/8 inch blade marked JOHN GILL WARRANTEED (London). LIET. COL. VIVIAN TO 2 MASTER GREENWOOD (both of 7th Light Dragoons) is also engraved on the blade. (Richard J. Dellar collection)

Blade presentation on pattern 1796 (variant) light cavalry officer's saber. (Richard J. Dellar collection)

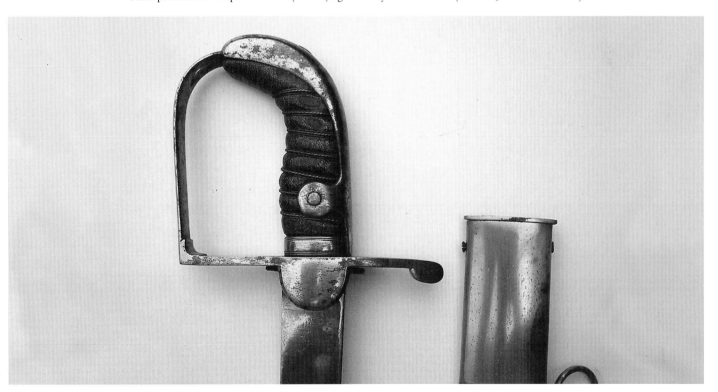

Hilt of pattern 1796 (variant) light cavalry officer's saber.
(Richard J. Dellar collection)

ENGLISH SWORD PHOTO SECTION

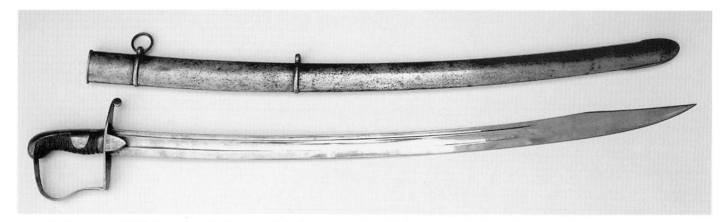

English pattern 1796 light cavalry officer's saber with leather grip covering with three strands of twisted silver wire. The 33 3/4 x 1 7/16 inch blade is a variant blade with two fullers and a long clipped point. Maker: Osborn & Gunby, Birmingham. (Richard J. Dellar collection)

Hilt of the Osborn & Gunby pattern 1796 light cavalry officer's saber. (Richard J. Dellar collection)

English pattern 1796 light cavalry officer's saber. Maker: Woolley (James) & Co. (Richard J. Dellar collection)

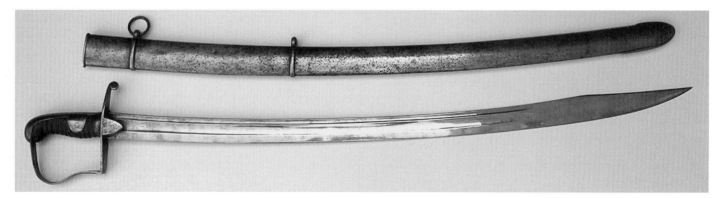

English pattern 1796 light cavalry officer's saber with leather grip covering with two strands of silver twisted wire. The etched 32 x 1 1/2 inch blade with cipher GR is marked BLD for the Berwickshire Light Dragoons. Maker: Woolley (James) & Co., Birmingham. (Richard J. Dellar collection)

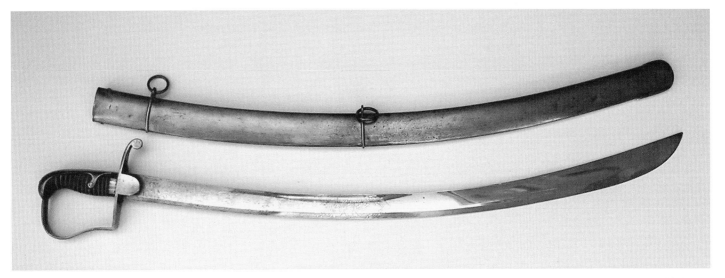

English pattern 1796 light cavalry officer's saber with leather grip covering with three strands of twisted silver wire. The 32 1/2 x 1 1/2 inch blade is etched with LBLHV for the Loyal Birmingham Light Horse Volunteers. Maker: Henry Osborn, Birmingham. (Richard J. Dellar collection)

Hilt of Osborn pattern 1796 light cavalry officer's saber marked LBLHV.
(Richard J. Dellar collection)

## 1796 pattern Loyal Birmingham Light Horse Volunteers

*Blade etching - left (reverse) side*

Left side of blade etching on Osborn pattern 1796 cavalry officer's saber on p. 303. (Richard J. Dellar collection)

### 1796 pattern Loyal Birmingham Light Horse Volunteers

*Blade etching - right (obverse) side*

Right side of blade etching on Osborn pattern 1796 cavalry officer's saber on p. 303. (Richard J. Dellar collection)

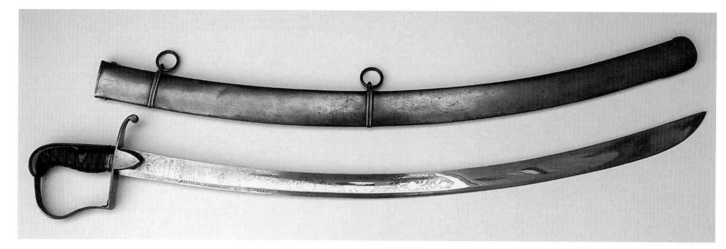

English pattern 1796 light cavalry officer's saber with leather grip covering with three strands of twisted silver wire. The 32 1/2 x 1 1/2 inch blade is engraved with the cipher GR and THO'S GILLS WARRANTED NEVER TO FAIL 1798. (Richard J. Dellar collection)

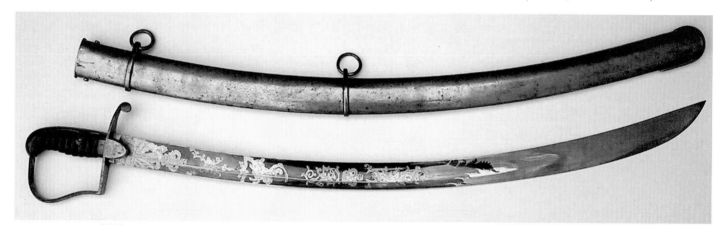

English pattern 1796 light cavalry officer's saber with leather grip covering with three strands of twisted silver wire. The 32 1/2 x 1 1/2 inch blued and gilded blade is etched with a crown over a harp mark and the badge of the 8th (Kings Royal Irish) Light Dragoons (regiment raised in Ireland from Irish Protestants in 1693). (Richard J. Dellar collection)

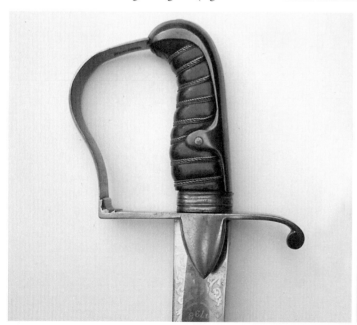

Hilt of the Gills pattern 1796 light cavalry officer's saber. (Richard J. Dellar collection)

Hilt of pattern 1796 8th Light Dragoons officer's saber. (Richard J. Dellar collection)

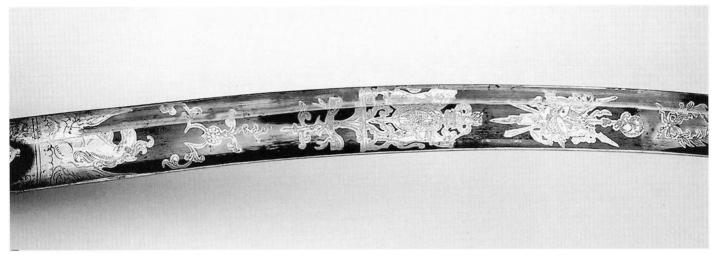

Front blade decoration on pattern 1796 8th Light Dragoons officer's saber on p. 306. (Richard J. Dellar collection)

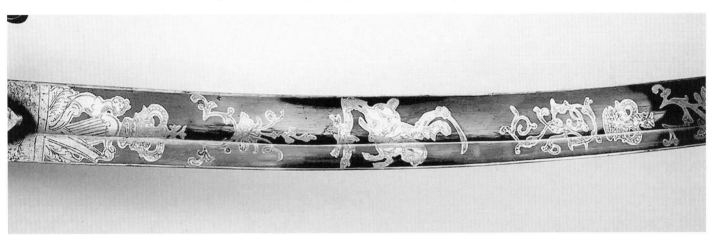

Reverse blade decoration on pattern 1796 8th Light Dragoons officer's saber on p. 306. (Richard J. Dellar collection)

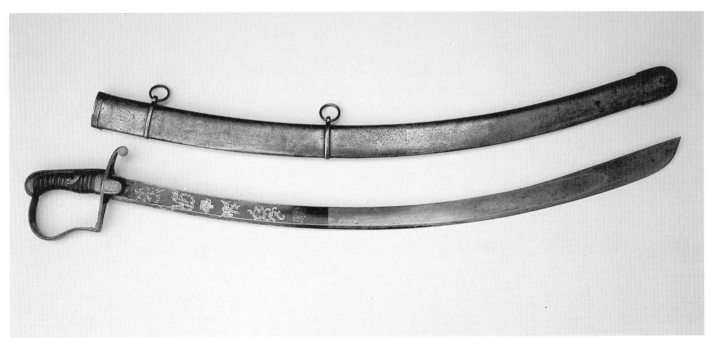

English pattern 1796 light cavalry officer's saber. The 34 3/4 x 1 3/8 inch blade is marked J.J. RUNKEL SOHLINGEN. The upper blade is blued and gilded. Maker: S. (Samuel) Brunn, London. (Richard J. Dellar collection)

Hilt of Brunn's pattern 1796 cavalry officer's saber on p. 307. (Richard J. Dellar collection)

English pattern 1796 heavy cavalry officer's saber. (C.M. Lewer-Allen collection)

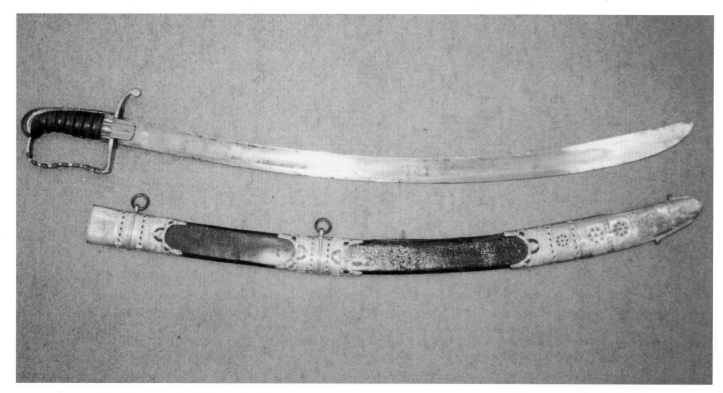

English pattern 1796 light cavalry officer's saber (presentation grade) with engraved and gilded blade and Hussar style leather scabbard with pierced brass mounts. Maker: S. (Samuel) Brunn, London. (Courtesy of West Street Antiques)

Hilt of the Brander & Potts pattern 1796 heavy cavalry officer's sword. (Richard J. Dellar collection)

Bottom of hilt of the Brander & Potts pattern 1796 heavy cavalry sword. (Richard J. Dellar collection)

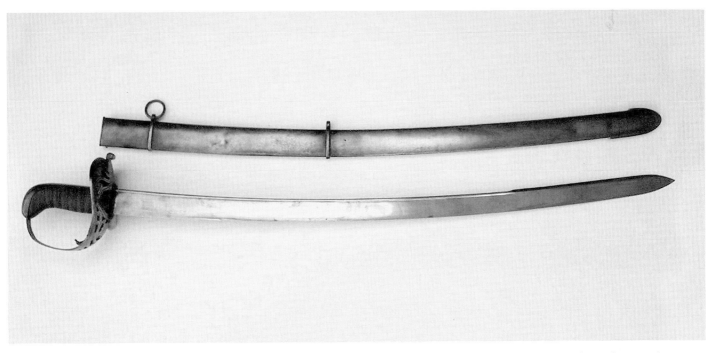

English pattern 1796 heavy cavalry officer's saber with sharkskin grip covering with two strands of twisted silver wire and one of copper wire and 34 1/4 x 1 3/8 inch pipe-backed blade with a spear point. Maker: Brander & Potts, London. (Richard J. Dellar collection)

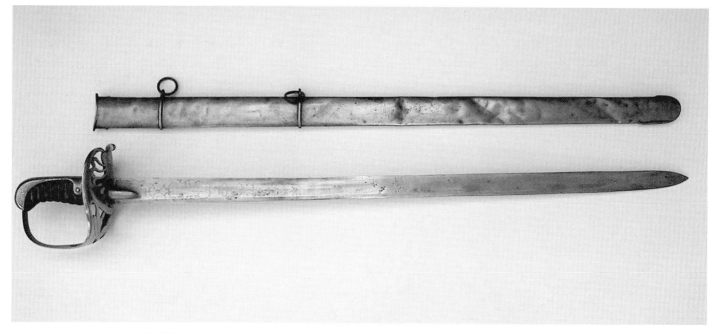

English pattern 1796 heavy cavalry officer's sword. The 34 1/2 x 1 1/4 inch spear point blade is etched with the GR cipher. Maker: J.J. Runkel, Solingen. (Richard J. Dellar collection)

Bottom hilt of Runkel pattern 1796 heavy cavalry officer's sword. (Richard J. Dellar collection)

Hilt of Runkel pattern 1796 heavy cavalry officer's sword. (Richard J. Dellar collection)

English pattern 1796 heavy cavalry officer's sword (dress pattern) with blued and gilded blade. (C.M. Lewer-Allen collection)

English pattern 1796 heavy cavalry officer's sword (dress pattern) with double-edged broadsword blade. (C.M. Lewer-Allen collection)

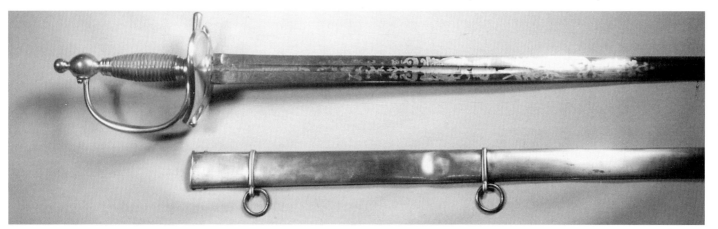
English pattern 1796 heavy cavalry officer's sword (dress pattern) with blued and gilded broadsword blade. (Sheperd Paine collection)

English horsehead pommel light cavalry officer's saber, c. 1805, with ivory grip, gilt brass hilt, and gilt brass mounted Hussar-style scabbard. (Courtesy of West Street Antiques)

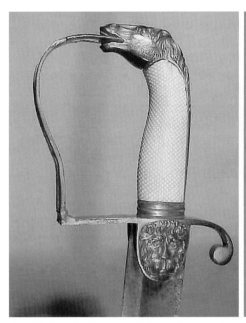 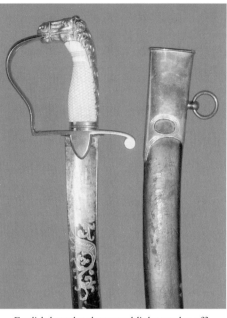 

| | | |
|---|---|---|
| English horsehead pommel light cavalry officer's saber, c. 1805, with checkered ivory grip and gilt copper hilt. Overall length is 34 1/2 inches. (Jan Zajac collection) | English horsehead pommel light cavalry officer's saber, c. 1805, with blued and gilded blade, gilt brass hilt, checkered ivory grip, and gilt brass mounted leather scabbard. (G.H. Cook collection) | English pattern 1807 light dragoon officer's saber (10th [Prince of Wales own] Light Dragoons) with gilded blade with hatchet point, fish-skin grip, and gilt brass hilt. (C.M. Lewer-Allen collection) |

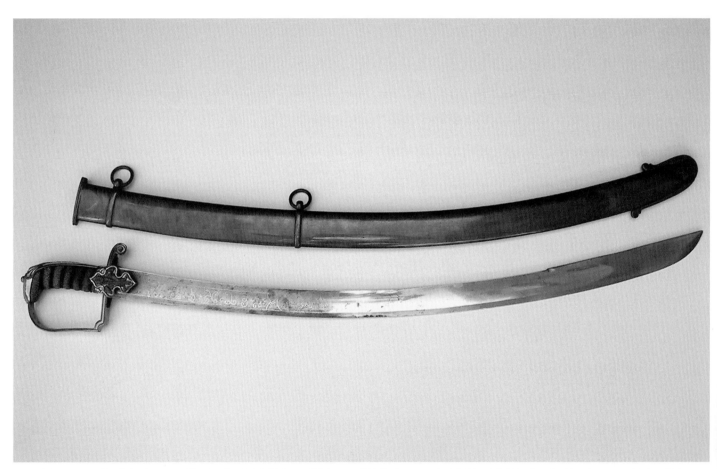

English pattern 1807 light dragoon officer's saber (10th [Prince of Wales own] Light Dragoons) with 32 x 1 1/2 inch etched hatchet point blade, gilt brass hilt, and fish-skin grip covering with two strands of plain and one strand of twisted bullion wire. Maker: Prosser (John), London. (Richard J. Dellar collection)

Hilt of Prosser pattern 1807 light dragoon officer's saber on p. 312. (Richard J. Deller collection)

Reverse hilt of Prosser pattern 1807 light dragoon officer's saber on p. 312. (Richard J. Deller collection)

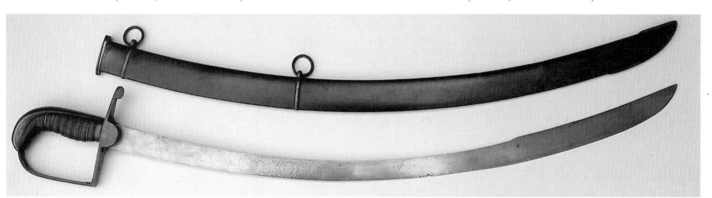
English light dragoon officer's saber (19th Light Dragoons), c. 1807, with steel hilt and scabbard and fish-skin grip covering with one silver and two copper strands of wire. The 30 7/8 x 1 1/4 inch kilij-type blade has a latch back point and etched panels. Maker: Prosser (John), London. (Richard J. Deller collection)

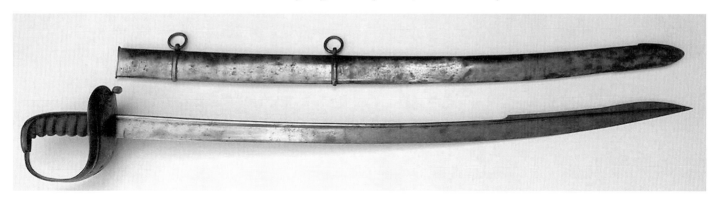
English heavy cavalry officer's saber, c. 1818, with fish-skin grip, pierced steel bowl hilt, and 34 3.4 x 1 5/16 inch pipe back blade. Maker: Prosser (John), London. (Richard J. Dellar collection)

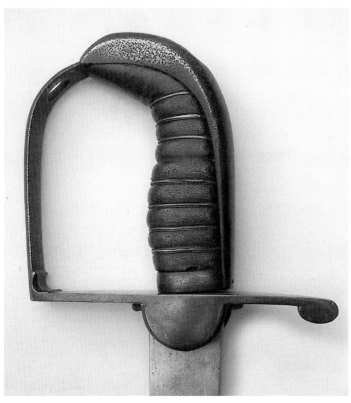

Hilt of Prosser 19th Light Dragoons officer's saber on p. 313.
(Richard J. Deller collection)

Hilt of Prosser c. 1818 heavy cavalry officer's saber on p. 313.
(Richard J. Dellar collection)

English heavy cavalry officer's saber, c. 1815,
with etched blade and scabbard. (C.M. Lewer-Allen collection)

Rear hilt of Prosser c. 1818 heavy cavalry officer's saber on p. 313.
(Richard J. Dellar collection)

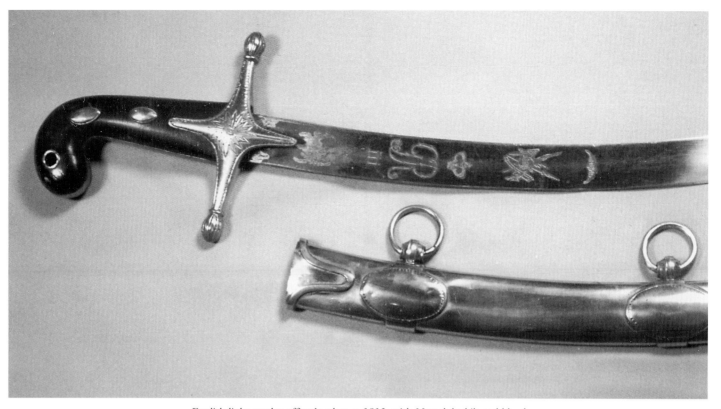

English light cavalry officer's saber, c. 1812, with Mameluke hilt and blued and gilded blade with cipher GRIII. (Sheperd Paine collection)

English pattern 1821 light cavalry officer's saber. The leather-covered branches are for a South African regiment. (C.M. Lewer-Allen collection)

English pattern 1821 light cavalry officer's saber with etched blade and gilt brass hilt. (C.M. Lewer-Allen collection)

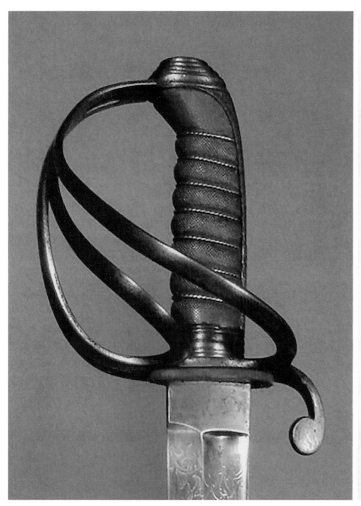

English pattern 1821 light cavalry officer's saber for the "Mysor Horse" Indian Regiment. The 32 1/2 x 1 1/2 inch blade is etched MYSOR HORSE. Maker: Robert Mole & Sons, Birmingham. (Jan H. Zajac collection)

English pattern 1821 heavy cavalry officer's saber with pipe-backed blade. (C.M. Lewer-Allen collection)

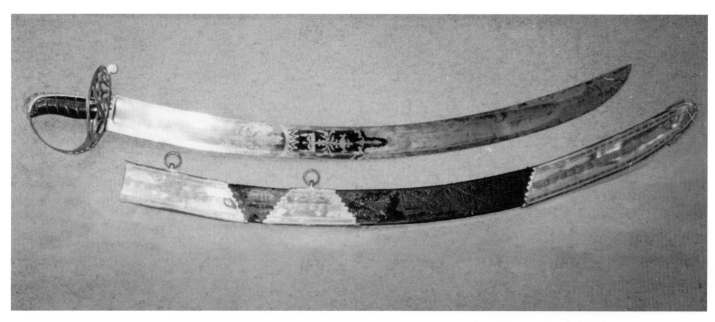

English presentation-grade pattern 1821 heavy cavalry officer's saber with huge blued and gilded 2 1/4 inch wide blade, Hussar-type leather scabbard with gilt brass mounts, and pierced gilt brass hilt. (Courtesy of West Street Antiques)

English 4th Dragoon Guards officer's saber, c. 1850.
(C.M. Lewer-Allen collection)

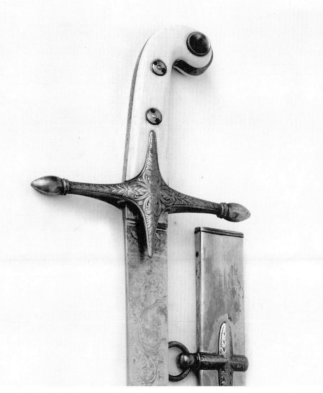

English 6th Dragoon Guards officer's saber, c. 1860 (for dress).
The blade is etched with the crest of the 6th Dragoon Guards.
(C.M. Lewer-Allen collection)

English 6th Dragoon Guards officer's saber, c. 1877.
(C.M. Lewer-Allen collection)

English pattern 1887 heavy cavalry officer's sword, South African
Police issue, with etched blade. Maker: Wilkinson, London.
(C.M. Lewer-Allen collection)

English pattern 1887 heavy cavalry officer's sword with etched blade. (C.M. Lewer-Allen collection)

English presentation light cavalry officer's saber with solid silver hilt, black fish-skin grip with silver wires, and profusely etched blade. Presented in July 1889 to the best swordsman in the regiment by the Right Honorable the Earl of Waddington, Lt. Colonel. Maker: Robert Mole & Sons, Birmingham. (Jan H. Zajac collection)

Reverse hilt of the Mole presentation sword. (Jan H. Zajac collection)

English pattern 1912 cavalry officer's sword with 35 inch blade. Maker: Henry Wilkinson. (Collection of Mervyn Emms)

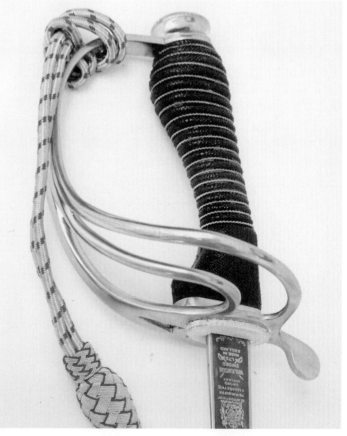

English pattern 1912 cavalry officer's sword, 10th Hussars. (C.M. Lewer-Allen collection)

# English Household Cavalry Officers Swords

English pattern 1814 household cavalry officer's sword, Royal Horse Guards. (C.M. Lewer-Allen collection)

English household cavalry officer's sword, c. 1832, Royal Horse Guards. (C.M. Lewer-Allen collection)

English pattern 1834 household cavalry officer's sword, 1st Life Guards. (C.M. Lewer-Allen collection)

English pattern 1874 household cavalry officer's sword, Royal Horse Guards. (C.M. Lewer-Allen collection)

English pattern 1874 household cavalry officer's sword, 1st Life Guards. (C.M. Lewer-Allen collection)

English pattern 1874 household cavalry officer's sword, 2nd Life Guards. (C.M. Lewer-Allen collection)

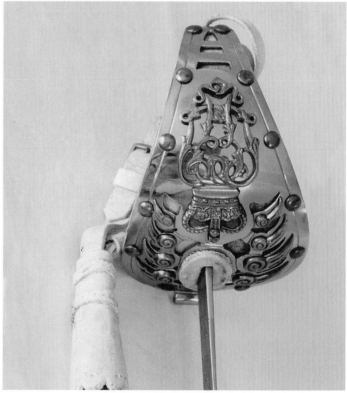

English pattern 1874 household cavalry officer's sword, Life Guards (amalgamated in 1922). (C.M. Lewer-Allen collection)

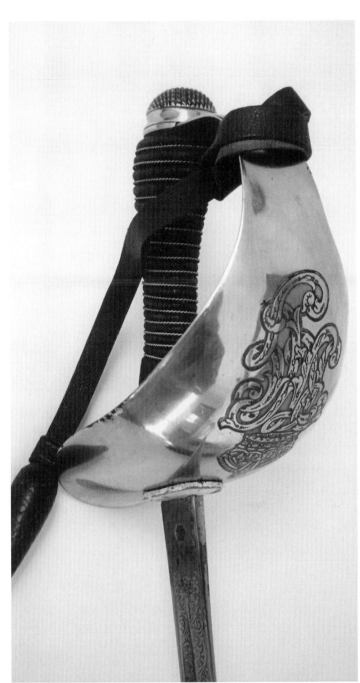

English pattern 1912 Royal Horse Guards officer's sword.
(C.M. Lewer-Allen collection)

Hilt logo on 1912 Royal Horse Guards officer's sword.
(C.M. Lewer-Allen collection)

# English Hussar Officers Swords

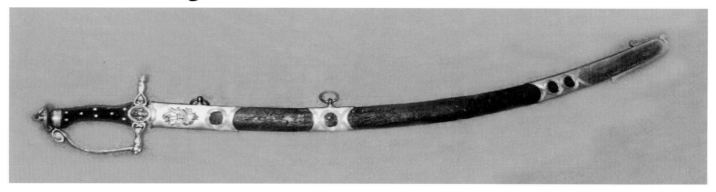

English Hussar officer's saber, c. 1800, with Indian "Tulwar" hilt. Maker: Henry Tatham, London. (Courtesy of West Street Antiques)

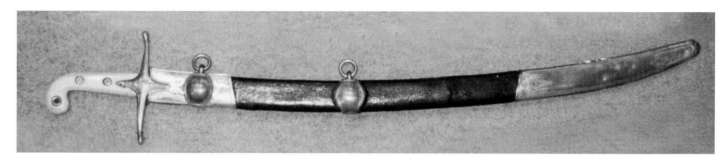

English Hussar officer's saber, c. 1805, with Mameluke hilt with ivory grip. (Courtesy of West Street Antiques)

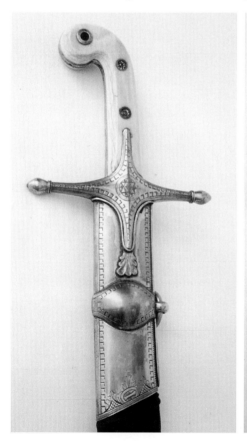

English Hussar officer's saber, c. 1805.
(C.M. Lewer-Allen collection)

English Hussar officer's saber, c. 1830, 4th Hussars Regimental scimitar.
(C.M. Lewer-Allen collection)

English Hussar officer's saber, c. 1870, 15th Hussars Regimental scimitar.
(C.M. Lewer-Allen collection)

English Hussar officer's saber, c. 1900,
18th Hussars Regimental scimitar.
(C.M. Lewer-Allen collection)

English Hussar officer's saber, c. 1910,
18th Hussars Regimental scimitar.
(C.M. Lewer-Allen collection)

# English Lancer Officers Swords

English pattern 1822 Lancer officer's saber, 9th Lancers Regimental scimitar, with Mameluke hilt. (C.M. Lewer-Allen collection)

English Lancer officer's saber, c. 1832, 16th Lancers Regimental scimitar. (C.M. Lewer-Allen collection)

English Lancer officer's saber, c. 1840, 16th Queens Lancers Regimental scimitar, with Mameluke hilt. (C.M. Lewer-Allen collection)

# English Infantry Hangers

English infantry hanger, c. 1680–1700, with brass hilt, gargoyle pommel, and 28 1/8 inch blade. (George C. Neumann collection)

English infantry hanger, c. 1725–1740, with brass hilt and 24 inch blade. (George C. Neumann collection)

English infantry hanger, c. 1727–1760, with 29 1/2 inch straight blade and mark of the House of Hanover on the guard.
(George C. Neumann collection)

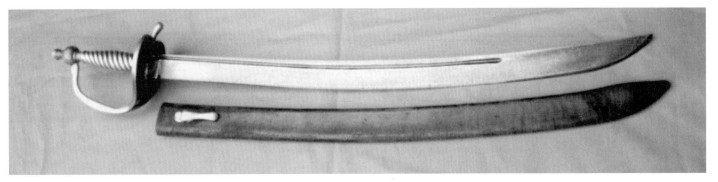

English infantry hanger, c. 1740, with brass hilt and 23 3/4 x 1 1/4 inch blade.
(Sam Saladino collection)

English pattern 1751 infantry hanger with brass hilt and 25 1/2 inch blade.
(G.H. Cook collection)

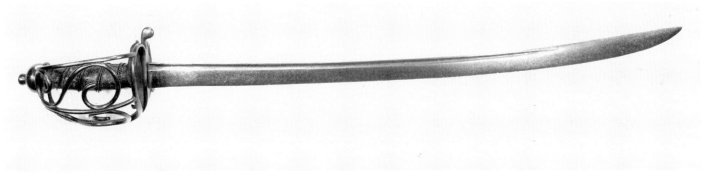

English infantry hanger, c. 1740–1760, for grenadier company.
(George C. Neumann collection)

# English Infantry Sergeants Swords

English sergeant's sword, c. 1850, English rifle regiment, 3rd Scots Fusilier Guards. (C.M. Lewer-Allen collection)

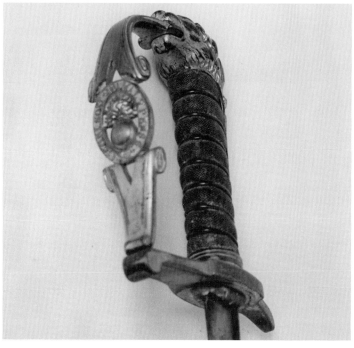

English sergeant's sword, c. 1850, 1st Grenadier Guards. (C.M. Lewer-Allen collection)

# English Infantry Officers Swords

English infantry officer's short saber, c. 1770, with split and divided hilt and 29 1/4 inch blade. Maker: Nathaniel Jeffreys Sr., London. (G.H. Cook collection)

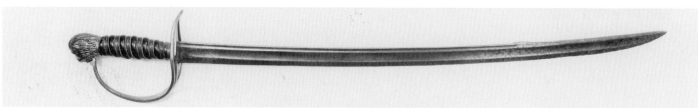

English infantry officer's short saber, c. 1770–1780, with 25 1/2 inch blade. (George C. Neumann collection)

English infantry officer's short saber, c. 1770–1780, with horn grip, split and divided hilt, and 26 3/4 inch blade.
(George C. Neumann collection)

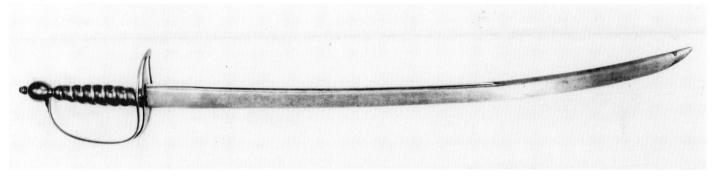

English infantry officer's short saber, c. 1770–1780, with horn grip with silver bands, split and divided hilt, and 26 1/8 inch blade.
(George C. Neumann collection)

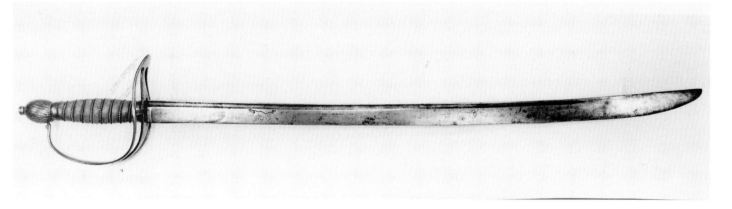

English infantry officer's short saber, c. 1775–1790, with split and divided hilt and 28 inch blade.
(George C. Neumann collection)

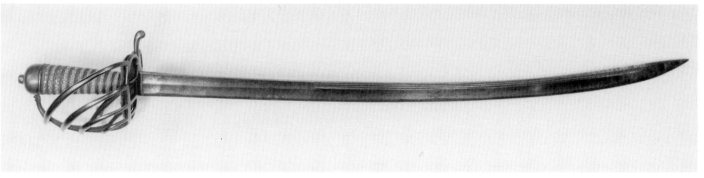

English half-basket infantry officer's short saber, c. 1775–1790, with fish-skin grips and 28 1/2 inch blade.
(George C. Neumann collection)

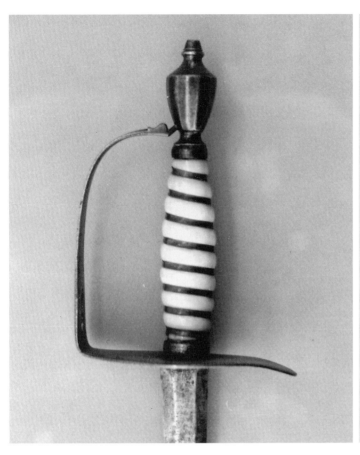

English infantry officer's sword (spadroon), c. 1775–1790, with 30 inch straight blade. (Collection of Mervyn Emms)

English infantry officer's sword, c. 1775–1790, with split and divided hilt. (C.M. Lewer-Allen collection)

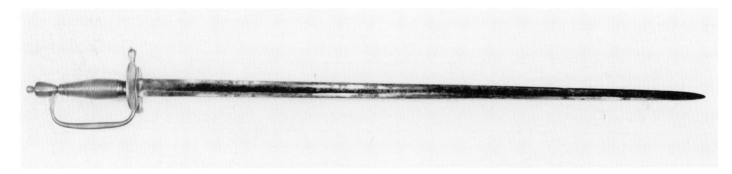

English infantry officer's sword (spadroon), c. 1775–1790, with 32 inch straight blade. (George C. Neumann collection)

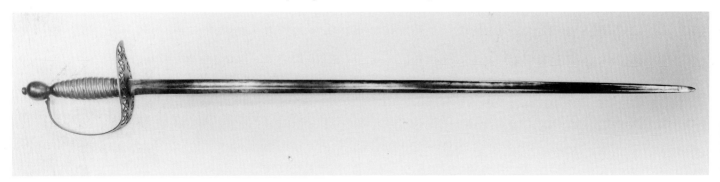

English infantry officer's sword (spadroon), c. 1775–1790, with cut steel hilt and 31 inch straight blade. (George C. Neumann collection)

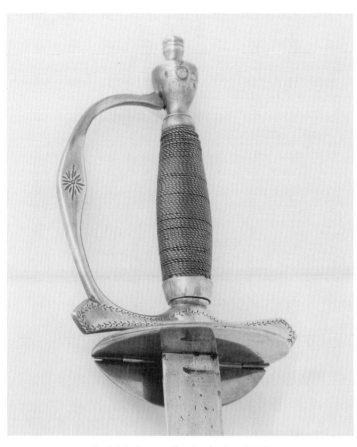

English infantry officer's sword, c. 1780.
(C.M. Lewer-Allen collection)

English infantry officer's sword (spadroon), c. 1780, with five-ball hilt and pillow pommel. (C.M. Lewer-Allen collection)

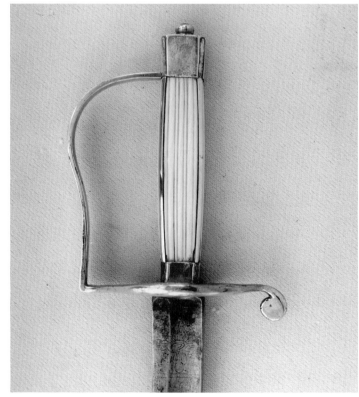

English infantry officer's sword (spadroon), c. 1790, with brass hilt and 31 1/2 inch straight blade marked GILL. Maker: Thomas Gill Sr., London. (Collection of Mervyn Emms)

English infantry officer's sword (spadroon), c. 1790, with pillow pommel and 32 1/2 inch blade. (Collection of Mervyn Emms)

English infantry officer's short saber, c. 1795, 23rd Welch Fusiliers. (Sheperd Paine collection)

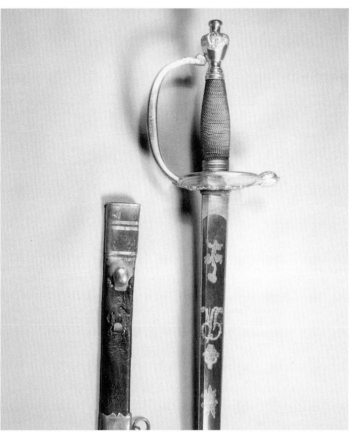

English pattern 1796 infantry officer's sword. Maker: T & S Goldney, London. (Sheperd Paine collection)

English pattern 1796 infantry officer's sword with one fold-down shell. (C.M. Lewer-Allen collection)

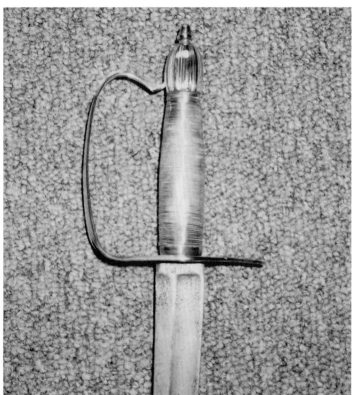

English silver-hilted infantry officer's sword with slotted hilt and 32 inch straight blade. Silver assay marks indicate 1799 fabrication. Maker: William Read, Portsmouth. (G.H. Cook collection)

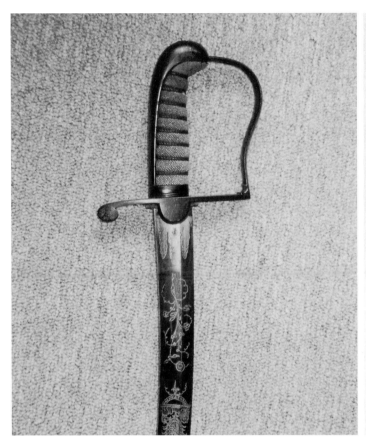

English flank officer's short saber, c. 1800, with iron hilt with fish-skin grip and 26 1/2 inch blued and gilded blade. (G.H. Cook collection)

English infantry officer's short saber, c. 1800, for grenadier company. (Marc Wesseler collection)

English infantry officer's sword with pillow pommel, c. 1800, Loyal St. James Volunteers. (C.M. Lewer-Allen collection)

English lion-head infantry officer's saber, c. 1800, with Mameluke brass hilt, brass scabbard, and 28 inch blade marked 2nd W. INDIA [Regiment]. (Collection of Mervyn Emms)

English pattern 1803 infantry officer's saber. (Sheperd Paine collection)

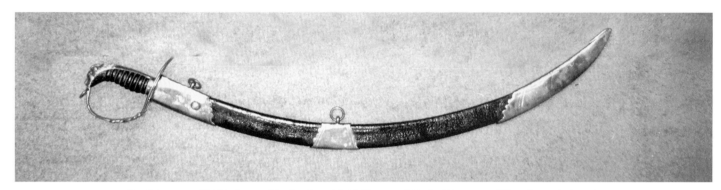

English pattern 1803 infantry officer's saber with highly curved blade. (Courtesy of West Street Antiques)

English pattern 1803 infantry officer's saber.
(C.M. Lewer-Allen collection)

English pattern 1803 infantry officer's saber, grenadier company
(C.M. Lewer-Allen collection)

English pattern 1803 infantry officer's saber with unusual flat blade. (Collection of Ulrich Dubi)

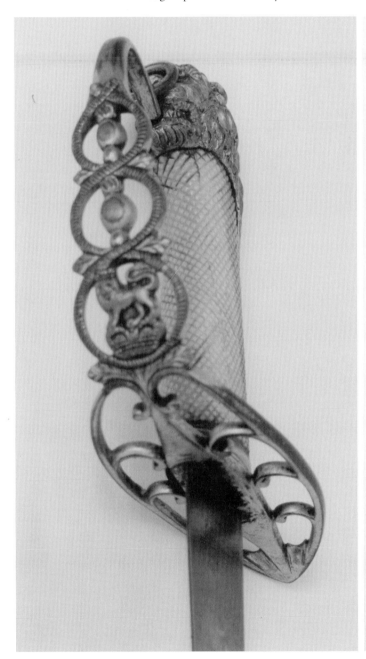

English pattern 1803 infantry officer's saber, grenadier company. (C.M. Lewer-Allen collection)

English c. 1803 light infantry officer's saber, 23rd Foot-Royal Welch Fusiliers. (C.M. Lewer-Allen collection)

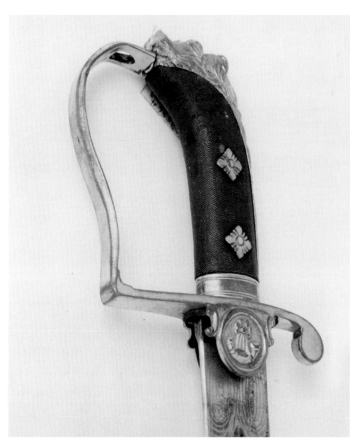
English c. 1803 light infantry officer's saber, 43rd Foot.
(C.M. Lewer-Allen collection)

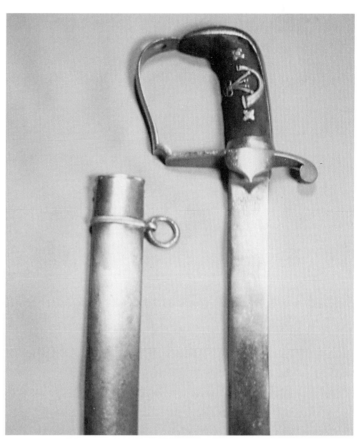
English c. 1803 light infantry officer's saber, 52nd Foot.
(Sheperd Paine collection)

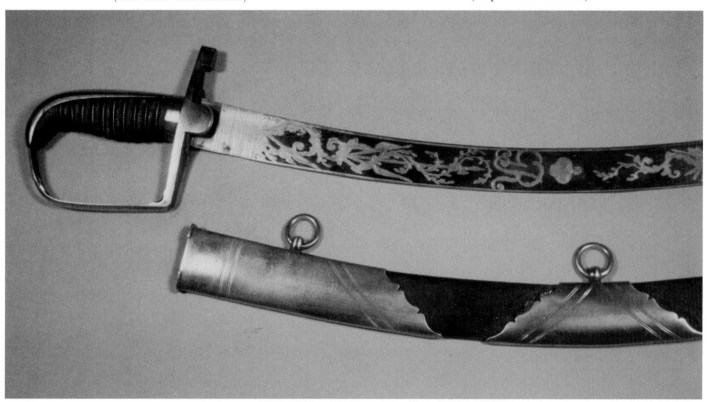
English pattern 1803 rifle [regiment] officer's saber, King's German Legion.
(Sheperd Paine collection)

English pattern 1822 infantry officer's sword, May 1823 presentation marked FROM COL KNOWLES AND OFFICERS OF THE MADRAS RIFLE CORPS TO THEIR BROTHER OFFICER CAPT. JAMES CROKAT. Maker: John Prosser, London. (Sheperd Paine collection)

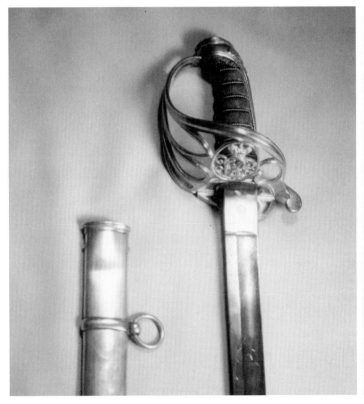

English pattern 1822 infantry officer's sword. Maker: Henry Wilkinson. (Sheperd Paine collection)

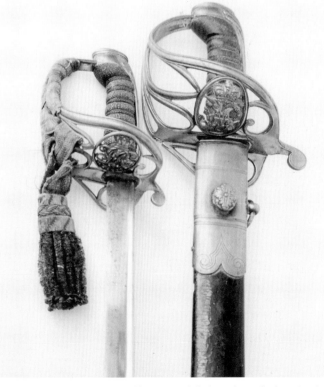

English pattern 1822 infantry officer's sword (light and standard versions). (C.M. Lewer-Allen collection)

English infantry officer's sword, c. 1824, 21st Foot-Royal Scots Fusiliers.
(C.M. Lewer-Allen collection)

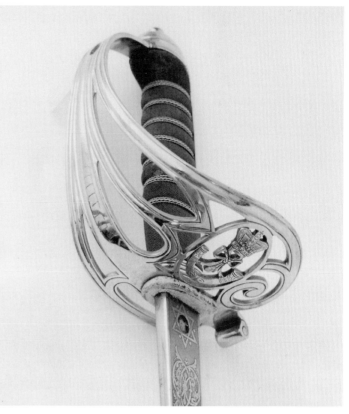

English pattern 1827 rifle [regiment] officer's sword.
(C.M. Lewer-Allen collection)

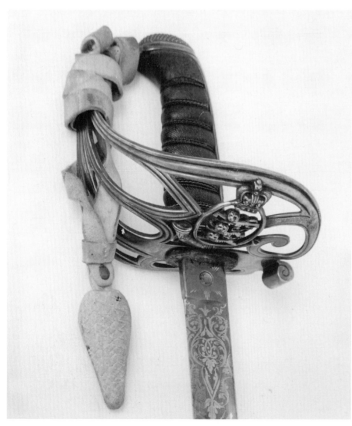

English pattern 1827 rifle [regiment] officer's sword, Lankeshire Rifles.
(C.M. Lewer-Allen collection)

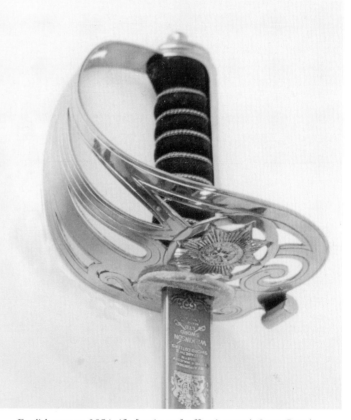

English pattern 1854 rifle [regiment] officer's sword, Scots Guards.
(C.M. Lewer-Allen collection)

English pattern 1854 rifle [regiment] officer's sword, Grenadier Guards. (C.M. Lewer-Allen collection)

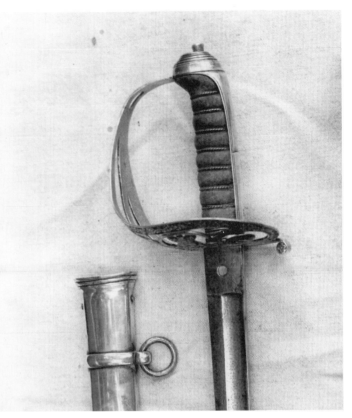

English pattern 1854 rifle [regiment] officer's sword, c. 1865, with custom silver hilt. Elginshire volunteers. (James Forman collection)

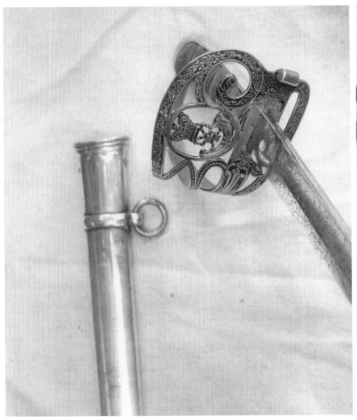

Hilt of Elginshire volunteers officer's sword, c. 1865. (James Forman collection)

English 1892 pattern rifle [regiment] officer's sword. (Collection of H. Coetzee; Rolf Duffur, photographer)

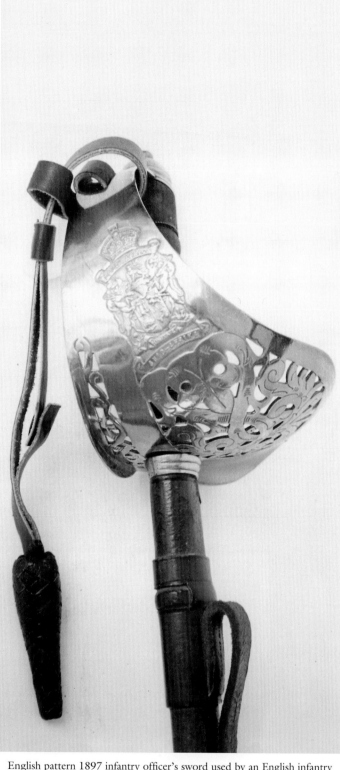

English pattern 1897 infantry officer's sword with 32 3/8 inch blade with cipher of George V (1910–1936). Maker: Robert Mole & Sons, Birmingham. (G.H. Cook collection)

English pattern 1897 infantry officer's sword used by an English infantry officer in a South American regiment during the Boer War, 1899–1902. (C.M. Lewer-Allen collection)

# English General Officers Swords

English general officer's sword, pattern 1796, Royal House guards.
(C.M. Lewer-Allen collection)

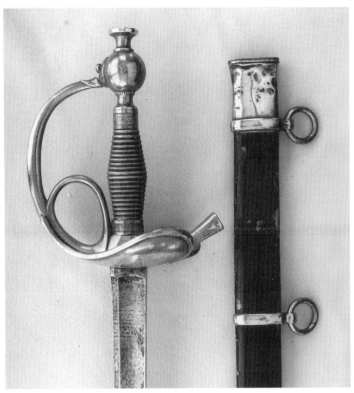

English general officer's sword, pattern 1814, with 32 inch blade.
(Collection of Mervyn Emms)

English general and staff officer's sword, pattern 1822.
(C.M. Lewer-Allen collection)

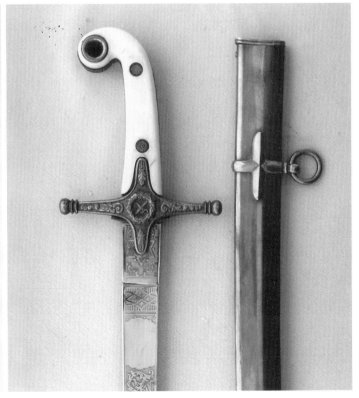

English general officer's sword, pattern 1831, with Mameluke hilt
and 33 1/2 inch blade. Maker: Manton & Co., London.
(Collection of Mervyn Emms)

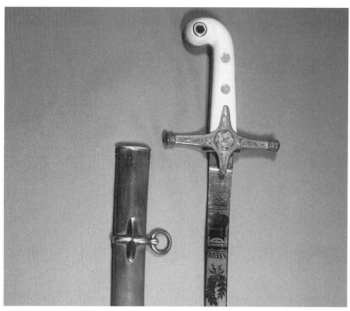

English general officer's sword, pattern 1831.
(Sheperd Paine collection)

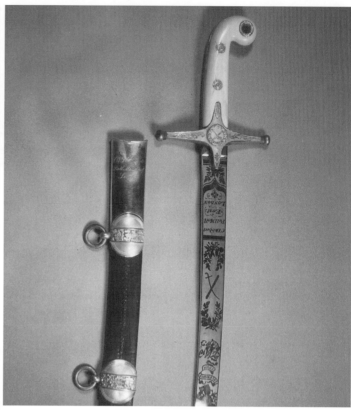

English general officer's sword, pattern 1831, with c. 1850 scabbard variation.
Maker: C. Hebbert, London. (Sheperd Paine collection)

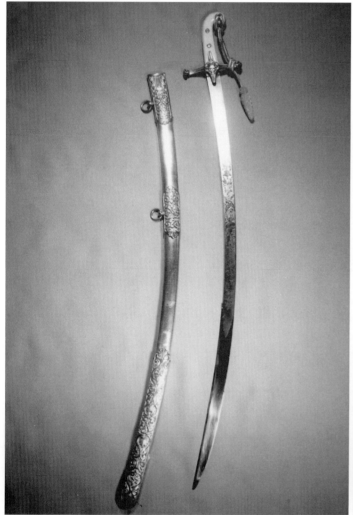

English general officer's sword, pattern 1831, c. 1860–1870, Kings 15th Hussars. Maker: Henry Wilkinson, London. (Sheperd Paine collection)

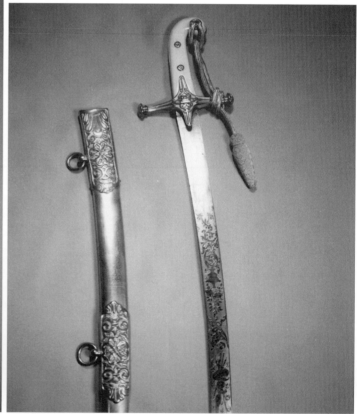

Hilt of the Wilkinson general officer's sword.
(Sheperd Paine collection)

# English Infantry Swords for Highland Regiments

English basket-hilted infantry backsword for Highland regiments (42nd Royal Highland regiment), c. 1760, with 32 inch blade. (Courtesy of Patrick Tougher)

English basket-hilted infantry backsword for Highland regiments, c. 1750, with 31 1/2 inch blade. (G.H. Cook collection)

English basket-hilted infantry backsword for Highland regiments (42nd Royal Highland regiment), c. 1760, with 31 inch blade. Maker: Dru Drury Sr., London. (James Forman collection)

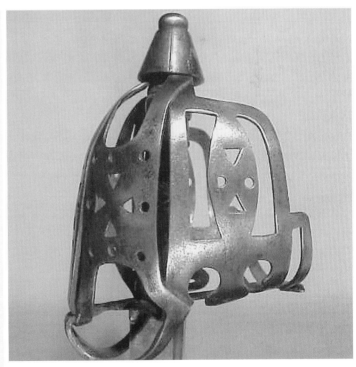

English basket-hilted infantry backsword for Highland regiments (42nd Royal Highland regiment), c. 1760, with leather-wrapped wooden grip wrapped with a single twisted wire and 32 inch blade. The hilt is marked 42 A 46 and the blade is marked with the cipher GR for George Rex. Overall length is 38 1/2 inches. Maker: Dru Drury Sr., London. (Jan H. Zajac collection)

# English Infantry Officers Swords for Highland Regiments

Very rare English half-basket infantry officer's backsword for "Loyal North Britons Volunteers," a regiment of ex-patriot Scots based in Knightsbridge, Middlesex, c. 1790. The blade is 30 inches in length. (James Forman collection)

English basket-hilted infantry officer's broadsword, c. 1790, with brass hilt and 32 inch blade. Sword used by the Gordon Fencibles, Highland Volunteer Battalion. (Courtesy of Neil McGregor)

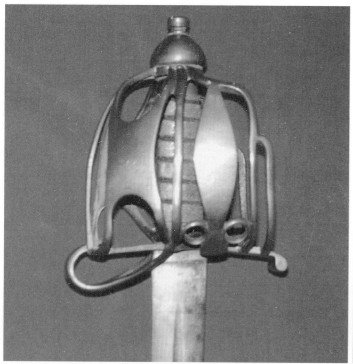

English pattern 1798 infantry officer's broadsword for Highland regiments with sharkskin grip, brass hilt, and 32 1/2 inch blade. (G.H. Cook collection)

English basket-hilted infantry officer's broadsword for Highland regiments (93rd Highlanders), c. 1820, with 33 inch blade with cipher of George III. (James Forman collection)

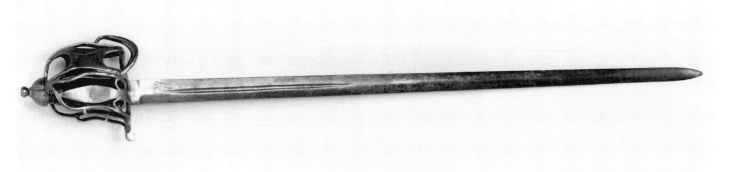

English pattern 1798 basket-hilted infantry officer's broadsword for Highland regiments with brass hilt.
(Courtesy of Patrick Tougher)

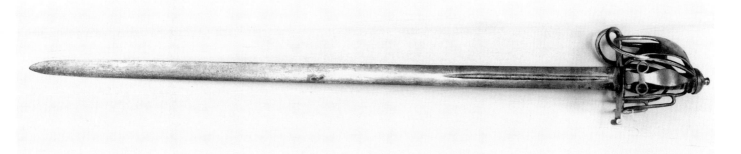

English pattern 1798 basket-hilted infantry officer's broadsword for Highland regiments with brass hilt.
(Courtesy of Patrick Tougher)

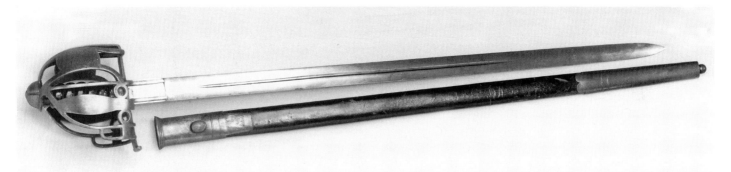

English basket-hilted infantry officer's broadsword for Highland regiments, c. 1820, with 31 1/2 inch blade marked A. & A.S. for August & Albert Schnitzler, Solingen, Germany. (Courtesy of Patrick Tougher)

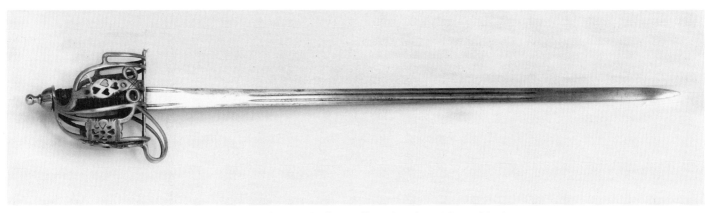

English pattern 1828 basket-hilted infantry officer's broadsword for Highland regiments.
(Courtesy of Patrick Tougher)

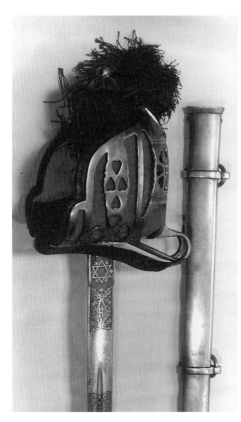  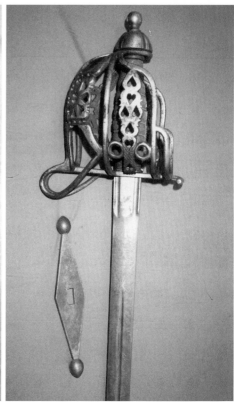

English pattern 1828 basket-hilted infantry officer's broadsword for Highland regiments, c. 1854, with 34 inch blade. Maker: Henry Wilkinson & Son, London. (James Forman collection)

English pattern 1828 basket-hilted infantry officer's broadsword for Highland regiments, January 11, 1856, presentation. Blade engraved, THE GIFT OF LIEUT. GENERAL SIR THOMAS BROTHERTON TO FRANCIS GEORGE COLERIDGE, JAN. 11, 1856. Maker: Hawkes & Co., London. (James Forman collection)

English 1828 pattern basket-hilted infantry officer's broadsword for Highland regiments (Gordon Highlanders), c. 1860–1880, with 31 3/4 inch blade. Alternative dress hilt cross-guard also shown. Maker: (Henry) Wilkinson & Son, London. (G.H. Cook collection)

Blade of the Gen. Sir Thomas Brotherton presentation sword. (James Forman collection)

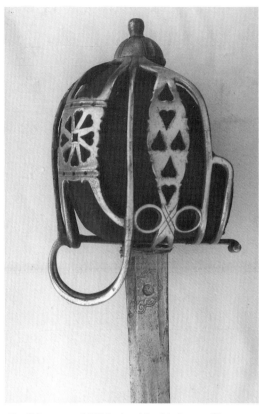
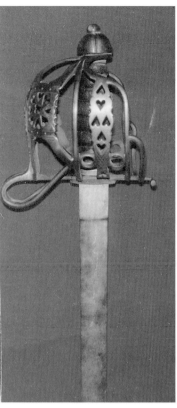

English pattern 1828 basket-hilted infantry officer's broadsword for Highland regiments (71st Highlanders), c. 1850–1880, with 32 inch blade. Retailer: Hamburger Rogers & Co., London. (Collection of Mervyn Emms)

English pattern 1828 basket-hilted infantry officer's broadsword for Highland regiments, c. 1880, with 33 1/4 inch blade. Maker: (Henry) Wilkinson & Son, London. (G.H. Cook collection)

English pattern 1828 basket-hilted infantry officer's broadsword for Highland regiments (Seaforth Highlanders), c. 1860–1880. Maker: (Henry) Wilkenson & Son, London. (Sheperd Paine collection)

English pattern 1828 basket-hilted infantry officer's broadsword for Highland regiments, c. 1860–1880. VR cipher and ABERDEENSHIRE HIGHLANDERS etched on the blade. Maker: Hawkes & Co., London. (Jan. H. Zajac collection)

Hilt of the pattern 1863 English officer's broadsword for Royal Scots Fusiliers on p. 346. (James Forman collection)

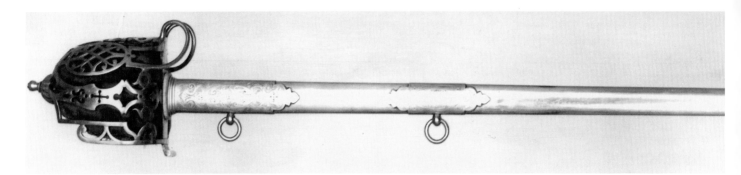

English pattern 1863 basket-hilted infantry officer's broadsword for Highland regiments (Royal Scots Fusiliers) with 32 inch blade. Maker: (Henry) Wilkinson & Son, London. (James Forman collection)

# English Artillery Officers Swords

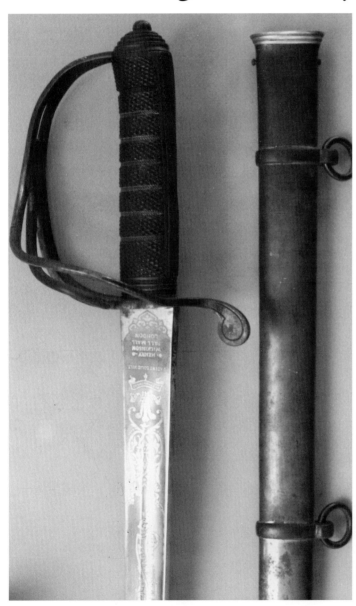

English Royal Artillery officer's sword, c. 1850, with 34 3/4 inch blade. (Collection of Mervyn Emms)

English Honorable Artillery Company officer's sword, c. 1875. (C.M. Lewer-Allen collection)

# English Royal Engineer Officers Swords

English pattern 1857 Royal Engineer officer's sword, 5th Northumberland Fusiliers. (C.M. Lewer-Allen collection)

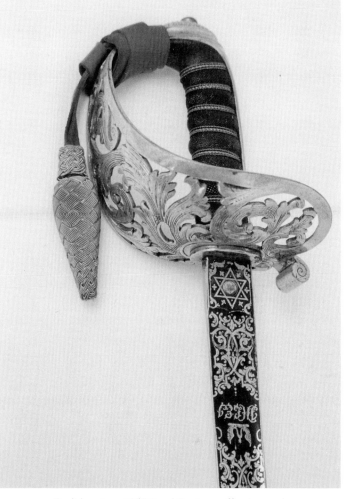

English pattern 1857 Royal Engineer officer's sword. (C.M. Lewer-Allen collection)

# English Naval Officers Swords

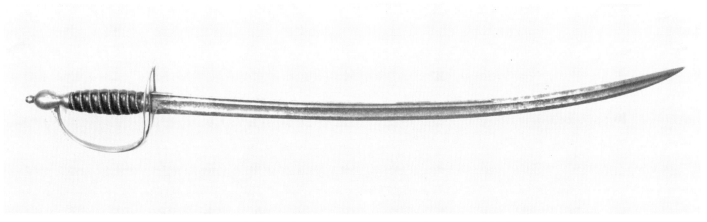

English naval officer's short saber, c. 1770-1785, with 25 1/8 inch blade. (George C. Neumann collection)

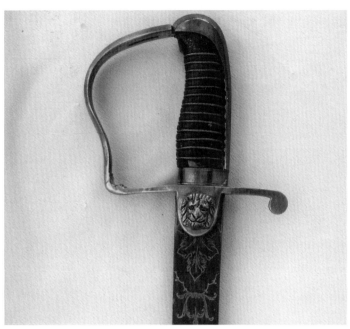

English naval officer's saber, c. 1800, with 31 inch blade. Retailer: J.J. Runkel, London. (Collection of Mervyn Emms)

Hilt of English pattern 1847 naval officer's sword on p. 349. (Collection of H. Coetzee; Rolf Duffur, photographer)

English pattern 1827 naval officer's sword, c. 1830, with 30 1/4 inch blade. Retailer: John Salter & Co., London. (Collection of Mervyn Emms)

English naval officer's saber, c. 1800–1820, with 32 inch blued and gilded blade. Cipher of George III on the blade. (Collection of Mervyn Emms)

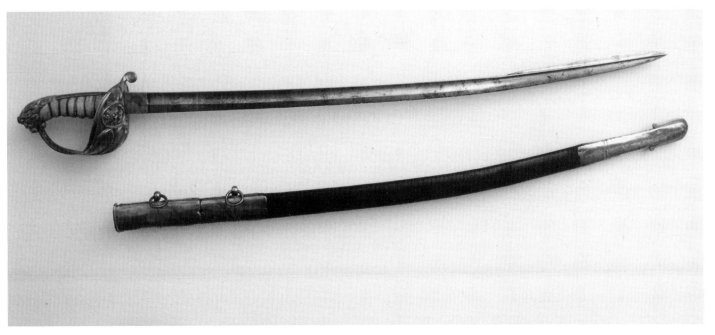

English pattern 1847 naval officer's sword. Retailer: Lawrence Phillips, London.
(Collection of H. Coetzee; Rolf Duffur, photographer)

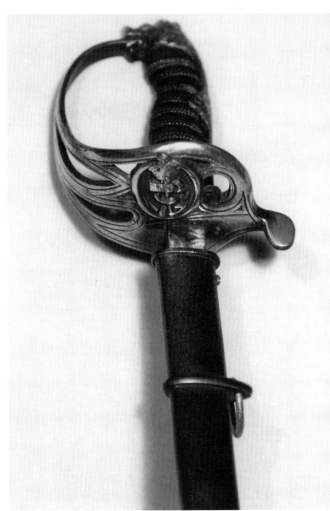

English naval officer's saber, c. 1870, with 26 inch blade.
(Jan Zajac collection)

English pattern 1847 naval officer's sword, c. 1900, with 31 1/2 blade.
Maker: Seagrove & Co., Portsea. (Jan Zajac collection)

# English Royal Company of Archers Swords

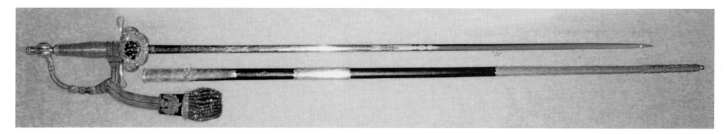

English Royal Company of Archers sword, c. 1830–1850.
(Collection of Ulrich Dubi)

English Royal Company of Archers sword, c. 1830, king's bodyguard in Scotland. (C.M. Lewer-Allen collection)

Hilt of Royal Company of Archers sword, c. 1830–1850.
(Collection of Ulrich Dubi)

English Royal Company of Archers sword, c. 1840–1860, queen's bodyguard in Scotland. (C.M. Lewer-Allen collection)

English Royal Company of Archers sword, c. 1910–1930, with 31 1/4 inch blade with cipher of George V. (Jan Zajac collection)

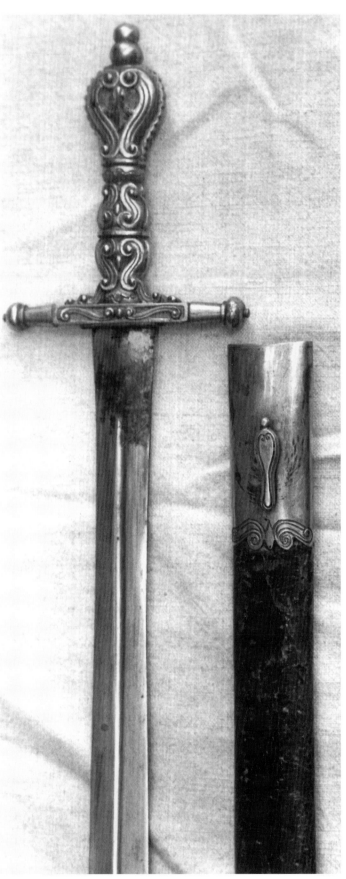

English Royal Company of Archers sword, c. 1860. (James Forman collection)

# English Drummer's Sword

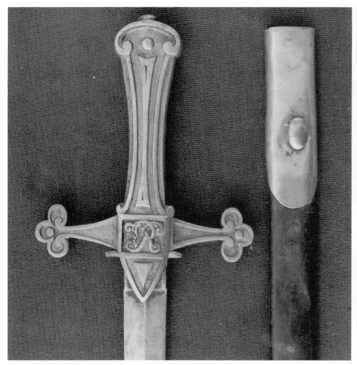

English drummer's sword, pattern 1856 Mark I, with brass hilt and 20 inch blade. (Collection of Mervyn Emms)

English drummer's sword, pattern 1895, with brass hilt and 13 inch blade. (Collection of Mervyn Emms)

# English Small Swords

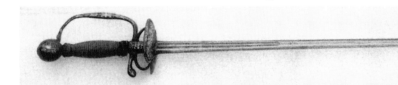

English silver-hilted small sword, c. 1750, with 27 3/4 inch blade. Blade maker: Tomas de Ayala, Toledo, Spain. (James Forman collection)

English silver-hilted small sword, c. 1770, with 31 3/4 inch Colichemarde blade. Maker: William Kersill, London. (James Forman collection)

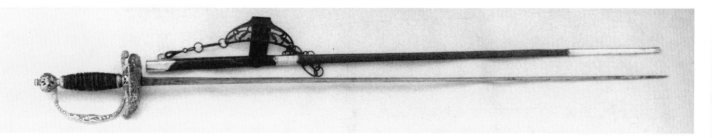

English silver-hilted small sword, c. 1770, with 31 inch blade. Maker: George Fayle, London. (James Forman collection)

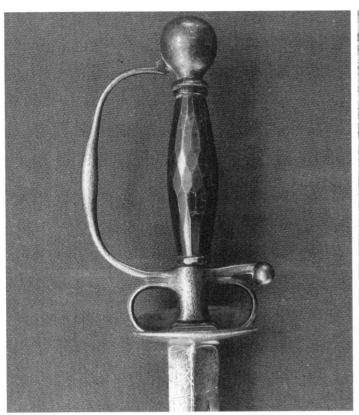

English small sword, c. 1720, with steel hilt and 28 inch blade. (Collection of Mervyn Emms)

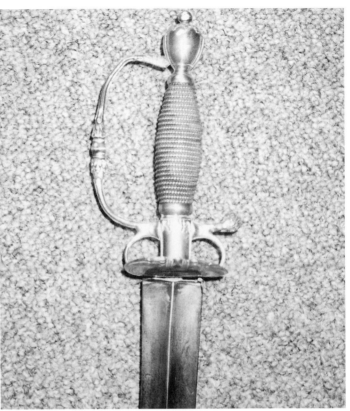

English silver-hilted small sword with 29 inch Colichemarde blade and assay marks of 1722. Maker: George Wickes, London. (G.H. Cook collection)

Hilt of silver-hilted small sword, c. 1750, on p. 352. (James Forman collection)

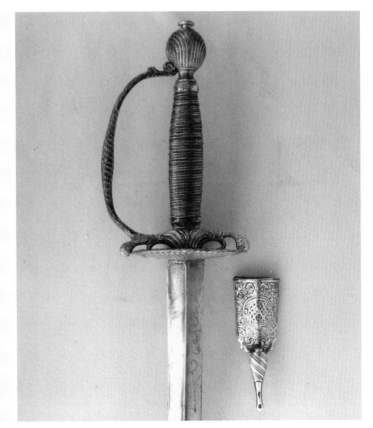

English silver-hilted small sword, c. 1760, with Colichemarde blade. Maker: Charles Bibb, London. (Collection of Mervyn Emms)

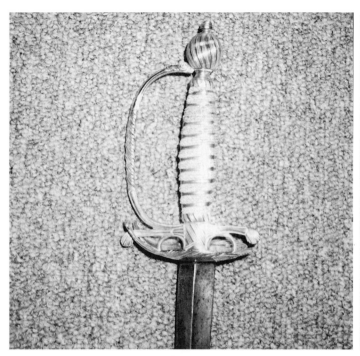

English silver-hilted small sword, c. 1770, with 33 1/4 inch Colichemarde blade. Maker: William Kersill, London. (G.H. Cook collection)

Hilt of English silver-hilted small sword by William Kersill, c. 1770, on p. 352. (James Forman collection)

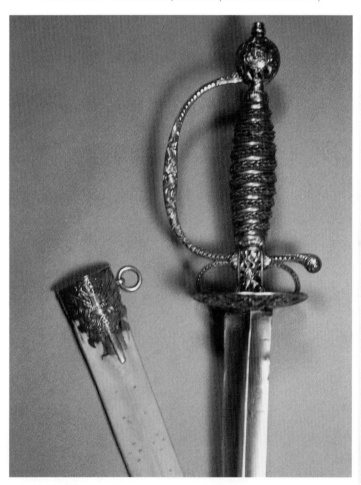

English silver-hilted naval officer small sword, c. 1750, with Neptune motif on hilt, Colichemarde blade, and sheepskin scabbard. Maker: William Loxham, London. (Sheperd Paine collection)

Hilt of silver-hilted small sword by George Fayle, c. 1770, on p. 352. (James Forman collection)

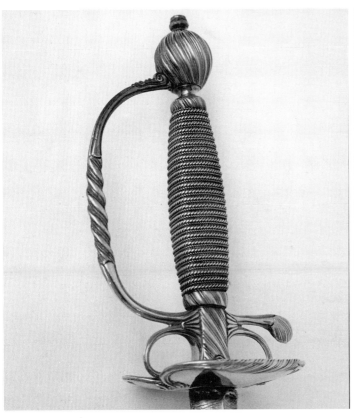

English infantry officer's silver-hilted small sword, c. 1780.
(C.M. Lewer-Allen collection)

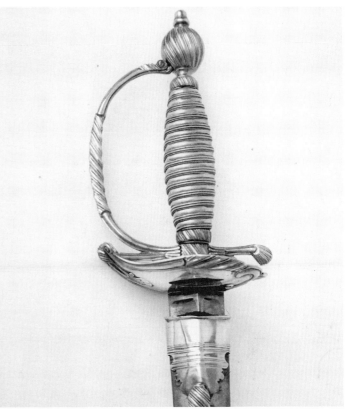

English silver-hilted small sword, c. 1780, with
Colichemarde blade. (C.M. Lewer-Allen collection)

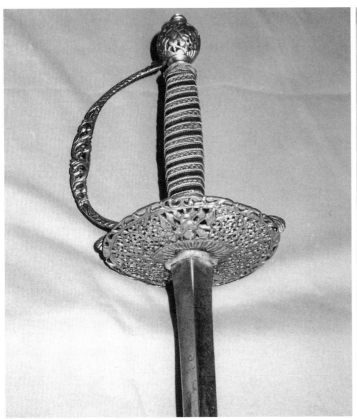

English small sword, c. 1780, with chiseled steel hilt, silver grip wire,
and 32 inch hollow ground blade. (Collection of Sam Saladino)

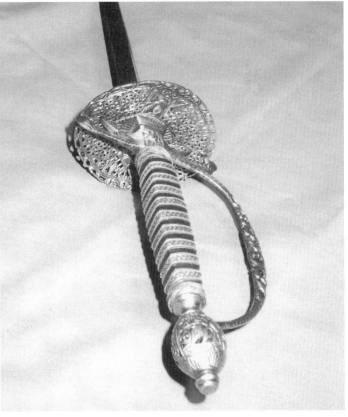

Hilt of chiseled-steel small sword.
(Collection of Sam Saladino)

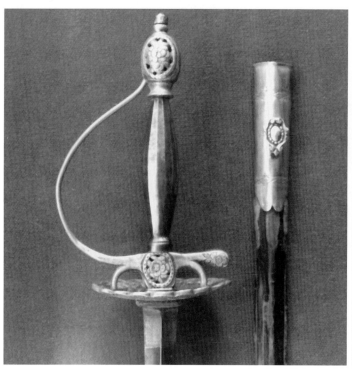

English small sword, c. 1780, with 34 inch blued and gilded blade.
(Collection of Mervyn Emms)

English small sword, c. 1780, with cut steel hilt. Maker: J.B. Wells, London.
(Collection of Mervyn Emms)

# English Court Swords

English regimental court sword, c. 1830, 62nd Foot.
(C.M. Lewer-Allen collection)

English court sword, c. 1840, with cut steel hilt.
Retailer: Stephen M. Silver & Co., London.
(Collection of Mervyn Emms)

English court sword, c. 1910, with 31 1/2 inch blade and Edward VII cipher on the shell. (Jan Zajac collection)

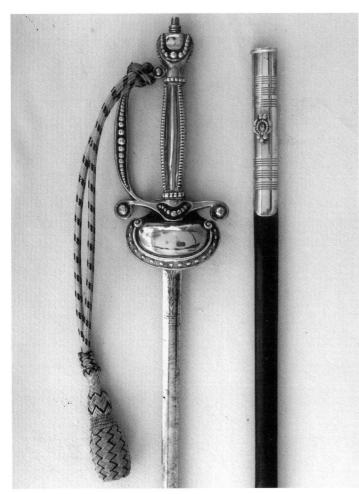

English court sword, c. 1936, with brass hilt and 32 inch blade and Edward VIII cipher on the blade. Retailer: Griffiths, McAlister, London. (Collection of Mervyn Emms)

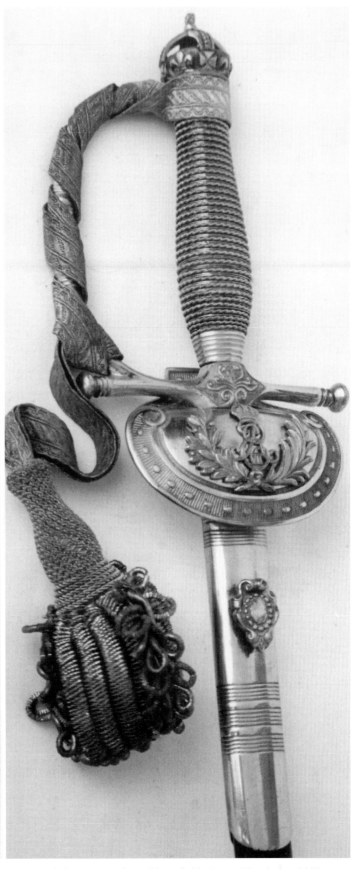

English court sword, royal household, George V period, c. 1920. (C.M. Lewer-Allen collection)

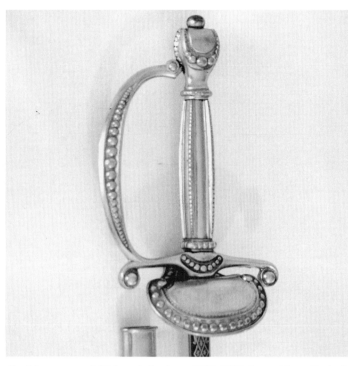

English court sword, Diplomatic Service, c. 1930. (C.M. Lewer-Allen collection)

English court sword, city of London sheriff, c. 1930. (C.M. Lewer-Allen collection)

English court sword, royal household, c. 1930. (C.M. Lewer-Allen collection)

# Miscellaneous English Swords

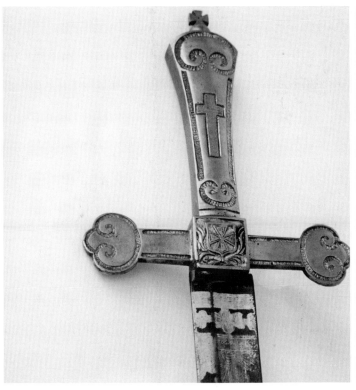

English Military Knights of Windsor Sword.
(C.M. Lewer-Allen collection)

English Royal Army Medical Corps officer's sword, pattern 1892.
(C.M. Lewer-Allen collection)

English Veterinary Corps officer's sword, South African Regiment, c. 1810.
(C.M. Lewer-Allen collection)

English infantry officer's hunting sword, c. 1776.
(C.M. Lewer-Allen collection)

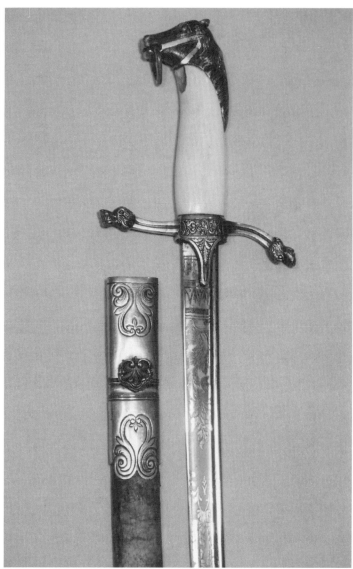

English Tower of London yeoman warder's sword, c. 1727–1760, with 35 5/8 inch blade. The gilt copper hilt has the mark of the House of Hanover on the guard. (C.M. Lewer-Allen collection)

English child's sword, c. 1840, with horsehead pommel, ivory grip, and 20 1/4 inch blade. Maker: H. Kettle, London. (Jan Zajac collection)

# Scottish Sword Photo Section

## Scottish Hand-and-a-Half Sword

Scottish hand-and-a-half broadsword, c. 1525, with a narwhale grip and 40 inch blade. (James Forman collection)

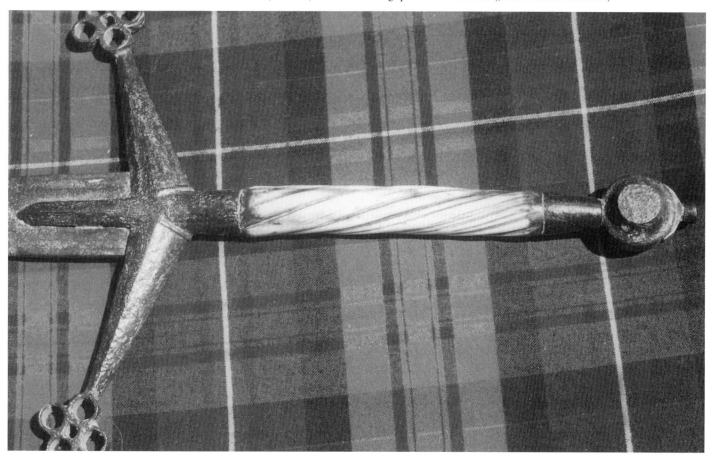

Hilt of the hand-and-a-half broadsword. (James Forman collection)

# Scottish Two-Handed Sword

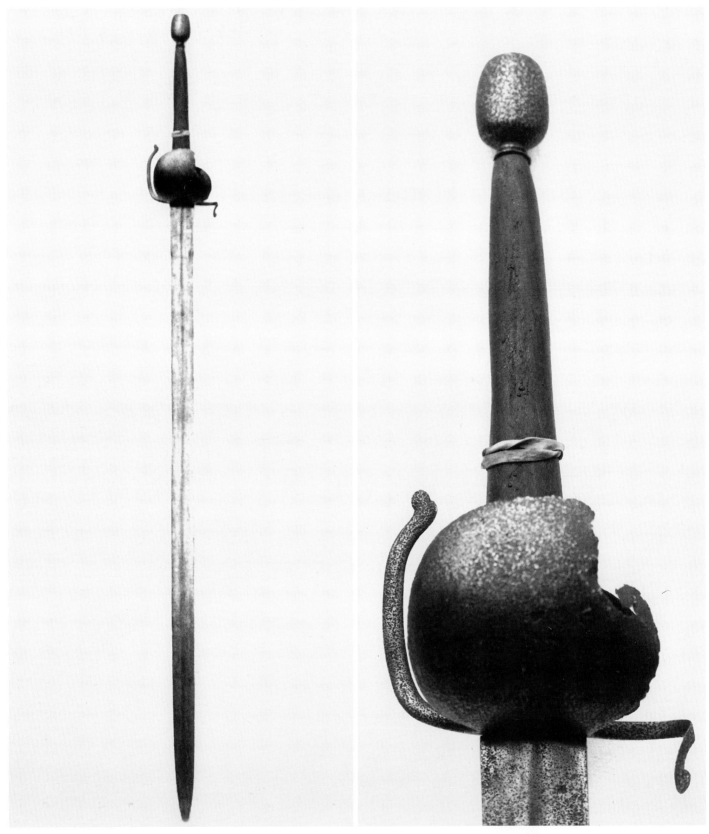

Scottish two-handed broadsword, c. 1600, with clamshell guard and 39 1/2 inch blade. (James Forman collection)

Hilt of the two-handed broadsword. (James Forman collection)

# Scottish Ribbon Hilt Swords

Scottish ribbon hilt broadsword, c. 1630–1650, with counter-curved quillons. The 32 inch blade is marked FERRARA. (Courtesy of Patrick Tougher)

Scottish ribbon hilt broadsword of West Highlands, c. 1660–1670, with 28 1/2 inch blade.
(James Forman collection)

Hilt of West Highlands broadsword. (James Forman collection)

Scottish ribbon hilt broadsword, c. 1600–1630, with an iron hilt and overall length of 38 1/4 inches. (Geoffrey Jenkinson collection)

Scottish ribbon hilt broadsword, c. 1630–1650, with iron hilt and fish-skin grip with silver wire. (Geoffrey Jenkinson collection)

Scottish ribbon hilt broadsword, c. 1665–1675, with iron hilt and overall length of 41 3/16 inches. Blade is marked ANDREA FARARA. (Geoffrey Jenkinson collection)

Scottish ribbon hilt broadsword, c. 1665–1675, with iron hilt and fish-skin grip. The French heavy cavalry blade is marked VIVE LE ROY ET MON SINNYOR LE DU. Sword slipper/maker: Thomas Gemmill (T.G.), Glascow. (Geoffrey Jenkinson collection)

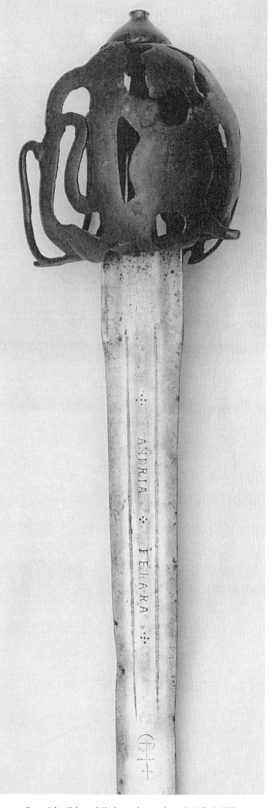

Scottish ribbon hilt broadsword, c. 1665–1675. The blade is marked ANDRIA FERARA. (Courtesy of Don F. Hamilton)

# Scottish Horseman's Swords

Scottish horseman's backsword, c. 1620–1640, with McAllister hilt and 35 1/2 inch blade. (James Forman collection)

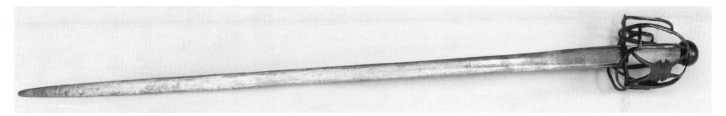

Scottish horseman's broadsword, c. 1640–1660, with bone grip and long 37 1/2 inch blade. (Courtesy of Patrick Tougher)

Scottish horseman's broadsword, c. 1660. The 35 1/2 inch blade is marked FERRARA. (Courtesy of Patrick Tougher)

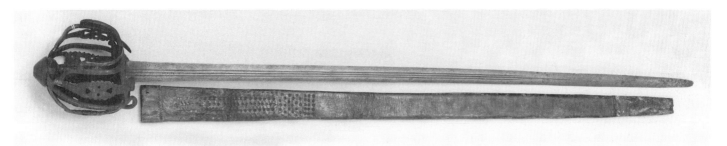

Scottish horseman's broadsword, c. 1725–1740, with 35 1/2 inch blade. (Courtesy of Patrick Tougher)

Scottish horseman's broadsword, c. 1710–1720, with Edinburgh-type hilt. The 35 inch blade is marked ANDREA FERRARA. (Courtesy of Patrick Tougher)

# Scottish Horseman's Swords for Officers

Scottish horseman's backsword for officers, c. 1690–1715, with iron hilt with horseman's ring. Overall length is 37 1/4 inches. Sword slipper/maker: John Simpson the Elder, Glascow (mark: I.S. over G). (Geoffrey Jenkinson collection)

Hilt of the John Simpson the elder's broadsword for officers, c. 1770. (James Forman collection)

Scottish horseman's backsword for officers, c. 1710–1720, with iron hilt and overall length of 40 1/16 inches. Sword slipper/maker: Thomas Gemmill, Glasgow (mark: T.G. over G.). (Geoffrey Jenkinson collection)

Scottish horseman's broadsword for officers, c. 1700, with 36 inch blade. Sword slipper/maker: John Simpson the Elder, Glascow. (James Forman collection)

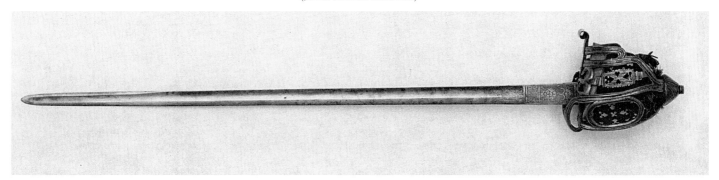

Scottish horseman's backsword for officers, c. 1715–1725. Sword slipper/maker: John Simpson the Younger, Glascow. (Courtesy of Don F. Hamilton)

Hilt of the John Simpson the Younger's horseman's backsword for officers, c. 1715–1725, on p. 366.
(Courtesy of Don F. Hamilton)

Scottish horseman's backsword for officers, c. 1730–1750, with iron hilt and black fish-skin grip. Overall length is 39 9/16 inches.
(Geoffrey Jenkinson collection)

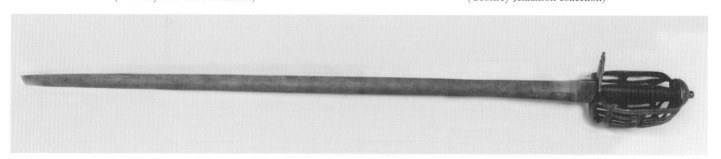

Scottish half-basket horseman's broadsword for officers, c. 1740–1750. Sword slipper/maker: Walter Allen, Stirling.
(Courtesy of Patrick Tougher)

Scottish horseman's broadsword for officers (English type), c. 1760–1780, with basket hilt with horseman's ring.
The 32 1/4 inch blade is marked FERRARA. (Courtesy of Patrick Tougher)

# Scottish Infantry Swords

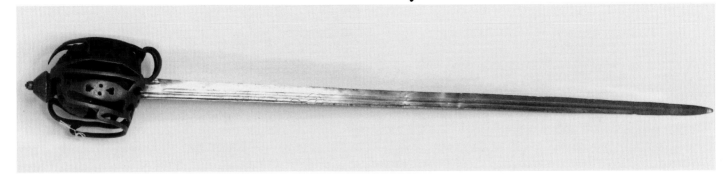

Scottish infantry broadsword, c. 1710, with 34 1/2 inch blade and sharkskin grip.
(Courtesy of Patrick Tougher)

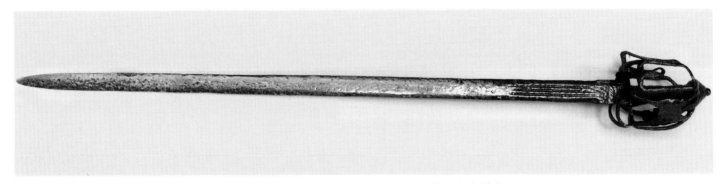

Scottish infantry broadsword, c. 1730–1740, with 34 inch blade.
(Courtesy of Patrick Tougher)

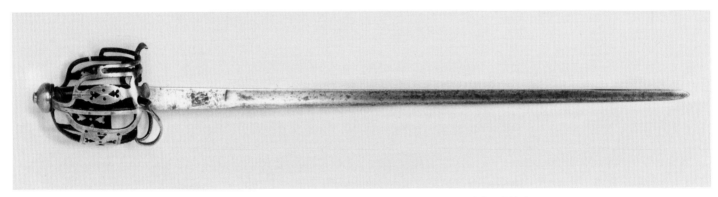

Scottish infantry broadsword, c. 1750–1760, with 33 inch wide-fullered blade.
(Courtesy of Patrick Tougher)

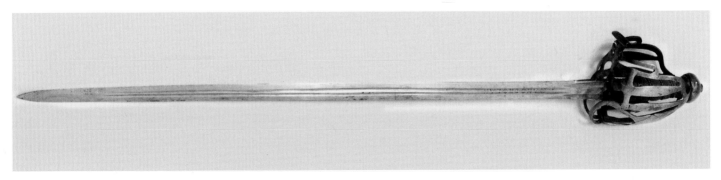

Scottish infantry broadsword, c. 1770, with 32 1/2 inch blade marked FERRARA.
(Courtesy of Patrick Tougher)

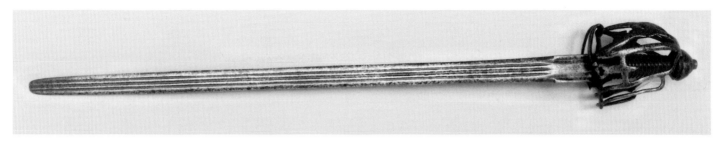

Scottish infantry broadsword, c. 1630–1640, with a wide 28 3/4 inch blade with fullers running the length of the blade.
(Courtesy of Patrick Tougher)

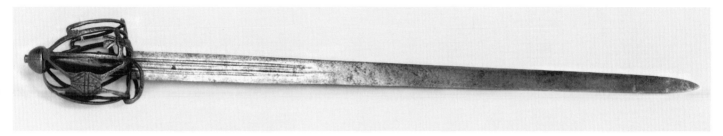

Scottish infantry broadsword, c. 1640, with wide 34 x 1 1/2 inch blade marked with a crown.
(Courtesy of Patrick Tougher)

Scottish infantry broadsword, c. 1670. The 30 inch blade is marked FERRARA.
(Courtesy of Patrick Tougher)

Scottish infantry broadsword, c. 1690, with 33 1/2 inch blade.
(Courtesy of Patrick Tougher)

Scottish infantry broadsword, c. 1710, with Edinburgh-type basket hilt and 34 1/2 inch blade marked FERRARA.
(Courtesy of Patrick Tougher)

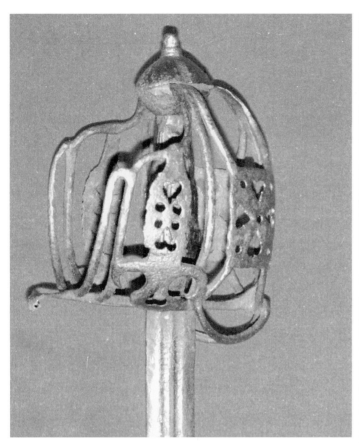

Scottish infantry broadsword, c. 1710. The 34 1/2 inch blade is marked ANDREA FARARA. (G.H. Cook collection)

Scottish infantry broadsword, c. 1725, with wood grip and 32 1/4 x 1 9/16 inch blade. (G.H. Cook collection)

Scottish infantry officer broadsword, c. 1690, with 34 1/2 inch blade. (James Forman collection)

# Scottish Infantry Officers Broadswords

Scottish infantry officer's broadsword, c. 1690–1710, with 32 incfh blade. (Courtesy of Patrick Tougher)

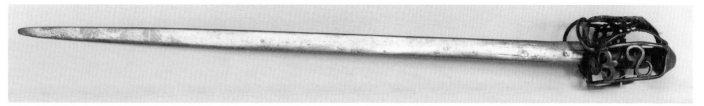

Scottish infantry officer's broadsword, c. 1690–1710, with S-bar basket hilt. The 30 1/2 inch blade with a king's head mark. Maker: Johannes Wundes III, Solingen, Germany. (Courtesy of Patrick Tougher)

Scottish infantry officer's broadsword, c. 1700–1710, with Edinburgh-style basket hilt and 32 inch blade marked FERRARA. (Courtesy of Patrick Tougher)

Scottish infantry officer's broadsword, c. 1700–1715, with iron hilt. Sword slipper/maker: John Simpson the Elder, Glascow (mark: I.S. over G). (Geoffrey Jenkinson collection)

Scottish infantry officer's broadsword, c. 1620, with 33 inch blade. (James Forman collection)

Scottish infantry officer's backsword, c. 1710. The 34 inch German blade is marked PROSPERITY TO SCHOTLAND AND NO UNION. (James Forman collection)

Scottish infantry officer's broadsword, c. 1710, with 29 1/2 inch blade marked PROSPERITY TO SCHOTLAND AND NO UNION. Sword slipper/maker: John Allan, Stirling. (James Forman collection)

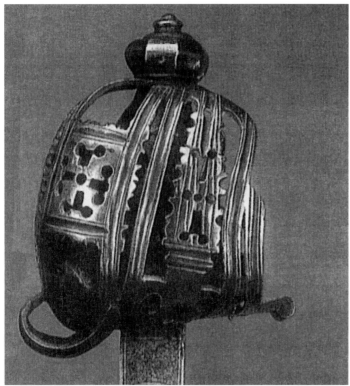

Scottish infantry officer's backsword, c. 1710, with brass hilt and white fishskin grip. Blade is marked on its outer side VIVAT JACABUS TERTIUS MAGNA BRITARMIA REX AS and on its inner side with WITH THIS GOOD SWORD THY CAUSE I WILL MAINTAIN AND FOR THY SAKE O JAMES WILL BREATHE EACH VIEN. Overall length is 39 5/8 inches. (Geoffrey Jenkinson collection)

Scottish infantry officer's broadsword, c. 1720, with a 33 inch cut-down blade. (G.H. Cook collection)

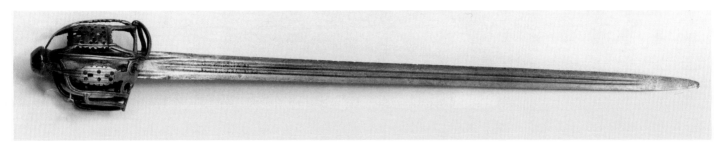
Scottish infantry officer's broadsword, c. 1700–1715, with Edinburgh-style basket hilt and 33 inch blade.
(Courtesy of Patrick Tougher)

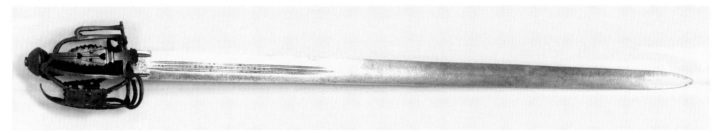
Scottish infantry officer's broadsword, c. 1710–1720.
(Courtesy of Patrick Tougher)

Scottish infantry officer's broadsword, c. 1710–1725. Its 33 1/4 inch blade is marked FERRARA.
(Courtesy of Patrick Tougher)

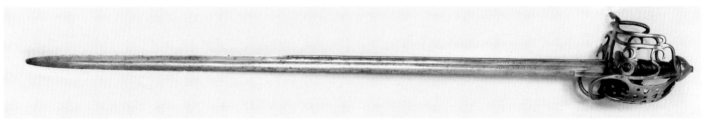
Scottish infantry officer's backsword, c. 1710–1725, with 34 inch blade marked FERRARA.
(Courtesy of Patrick Tougher)

Scottish infantry officer's broadsword, c. 1720–1730, with checkered fruitwood grip and 32 inch blade.
(Courtesy of Patrick Tougher)

Scottish infantry officer's broadsword, c. 1720, with gilt brass hilt and white fish-skin grip. The blade is marked ANDREIA FERRARA. Overall length is 38 1/4 inches. (Geoffrey Jenkinson collection)

Scottish infantry officer's broadsword, c. 1720, with gilt brass hilt. The blade is marked FERARA and SAHAGUN EL VIEJO. Overall length is 39 7/8 inches. Maker: Alonzo de Sahagun the Elder, Toledo, Spain. (Geoffrey Jenkinson collection)

Scottish infantry officer's broadsword, c. 1720, with gilt brass hilt and black fish-skin grip. The blade is marked ANDRI FERARA. (Geoffrey Jenkinson collection)

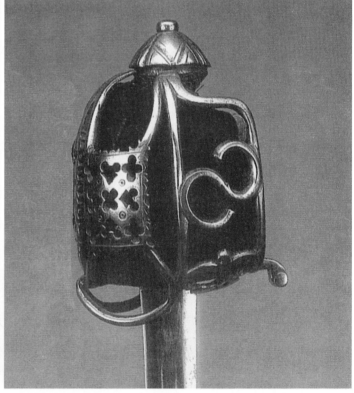

Scottish infantry officer's broadsword, c. 1720–1730, with black fish-skin grip. The blade is marked (AND)RIA. Overall length is 38 5/16 inches. (Geoffrey Jenkinson collection)

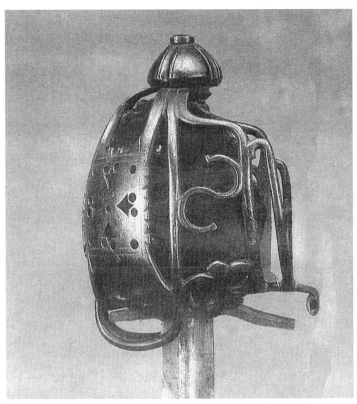

Scottish infantry officer's broadsword, c. 1720–1730, with iron hilt and overall length of 39 7/16 inches. The blade is marked ANDREIA FARARA. (Geoffrey Jenkinson collection)

Scottish infantry officer's broadsword, c. 1720–1730, with iron hilt. The blade is marked ANDRIA FARARA. Overall length is 41 3/4 inches. Sword slipper/maker: John Allan the Elder, Stirling (mark: J.A. over S.). (Geoffrey Jenkinson collection)

Scottish infantry officer's backsword, c. 1720–1740, with iron hilt and grey fish-skin grip. The blade is marked (A)NDR(EA) (F)ARA(RA) and, in script, L. over L.G. Overall length is 38 1/8 inches. (Geoffrey Jenkinson collection)

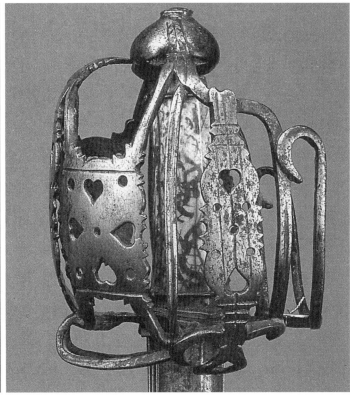

Scottish infantry officer's backsword, c. 1720–1740, with iron hilt and marbled bone grip. The blade is marked ANDRIA FARARA and has the king's head mark of Peter Wundes III, Solingen, Prussia. Overall length is 39 11/16 inches. (Geoffrey Jenkinson collection)

Scottish infantry officer's broadsword, c. 1720–1740, with iron hilt. Blade has the king's head mark of Peter Wundes III, Solingen, Germany. Overall length 38 1/4 inches. (Geoffrey Jenkinson collection)

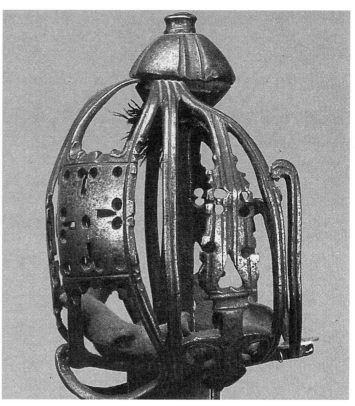

Scottish infantry officer's broadsword, c. 1725–1740, with iron hilt and an overall length of 39 5/16 inches. The blade is marked ANDREA FARARA and has the king's head mark of Peter Wundes III, Solingen, Germany. (Geoffrey Jenkinson collection)

Scottish infantry officer's broadsword, c. 1725–1750, with iron hilt and black fish-skin grip. The blade is marked ANDREA FERARA. Overall length is 38 1/8 inches. (Geoffrey Jenkinson collection)

Scottish infantry officer's broadsword, c. 1725–1750, with iron Stirling-type hilt and grey fish-skin grip. Blade is marked ANDRIA FARARA. Overall length is 39 13/32 inches. (Geoffrey Jenkinson collection)

Scottish infantry officer's backsword, c. 1725–1750, with iron Stirling-type hilt and white fish-skin grip. Blade is marked ANDRIA FARARA. Overall length is 40 15/16 inches. (Geoffrey Jenkinson collection)

Scottish infantry officer's backsword, c. 1725–1750, with iron hilt and overall length of 39 15/16 inches. Blade is marked ANDREA FARARA and has the black Moor's head mark of Peter Wundes III, Solingen, Prussia. (Geoffrey Jenkinson collection)

Scottish infantry officer's broadsword, c. 1730, with gilt brass hilt and overall length of 41 1/8 inches. The German blade is marked with a crowned AR for Augustus I, Elector of Saxony. (Geoffrey Jenkinson collection)

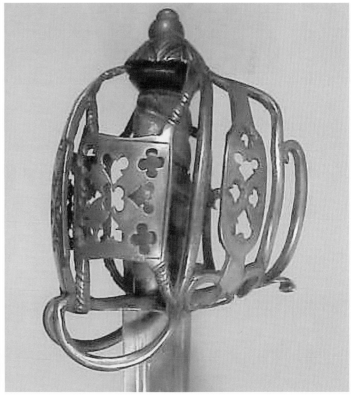

Scottish infantry officer's broadsword, c. 1730, with sharkskin grip. The 28 inch blade is marked ANDREA FERARA. (Jan H. Zajac collection)

Blademark of the c. 1730 Scottish broadsword on p. 377.
(Jan H. Zajac collection).

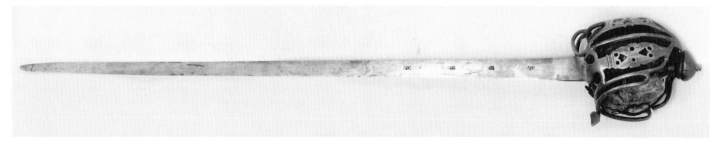

Scottish infantry officer's broadsword, c. 1730. It has an earlier blade marked with the king's head mark of Johunnes Wundes the Younger (1630–1685).
(Courtesy of Patrick Tougher)

Scottish infantry officer's broadsword, c. 1730–1740. Sword slipper/maker: John Allan the Elder, Stirling.
(Courtesy of Patrick Tougher)

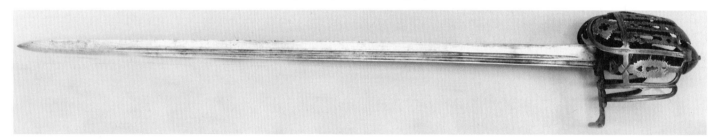

Scottish infantry officer's backsword, c. 1730–1760, with 31 1/2 inch blade. Sword slipper/maker: Walter Allan, Stirling.
(Courtesy of Patrick Tougher)

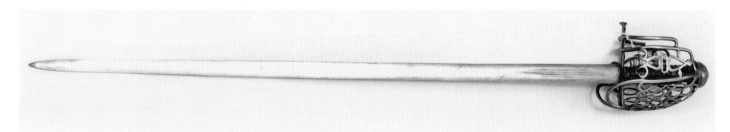

Scottish infantry officer's broadsword, c. 1740–1760, with 34 1/2 inch blade marked FERRARA.
The pommel has been replaced. Sword slipper/maker: Walter Allan, Stirling. (Courtesy of Patrick Tougher)

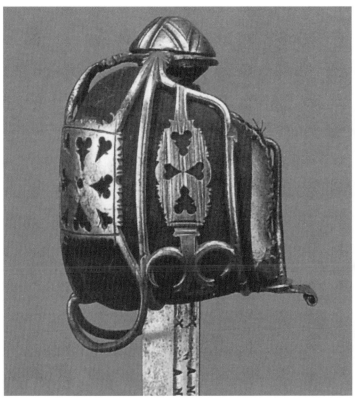

Scottish infantry officer's backsword, c. 1740–1760, with iron hilt and overall length of 37 1/2 inches. Blade is marked ANDRIA FARARA. Sword slipper/maker: Walter Allan, Stirling. (Geoffrey Jenkinson collection)

Scottish infantry officer's backsword, c. 1740–1760, with iron hilt. The blade is marked ANDRIA FARARA. Overall length: 38 inches. Sword slipper/maker: Walter Allan, Stirling (mark: W.A. over S). (Geoffrey Jenkinson collection)

Scottish infantry officer's backsword, c. 1740–1760, with iron hilt and black fish-skin grip. Blade is marked ANDRIA FARARA. Sword slipper/maker: Walter Allan, Stirling (mark: W.A. over S). (Geoffrey Jenkinson collection)

Scottish infantry officer's backsword, c. 1740–1760, with iron hilt and overall length of 38 21/32 inches. The blade is marked ANDREA FERARA. Sword slipper/maker: Walter Allan, Stirling (mark: W.A. over S). (Geoffrey Jenkinson collection)

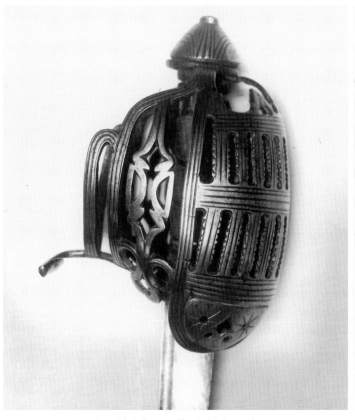

Scottish infantry officer's backsword, c. 1740, with 33 1/2 inch blade. (James Forman collection)

Scottish infantry officer's broadsword, c. 1740–1750, with 35 inch blade. (James Forman collection)

Scottish infantry officer's broadsword, c. 1740–1750. (James Forman collection)

Scottish infantry officer's backsword with 32 inch blade. (James Forman collection)

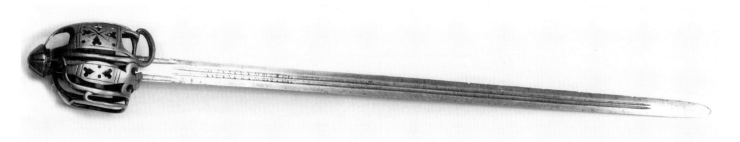

Scottish infantry officer's broadsword, c. 1740. The 33 1/4 inch blade is marked ANDREA FERRARA.
(Courtesy of Patrick Tougher)

Scottish infantry officer's backsword, c. 1740–1750, with brass hilt and 32 inch blade.
(James Forman collection)

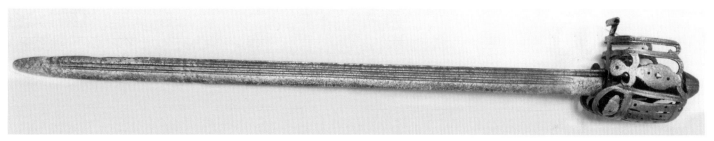

Scottish infantry officer's backsword, c. 1750–1760. The 32 inch blade is marked FERRARA.
(Courtesy of Patrick Tougher)

Scottish infantry officer's backsword, c. 1760, with 31 1/2 inch blade.
(Courtesy of Patrick Tougher)

Scottish infantry officer's broadsword, c. 1790, with thistle-pierced basket and 34 inch blade.
(Courtesy of Patrick Tougher)

SCOTTISH SWORD PHOTO SECTION

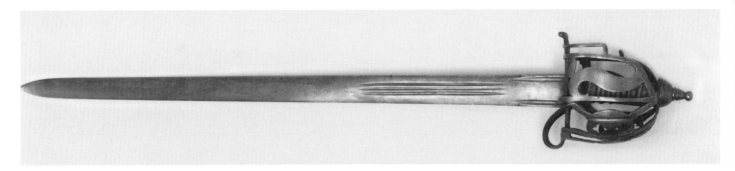

Scottish infantry officer's broadsword, c. 1800, with a wide 30 1/4 x 1 5/8 inch blade.
(Courtesy of Patrick Tougher)

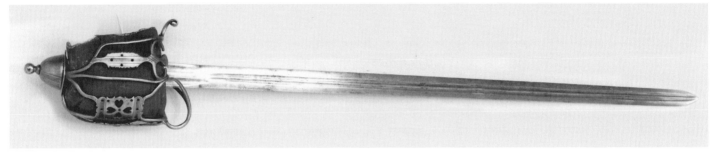

Scottish infantry officer's broadsword, c. 1815–1830, with 31 inch blade. Sword slipper/maker: John Macleod, Edinburgh.
(Courtesy of Patrick Tougher)

## Scottish Musician's Sword

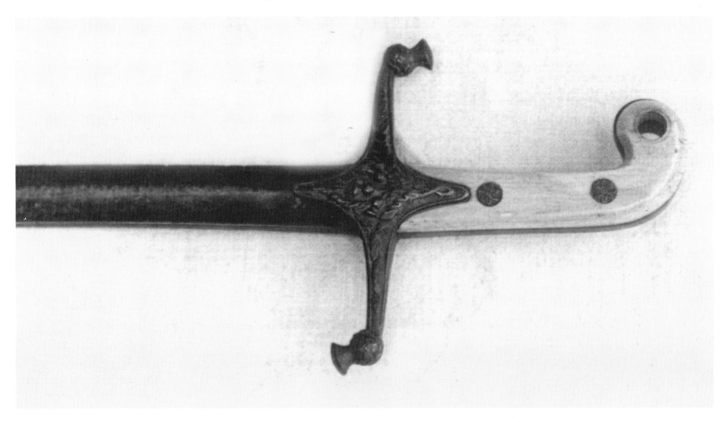

Scottish musician's sword, c. 1850, with Mameluke hilt and 27 inch blade.
(James Forman collection)

# Scottish Silver-Hilted Swords

Scottish silver-hilted horseman's broadsword with 35 inch blade. Hall marked 1715–1716 (silversmith's date mark). Maker: Colin Mackenzie, Edinburgh. (James Forman collection)

Scottish silver-hilted infantry officer backsword with 34 inch blade marked 1743 W.M. and LONG LIVE PRINCE CHARLES. (James Forman collection)

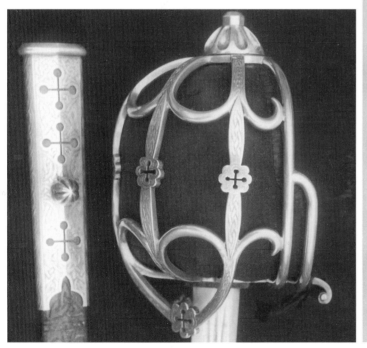

Scottish silver- and basket-hilted broadsword, c. 1884, with 34 inch blade. (James Forman collection)

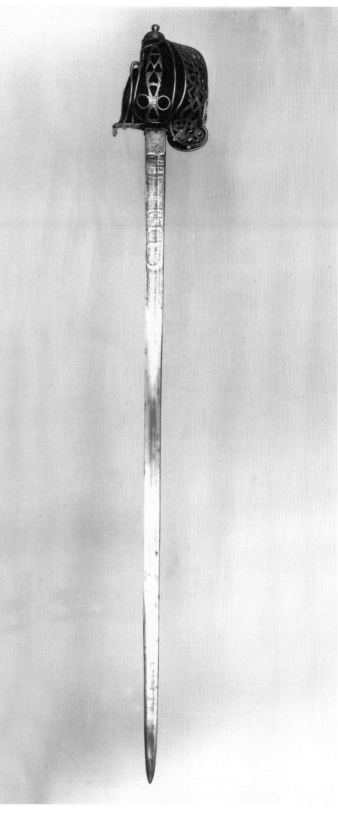

Scottish silver-hilted broadsword of Scottish nobleman Kennedy of Garvan Mains. Blade marked PROSPERITY TO SCHOTLAND GOD SAVE KING JAMES. Hilt c. 1875–1890, blade c. 1807. (James Forman collection)

# Scottish Family Swords

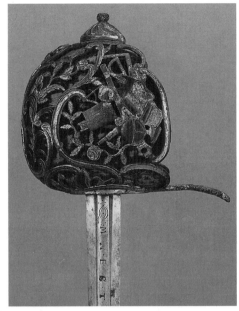

Basket-hilted broadsword of William Murray, Marguis of Tullibardine and Duke of Atholl, c. 1720. Features a chiseled and pierced iron basket hilt, russeted and fire gilded. The black leather grip is bound with a herringbone of silver wire. The pasteboard liner is covered with crimson velvet. Blade marked DOMINE MAESTRE. Overall length is 40 11/16 inches. (Geoffrey Jenkinson collection)

Basket hilt of the Robertson clan broadsword showing the Robertson clan family crest. (James Forman collection)

Basket hilt of the Mar family broadsword. (James Forman collection)

Right: Scottish broadsword, c. 1740, with 32 inch blade presented to the Robertson clan by Charles Stuart. (James Forman collection)

Scottish silver-hilted broadsword, c. 1822, presented to the Mar family by Prince Charles. (James Forman collection)

# Scottish Sinclair Swords

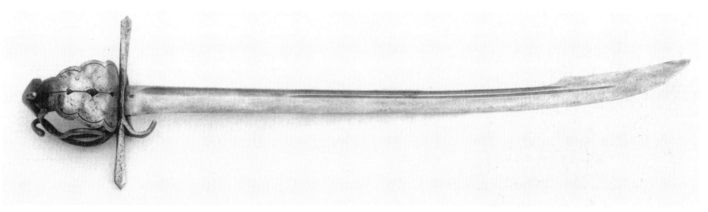

Seventeenth century Sinclair saber with 36 inch blade carried by Scottish mercenaries. (James Forman collection)

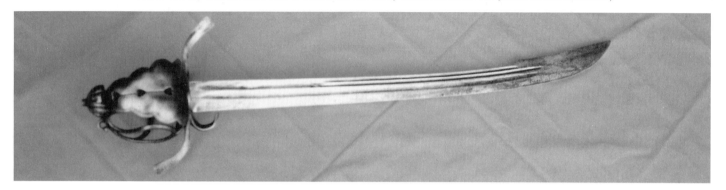

Seventeenth century Sinclair saber with 31 inch blade carried by Scottish mercenaries. The blade has the unicorn mark of sword maker Clemens Horn of Solingen, Prussia. (Sam Saladino collection)

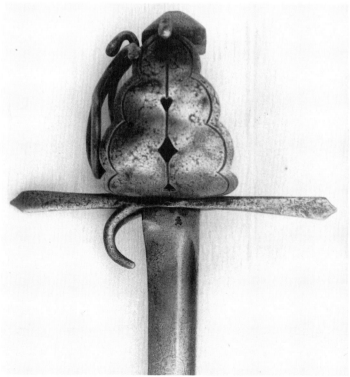

Hilt of the seventeenth century Sinclair saber.
(James Forman collection)

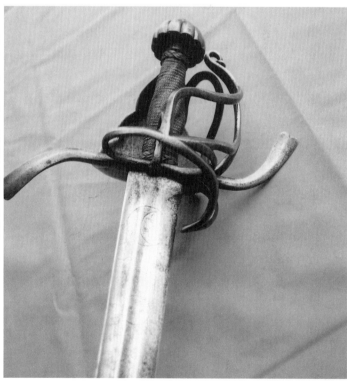

Reverse view of the Clemens Horn Sinclair saber hilt.
(Sam Saladino collection)

# Bibliography and Reference Material on English and Scottish Swords and Sword Makers

**Books and Catalogs**

*American and European Swords in the Historical Collections of the United States National Museum*
by Theodore T. Belote, 1932

*The American Eagle-Pommel Sword: The Early Years, 1794–1830*
by E. Andrew Mowbray, 1988

*Ancient Scottish Weapons*
by James Drummond, 1881

*The Armoury of Windsor Castle*
by Sir Guy Francis Laking, 1904

*Arms and Armour in Tudon Stuart London*
by M.R. Holmes, 1970

*Arms and Armour of the English Civil Wars*
by David Blackmore, 1990

*Battle Weapons of the American Revolution*
by George C. Neumann, 1998

*Blades of Glory: Swords of the Scottish Infantry, 1756–1900*
by Stephen Wood
Keeper of the Castle, Edinburgh Castle
Scottish United Services Museum

*Boarders Away*
by William Gilkerson, 1991

*British and Commonwealth Bayonets*
by Ian D. Skennerton and Robert Richardson, 1986

*British Cut and Thrust Weapons*
by John Wilkinson-Latham, 1971

*British Military Swords*
by John Wilkinson-Latham, 1966

*Burgh Laws of Dundee*
by A.J. Warden, 1872

*A Catalogue of the Sword Collection at York Castle Museum*
by P.R. Newman, 1985

*Cut and Thrust Weapons*
by Edward Wagner, 1967

*The Dewar Manuscript*
by J. MacKenzie, 1963

*Die Geschichte Der Solinger Industrie*
by Franz Hendrich, 1933

*The Dunfermline Hammerman*
by D. Thompson, 1909

*English Cutlery—Sixteenth to Eighteenth Century in the Victoria and Albert Museum*
by J.F. Hayward, 1956

*Europaisehe Hieb and Stich-Waffen*
by Heinrich Muller and Hartmut Kolling, 1986

*European and American Arms*
by Claude Blair, 1962

*European Arms and Armour at Kelvingrove, Glasgow*
by J.G. Scott, 1980

*European Swords (Victoria & Albert Museum)*
by Anthony North, 1982

*European Swords and Daggers in the Tower of London*
by Arthur Richard Duffy (Master of Armouries), 1974

*European Weapons and Armour*
by Ewart Oakshott, 1980

*Five Centuries of Gunsmiths, Swordsmiths, and Armourers, 1400–1900*
by Robert Edward Gardner, 1948

*The Goldsmiths of Aberdeen*
by I.E. James, 1981

*Gunmakers of London (1350–1850)*
by Howard L. Blackmore, 1986

*A Handbook of Court and Hunting Swords, 1660–1820*
by P. Carrington-Peirce, 1937

*Highland Dress, Arms and Ornament*
by Lord Archibald Campbell, 1899

*History of Consett and District (Shotley Bridge)*
by G. Lister, 1946

*History of Highland Dress*
by John Telfer Dunbar, 1962

*History of the Cutlers Company of London and the Minor Cutlery Crafts*
by Charles Welch, 1916 (Vol. I) and 1923 (Vol. II)

*History of the Hammerman of Glascow*
by Harry Lumsden and P. Henderson Aitken, 1912

*The Hollow Sword Blade Company and Sword Making at Shotley Bridge*
(North of England Institute of Mining and Mechanical Engineers, Newcastle Upon Tyne)
by Rhys Jenkins, 1935

*Hunting Weapons*
by Howard L. Blackmore, 1971

*An Introduction to European Swords*
by Anthony R.E. North, 1971

*Jacksons Silver and Gold Marks of England, Scotland and Ireland*
by Ian Pickford, editor, 1996, 3rd edition

*Jacobites of Aberdeenshire and Banffshire in the "45"*
by A. & H. Taylor, 1938

*London Goldsmiths, 1697–1897: Their Marks and Lives*
by Arthur G. Grimwade, 1976

*The Lyle Official Arms & Armour Review*, 1976–1983

*A Military History of Perthshire, 1660 to 1902*
by the Marchioness of Tullibardine

The Museum of Historical Arms catalogs
by Marvin E. Hoffman, 1952–1998

N. Flayderman & Co. Inc. catalogs
(Military and Nautical Antiquities), 1955–1998

*The Naval Officers Sword*
by Capt. Henry T.A. Bosanquet, 1955

*Naval Swords*
by P.G.W. Annis, 1970

*Naval Swords and Firearms*
by Com. W.E. May, R.N., 1962

Old Glascow Exhibition catalogue
Institute of Fine Arts, Glascow, 1894

*Old Irish and Highland Dress*
Dundalk, 1950

*1,000 Marks of European Blademakers*
by Zygmunt S. Lenkiewicz, 1991

*The Perth Hammermen Books*
by C.A. Hunt, 1889

*The Price Guide to Antique Edged Weapons*
by Leslie Southwick, 1982

*The Rapier and Small Sword, 1460–1820*
by A.V.B. Norman, 1980

*Rapiers*
by Eric Valentine, 1968

*Regulation Military Swords*
by J. Wilkinson-Latham, 1970

*Scottish Arms Makers*
by Charles E. Whitelaw, 1977

*Scottish Gold and Silver Work*
by Ian Finlay, 1991

*The Scottish National Dictionary*, 10 volumes, 1931–1975

*Scottish National Memorials*
"A Record of the Historical and Archaelogical Collection of the Bishop's Castle, Glascow"
by James Paton, 1890

*Scottish Swords and Dirks*
by John Wallace, 1970

*The Sheffield Knife Book*
by Geoffrey Tweedale, 1996

Shotley Bridge "Preface" (unpublished paper)
by John G. Bygate

*Small Arms of the Sea Service*
by Col. Robert H. Rankin, USMC, 1972

*Small Arms Makers*
by Col. Robert E. Gardner, 1963

*The Small Sword in England*
by J.D. Aylward, 1945

*Small Swords and Military Swords*
by A.V.B. Norman, 1980

*Sword, Lance, and Bayonet*
by Charles Ffoukles and E.C. Hopkinson, 1938

*Swords and Blades of the American Revolution*
by George C. Neumann, 1973

*Swords and Daggers (Victoria & Albert Museum)*
by John Hayward, 1951

*The Swords and the Sorrows: An Exhibition to Commemorate the Jacobite Rising of 1745 and the Battle of Culloden 1746*
The National Trust for Scotland Trading Company Ltd., 1996

*Swords for Sea Service* (two volumes)
by Com. W.E. May R.N. and P.G.W. Annis, 1970

*Swords for the Highland Regiments, 1757–1784*
by Anthony D. Darling, 1988

*Swords of the British Army*
by Brian Robson, 1996

*Wallace Collection Catoloques (European Arms & Armour)*
(two volumes)
by Sir James Mann, 1962

*Weapons of the American Revolution*
by Warren Moore, 1967

*Wilkinson Sword: A Short History*
by John Arlett

**Miscellaneous Magazine Articles and Publications**

"And So Make A City Here"
1948 Thamason Transactions (paper)
by C.E. Bate

"Andrew Fogelberg and the English Influence on Swedish Steel"
*Appollo*, June 1947
by Charles Oman

"Basketed Hilted Swords of Glascow Make"
*Scottish Review*, Vol. 4, No. 1, 1963
by J.G. Scott

"Beauties and the Beast"
*Arms Collecting*, Vol. 30, No. 2
by James D. Forman

"Benjamin Stone (Hounslow)"
*Dictionary of National Biography*, 1897

"British and Canadian Honeysuckle Hilts"
*Arms Collecting*, Vol. 31, No. 3
by Norman J. Crook

"The British Basket Hilted Cavalry Sword"
*The Canadian Journal of Arms Collecting*, Vol. 7, No. 3
by A.D. Darling

"The British Heavy Cavalry Sword of 1788"
*Classic Arms & Militaria*, Nov. 1994
by Geoff R. Worrall

"The British Infantry Hangers"
*The Canadian Journal of Arms Collecting*
by A.D. Darling

"British Infantry Officer Swords, Patterns 1786–1796"
*Arms Collecting*, Vol.32, No. 4
by David Patten

"British Light Dragoon Sword of 1788, Parts 1 and 2"
*Classic Arms & Militaria*, July and Aug. 1994
by Geoff R. Worrall

"The British Pattern 1908 Cavalry Sword"
*The Canadian Journal of Arms Collecting*
by Norman J. Crook

"British Presentation Swords"
*Connoisseur*, Vol. CLXVI, 1967
by A.V.B. Norman

"The Burgesses and Guild Brethren of Glasgow, 1751–1846"
Scottish Record Society, LXVI, 1935
by J.R. Anderson

"Court and Hunting Swords"
*Antique Collector*, VI, No. 8, Aug. 1935
by P. Carrington-Peirce

"The Earliest Scottish Basket-Hilted Swords"
*Man at Arms*, July/Aug. 1979
by Anthony D. Darling

"The Early Basket Hilt in Britain"
*Scottish Weapons and Fortifications, 1100–1800*
(David H. Caldwell, Editor), 1981
by Claude Blair

"Early Scottish Edged Weapons and Related Militaria"
*American Society of Arms Collectors Bulletin*, Sept. 1987
by Howard Mesnard

"The 1814 Household Cavalry Officers Dress Sword"
*Antique Arms & Militaria*
by Geoff R. Worrall

"English Signed Swords in the London Museums"
*Appollo*, Vol.XXIX, May 1939
by Clement Milward

"English Swords, 1600–1650"
*Arms and Armour Annual*, #1, 1973
by John F. Hayward

"Four Fragments of Grindstones Found in the Ruins of a Sword Mill at Shotley Bridge" (paper)
by R.G. Barclay

"From Rapier to Small Sword"
*Swords and Hilt Weapons*, 1989
by Anthony North

"A Further Note on Scottish Regimental Dirk"
*Arms Collecting*, Vol. 25, No.2
by James D. Forman

"Further Notes on London and Hounslow Swordsmiths"
*Appollo Magazine*, Vol. XXXV, April 1942
by Clement Milward

"The German Sword Makers of Shotley Bridge"
University of Durham
School of Education
Occasional paper #2, 1991
by David Atkinson

"German Swordsmiths in England during the 17th Century"
*The Antique Collector Magazine*, Vol. V, 1934
by C. Trenchard

"Glasgow and the Jacobite Rebellion of 1715"
*Scottish Historical Review*, XIII, No. 50, Jan. 1916
by T.F. Donald

"The Greatest of Basket Hilts"
*Monthly Bugle*, No. 108, Sept. 1978
(The Pennsylvania Antique Gun Collectors Association)
by Howard W. Mesnard

"Highland Broadswords"
*Proceedings of the Society of Antiquaries of Scotland*,
LXXXIV, 1949–1950
by R.L. Hunter

"Highland Weapons at the Royal Academy"
*Connoisseur*, CIII, 1939
by I. Findlay

"The Hollow Sword Blade Company"
Notes and Queries, Literary and Historical Notes
Oxford University Press, Academic Division
Vol. 193, Nos. 18 and 19, Sept. 1948
by J.D. Aylward

"Hounslow Blades and their Makers"
*Arms Fair Spring 1979 Guide* (Royal Lancaster Hotel)
by John Tofts White

"The Hounslow Sword Blade Industry"
*Hounslow Guide to Local Industry and Commerce*, 1978
by John Tofts White

"The Hounslow Swordsmiths"
*The Honeslaw Chronicle*, Vol. 1, No. 2, Sept. 1978
by John Tofts White

"The Infantry Sergeants Sword"
*The Bulletin of the Military Historical Society*, Vol. 5, No. 17, 1954
by J.F.R. Winsbury

"The Jacobite Relics"
*Connoisseur*, LXVII, No. 265, Sept. 19–20
by Charles R. Beard

"Jacobite Sword Blades, Parts I and II"
*The Antique Collector*, Vol. VII, February and April 1936
by C. Trenchard

"The Jenks Family of England"
*The New England Historical and Genealogical Register*,
Vol. CX, 1956
by Meredith B. Colket Jr.

"The John Simpsons of Glasgow"
*Journal of the Scottish Military Collectors Society*,
No. 131, Spring 1993
by A.V.B. Norman

"Joseph Jenckes, Sword Cutler of Hounslow"
Powysland Club, 1938 (British society paper)
by Richard Williams

"Joseph Jenckes and the Hounslow Sword Blade Industry"
*English Civil War Notes and Queries*, Issue 18, 1986
by John Tofts White

"Letters from Walter Allan, Armourer of Stirling,
to Colin Mitchell, Goldsmith of Cannongate"
*Scottish Weapons and Fortifications, 1100–1800*
(David H. Caldwell, Editor), 1981
by Stuart Maxwell

"A Mid-18th Century British Grenadier's Sword"
*Man at Arms*, Nov/Dec 1983
by Anthony D. Darling

"Neo-Classical Ornament and Design
on British Presentation Swords"
*Man at Arms*, July/Aug. 1990
by Lelsie Southwick

"A Note on English Sword Blades"
Notes and Queries, Literary and Historical Notes
Oxford University Press, Academic Division
Vol. 184, No. 11, May 1949
by J.D Aylward

"A Note on Some Hangers Possibly of Scottish Origin"
*Proceedings of the Society of Antiquaries*, CVI
by A.V.B. Norman

"Notes on Swords with Signed Basket Hilts of Glasgow
and Stirling Makers"
Transactions of the Glasgow Archaelogical Society,
New Series, Vol. VIII Supplement, 1934
by Charles E. Whitelaw

"Notice of Armour and Arms at Eqlinton Castle"
Transactions of the Glasgow Archaelogical Society,
New Series, IV, 1899–1901
by Robert Brydall

"Observations on the Article 'Benjamin Stone'"
*The Dictionary of National Biography*, 1897
*New Dictionary of the National Biography*, 1994
by John Tofts White

"Observations on the Dating of Scottish Basket Hilted Swords"
The Seventh Parklane Arms Fair Catalogue, February 1990
by Colin Rolland

"The Oldest Minute Book of Stirling Incorporation of Hammermen, 1596–1621"
Transactions of the Stirling National Historical and Archaeological Society, 1901–1902
by W.C. Cook

"The Origin and Development of the Highland Dirks"
Transactions of the Glasgow Archaeological Society, New Series, Vol. V, 1908
by Charles E. Whitelaw

"The Pipe Backed Heavy Cavalry Officers Sword, circa 1815–1820"
*Classic Arms & Militaria*, Sept./Oct. 1995
by Richard Deller

Proceedings of the Society of Antiquaries of Newcastle upon Tyne
Series 3, Vol. IV, No. 26, 1910

"Proof Marks on Officers Blades"
*The Canadian Journal of Arms Collecting*, Vol. 17, No. 2
by Norman J. Crook

"The Ray & Montaque London Presentation Swords"
*Man at Arms*, July/Aug. 1994
by Leslie Southwick

"Register of Edinburgh Apprentices"
1666–1700 *Scottish Record Society*, LX, 1929
1701–1755 *Scottish Record Society*, LXI, 1919
by Charles B. Boog Watson

"Roll of Edinburgh Burgesses and Guild Brethren, 1701–1760"
*Scottish Record Society*, LXII, 1930
by Charles B. Boog Watson

"Scots or Still English"
*Scottish Art Review*, Special Number (issue)
Scottish Weapons, Vol. IX, 1963
by C. Blair and J.M. Wallace

"Scottish Claymores"
*Man at Arms*, 1998, Volume 1
by James Forman

Scottish Exhibition of National Art and Industry
Catalogue of Exhibits, 2 Volumes (museum exhibit catalog)
Glasgow–Palace of History, 1911

"Scottish Officers Swords, Pattern 1865"
*The Canadian Journal of Arms Collecting*, Vol. 19, No. 4
by Norman J. Crook

"The Scottish Regimental Dirk"
*Arms Collecting*, Vol. 25, No. 1
by James Forman

"Scottish Silver Hilts"
*Man at Arms*, Sept./Oct. 1983
by James Forman

"Selections from the Records of Kirk of Aberdeen"
Spalding Club, Vol. XV, 1846 (British society paper)

"The 1796 Light Cavalry Sword, Parts 1 and 2"
*Classic Arms & Militaria*, Sept. and Oct. 1994
by Geoff R. Worrall

"Seventeenth Century Europe"
*Swords and Hilt Weapons*, 1989
by Anthony North

"A 17th Century Hounslow 'Mortuary' Sword in Gunnerbury Park Museum"
London Middlesex Archeological Society transactions, Vol. 35, 1984
by Phil Philo

"The Shotley Bridge Sword Blade Factory and the First Manufacture of Steel in the North Countrie"
*The Edgar Allen News*, Vol. 22, No. 253, June 1943
by R.N. Appleby-Miller

"The Shotley Bridge Swordmakers—Their Strange History"
*Northern History Booklet*, No. 37, 1973
by Richard Richardson

"Some 18th Century Civilian Swords"
*Scottish Art Review*, Vol. XII, No. 2, 1969
by A.V.B. Norman

"Some Notes on the Scottish Army in the First Half of the Sixteenth Century"
*Scottish Historical Review*, XXVIII
by H. Dickinson

"Staff Weapons of the British Army"
*The Canadian Journal of Arms Collecting*, Vol. 9, No. 1
by Anthony D. Darling

"Standardization of British Army Swords"
*Journal of the Society for Army Historical Research*, Vol. 39, 1961
by Brian Robson

"The Story of a Sword"
*Appollo*, Vol. 194
by J.D. Aylward

"The Strange Case of the Hollow Sword Blade Company"
*The Edgar Allen News*, Vol. 41, No. 476, Feb. 1962
and Vol. 41, No. 477, March 1962
by J.D. Aylward

"The Stylish Evolution of the Scottish Dirk"
*Arms Collecting*, Vol. 35, No. 3
by James D. Forman

"Survey of the Hundred of Isleworth"
(Hounslow Heath included)
The Syon Map, 1635
by Moses Glover

"The Sword Blade Makers at Hounslow Sword Mill, Part I"
*The Honeslaw Chronicle*, Vol. 3, No. 2, Sept. 1980
by John Tofts White

"The Sword Blade Makers at Hounslow Sword Mill, Part II"
*The Honeslaw Chronicle*, Vol. 6, No. 1, 1983
by John Tofts White

"A Sword from Cromwell and Fairfax"
*The Canadian Journal of Arms Collecting*, Vol. 12, No. 3
by James D. Forman

"Swords of Honour and Glory"
*American Society of Arms Collectors Bulletin*, Sept. 1979
by Peter Dale

"Swords of the Scottish Highlanders"
*Military Modeling*, Dec. 1996
by Major A. Logan-Thompson

"Thomas Gill: Improvements to the Manufacture of Sword Blades, Parts 1 and 2"
*Classic Arms & Militaria*, Feb.–Mar. 1994

"Walter Allan—Armourer of Stirling"
*Scottish Review*, Vol. 9, No. 1, 1963
by William Reid

"Weapons of the British Dragoon (1750)"
*The Gun Report*, Oct. 1972
by Anthony D. Darling

"Weapons of the Highland Regiments, 1740–1780"
*The Canadian Journal of Arms Collecting*, Vol. 8, No. 3
by Anthony D. Darling

### Articles from the *Journal of the Arms & Armour Society*

Volume I, No. 6
"A Royal Cutler's Bill of 1547"
by C. Blair

Volume IV, No. 8
"The 5 Ball Type of Sword Hilt"
by W.E. May

Volume V, No. 1
"Notes on Some Scottish Infantry Swords in the Scottish United Services Museum"
by A.V.B. Norman

Volume VI, No. 3
"Contracts for the Supply of Equipment to the 'New Model' Army in 1645"
by Gerald I. Mungeam

Volume VI, No. 7
"Some Notes on London Made Sword Hilts of the Seventeenth Century"
by Gerald I. Mungeam

Volume VII, No. 7/8
"Sword Cutlers to the Board of Ordnance"
by W.E. May

Volume VIII, No. 6
"Some References to Arms, Sword Cutlers and Gunsmiths in English Newspapers, 1660–1727"
by A.V.B. Norman

Volume IX, No. 6
"The Dating and Identification of Some Swords in the Royal Collection at Windsor Castle"
by A.V.B. Norman

Volume X, No. 4
"A List of Birmingham Makers of Arms and Their Accessories from Wrightsons New Triennial Directory, 1818"
by Arthur G. Credland

Volume X, No. 5
"Silver Sword Hilts Assayed at Birmingham"
by Norman Dixon

Volume X, No. 6
"Some Swords of the British Civil War with Notes on the Origin of the Basket Hilt"
by A.G. Credland

Volume XII, No. 1
"Notes on Some Early Basket Hilted Swords"
by G.M. Wilson

Volume XII, No. 1
"The Armours Bill of 1581: The Making of Arms and Armour in Sixteenth Century London"
by Claude Blair

Volume XII, No. 4
"Patriotic Fund Swords, Part 1"
by Leslie Southwick

Volume XII, No. 5
"Patriotic Fund Swords, Part 2"
by Leslie Southwick

Volume XIII, No. 3
"The Recipients, Goldsmiths and Cost of the Swords Presented to the Corporation of the City of London"
by Leslie Southwick

Volume XIII, Supplement
"Further Notes on Early Basket Hilted Swords"
by Guy Wilson

Volume XIII, Supplement
"A Basket Hilted Sword Marked A.C. in the Royal Armouries"
by Philip S. Lankester

Volume XIV, No. 1
"Sword Cutlers in the London Directories, 1736–1811"
by David Wright

Volume XIV, No. 5
"Notes on Some Scottish Silver Hilted Swords and Related Swords"
by Philip J. Lankester

Volume XV, No. 1
"The Fourteenth Century Scottish Sword"
by Tony Willis

Volume XV, No. 3
"Notes on Two British Light Cavalry Sabres"
by A.V.B. Norman and J.P. Puype

Volume XV, No. 6
"New Facts about James Morisset and a Revised List of his Known Works with Others his Successors Ray & Montaque"
by Leslie Southwick

Volume XV, No. 7
"Additional Material Towards a History of the Basket Hilt"
by A.V.B. Norman